Governing Through Markets

BENJAMIN CASHORE
GRAEME AULD
DEANNA NEWSOM

Governing Through Markets

FOREST CERTIFICATION
AND THE EMERGENCE
OF NON-STATE
AUTHORITY

Yale University Press
New Haven &
London

Set in Sabon type by Keystone Typesetting, Inc. Printed in the United States of America by Sheridan Books.

Library of Congress Cataloging-in-Publication Data
Cashore, Benjamin William, 1964–
    Governing through markets : forest certification and the emergence of non-state authority / Benjamin Cashore, Graeme Auld, Deanna Newsom.
        p. cm.
    Includes bibliographical references (p.   ).
    ISBN 0-300-10109-0 (alk. paper)
    1. Forest management — Standards. 2. Forest products — Certification. 3. Forest policy. I. Auld, Graeme. II. Newsom, Deanna. III. Title.
SD387.S69C37 2004
634.9′2′0218 — dc22
2004041288

A catalogue record for this book is available from the British Library.

The paper in this book meets the guidelines for permanence and durability of the Committee on Production Guidelines for Book Longevity of the Council on Library Resources.

10 9 8 7 6 5 4 3 2 1

To
Donna
Volker
Ed and Joyce

Contents

Preface   ix

Acknowledgments   xii

List of Abbreviations   xvii

I. The Transformation of Global Environmental Governance

1. The Emergence of Non-State Market-Driven Authority   3

2. The Research Design: Toward an Analytical and
    Explanatory Framework   31

II. North America

3. British Columbia, Canada   59

4. The United States   88

III. Europe

5. The United Kingdom   129

6. Germany   160

7. Sweden   189

IV. Private Authority and Sustainability

8. Competing for Legitimacy   219

Appendixes   249

Notes   263

Glossary of Terms   286

References   290

Index   317

Preface

Those policy scholars who had spent most of their careers studying *public policy* were blindsided in the 1990s by the widespread use of private authority as a means to address environmental deterioration and social issues. New arenas began to emerge at this time in which traditional Westphalian state authority was either diminished or non-existent, and political scientists, sociologists, legal scholars, and economists have been scrambling to make sense of what this all means for addressing matters of concern to civil society. Can these new initiatives address key policy problems in ways that traditional public policy processes have been unable? Or are these new private systems simply efforts to reduce what would have been more forceful governmental regulations?

The origins of this book can be traced back to an interest in these broad questions by its senior author, as he moved in the summer of 1998 from the Canadian province of British Columbia to begin a new faculty position at Auburn University's School of Forestry in Alabama. Within forestry, private authority, commonly known as "forest certification," was beginning to emerge as a key issue, both in the United States and globally. Yet little was known about what certification would do to enhance sustainable forestry on the ground or how certification might influence the socially constructed, and highly contested, debates about the appropriate definitions of sustainable forestry.

Recognition that forest certification was emerging *globally* as the most advanced case of non-state market-driven private authority pushed Cashore's interest in exploring and comparing domestic policy responses to this global phenomenon. He quickly began an international search to recruit graduate students to work with him. He found two peers. Graeme Auld and Deanna Newsom enthusiastically and rigorously pounced on the ambitious theoretical and empirical efforts that were needed to conduct such a study. As a team, we pushed forward both the theoretical and conceptual issues, consciously choosing Canada, the United States, Germany, the United Kingdom, and Sweden as cases in which forest certification was quickly emerging as an arena of private authority but in what appeared to be significantly different ways.

Our research and writing moved back and forth in an iterative process (author order reflects Cashore first as senior author; Auld and Newsom are listed alphabetically). However, we did divide up initial tasks — Auld was given the task of researching the United Kingdom and the US Pacific Coast, Newsom focused on the US Southeast and Germany (where she ended up living for a year and by the summer of 2001 had managed to entice the entire team there), and Cashore developed the framework, guided the overall research, participated in key informant interviews in all the cases, and took the lead in researching the Swedish case. We divided research on the British Columbia case. The writing and analysis of the results followed an intense iterative process over time and among the three authors. Cashore first wrote a key piece (with strong input from Auld and Newsom) that began our theoretical and conceptual journey. We then wrote an initial paper designed to both summarize the empirical cases and to develop initial explanations of the noted patterns of divergence that we had uncovered. In the summer of 2001 Auld and Newsom each wrote their master's theses that compared their three cases. These theses became key steps that well positioned the manuscript.

Cashore took up a new position at this time as assistant professor, Sustainable Forest Policy, Yale School of Forestry and Environmental Studies, and headed up the school's new Program on Forest Certification. At this stage interactive writing began in earnest. We worked hard to make sure our framework and analysis fit across cases and significantly revised our initial explanatory framework.

Owing to our strong theoretical and empirical focus, an array of practitioner and scholarly audiences have been strongly interested in this project. Professional foresters continue to show strong interest because they want to know whether forest certification will become a key arena in which forestry regulations develop, and if so, the ways in which policy choices are made within these systems. Will the views of forestry professionals be heard? If so, how? While our effort is to explain rather than to prescribe, those environmental, economic, and social groups promoting markets and sustainable forestry find this project highly important since it points to those factors that explain how support might occur and the obstacles to support. Likewise, those international and domestic governmental agencies who have long focused on the environmental and social problems will want to know what these new arenas of private authority can actually do and whether they are viewed as a threat, or complement, to traditional public policy approaches.

The scholarly world is equally interested in this project because it sheds considerable light on how these systems of private authority emerge in the forest sector, arguably the most advanced case of non-state, market-driven governance systems globally. Because this is a highly complex and understudied phenomenon, we were not content to apply existing (and what we felt to be inappropriate) lenses on these issues. Instead, we searched across sociology, economics, political science, law, and policy studies literature to develop a unique and important conceptual framework with which to address this phenomenon and advance the accumulation of knowledge. As a result, scholars interested in the formulation and implementation of policy choices from all social science disciplines and those natural science disciplines affected by these new systems will find this contribution invaluable.

For all these reasons this book is needed and useful. It speaks to the importance in grounding any theory of private governance with well-researched and exhaustive empirical accounts. The changing and complex nature of these systems also speaks to the need to conduct systematic and rigorous collaborative research in matters of private governance, especially if results are to be produced in a timely manner. Indeed, what we find in this study matters to ongoing struggles to address environmental, social, and economic problems facing the world's biosphere. We are happy to offer this book as an example of the best kind of collaboration — an exhaustive, innovative, and ultimately exhilarating read.

Acknowledgments

We have incurred numerous debts collectively and individually through-out this three-and-a-half-year project. The bulk of the research was carried out under a competitive grant from the USDA's National Research Initiative Markets and Trade Program. Other significant support has come from the USDA's Southern Experiment Station, the Bradley/Murphy Forestry and Natural Resources Extension Trust, the Washington, DC, Canadian Embassy's Canadian Studies Faculty Research Grant Program, Auburn University's internal competitive grant-in-aid program, Auburn University's Center for Forest Sustainability, the German Academic Exchange Service, the Yale School of Forestry's Program on Forest Certification, the Ford, Merck, and Kohlberg foundations, the Rockefeller Brothers Fund, and the Doris Duke Charitable Trust.

We were fortunate to be able to present preliminary results to a number of scholarly and practitioner conferences. The Association for the Canadian Studies in the United States biennial meeting in Minneapolis in November 1999 provided an initial arena in which to advance this project. Auburn University's Forest Policy Center's conference on private forestry in March 2001 provided a key arena for feedback from forestry practitioners and policy scholars. Our own NRI seminar in Freiburg, Germany, in May 2001, Michele Micheletti's May 2001 Stockholm conference on political consumerism, and Errol Meidinger and Chris Elliott's June 2001 workshop in Freiburg, Germany, on the social and political impacts of forest certification all provided important venues for further refinement and reflection, and we are grateful to the participants in those conferences. Laura McDonald's "Visioning North America" conference held at Carleton University in October 2001 provided important input in our theoretical framework. Similarly, the Consultative Group on Biological Diversity's Charlotte meeting in December 2001 provided feedback from the funding community, and the seventh annual Colloquium on Environmental Law and Institutions sponsored by the Ford Foundation and Duke University's Center for Environmental Solutions provided an enormous arena for intellectual feedback and stimulation. We are grateful to the organizers of the Forest Leadership Forum in Atlanta in April 2002, who gave us a chance to present and gain feedback from forest company executives, environmental groups officials, and private forest owners. The Canadian Political Science Association's 2002 annual meeting afforded us time to explore our theoretical argument with other scholars working on private authority, and we are likewise grateful to them. Thanks to Richard Donovan for asking Cashore to speak and gain feedback at the market linkages workshop, which preceded the FSC General Assembly in Oaxaca, Mexico, in November 2002. The Forest Dialogue's meeting on forest certification in Geneva in October 2002 and the World Business Council on Sustainable Development's Global Forest Industry CEO Forum Action Team meetings in London, May 2003, and New York, June 2003, gave us an unparalleled opportunity to present results to and gain feed back from those engaged in deciding the future of forest certification.

We owe a great deal to the individuals who have nurtured us, given us advice, and commented on our papers. Their individual contributions in our development are deserving of much more than a mention in this book, and we hope in the years to come to give adequate and proper thanks. Cashore's approach to this project was the direct result of those who trained him in comparative public policy, including Grace Skogstad, Richard Simeon, Carolyn Tuohy, Ron Manzer, and Jeremy Wilson. His collaborations with George Hoberg, Jeremy Rayner, Michael Howlett, and Jeremy Wilson proved an important step in Cashore's attention to private authority. Cashore's ongoing collaboration with Steven Bernstein (who came up with the book's final title) has been a source of intellectual stimulation and fulfillment, and Bernstein's comments on early versions of this manuscript have pushed the theoretical framework in important directions. Auld is grateful to John Worall, who planted the seeds of his interest in forest management, and David Haley, Peter Marshall, Tony Kozak, Valerie LeMay, George Hoberg, Tim Ballard, and Sandra Schinnerl, who gave him direction. Knowing that her shift from ecology to forest policy research would not have occurred without the five years of field research in Clayoquot Sound, BC, preceding this project, Newsom would like to thank Barbara Beasley, Alan Burger, and Trudy Chatwin for providing those opportunities.

Our work at Auburn University benefited greatly from a number of individuals, including Larry Teeter, whose attention to Cashore's career choices and mentorship was selfless and invaluable, and Dean Brinker, who provided support and resources to all three authors. Important comments and advice on research designs and drafts while at Auburn came from David Laband, John Schelhas, Jill Crystal, Mark McKenzie, Mark Dubois, Jamie Lawson, Conner Bailey, and Daowei Zhang. We are fortunate to have benefited from additional support from the Auburn community, including Graeme Lockaby, Murray Jardin, John Kush, Changyou Sun, Beverly Edwards, Marjorie Gentry (without whom the NRI grant would not have been handed in on time!), Lenore Martin, Pat Chiroux, and Harry Murphy. We are grateful to research assistance while at Auburn from Sarah Day, Sheila Saunders, Gina Glaze, and Anthony Stewart.

At Yale, those who took the time to comment on our manuscript or otherwise provide important support include Dean Gus Speth, Jacob Hacker, Chad Oliver, Dan Esty, Rob Mendelsohn, Marian Ivanova, Monica Araya, Mary Tyrrell, Michael Washburn, Gary Dunning, and Barbara Bamberger. Research support from Yale graduate students was invaluable in getting this manuscript out. We are in debt to Margaret Francis and Emily Noah for incredible research support during the final year of manuscript writing and for excellent comments on previous drafts. We also wish to thank for their research assistance at Yale Johannes Wurm, Elizabeth Shapiro, Beth Egan, James Lucas, Ho Way Wang, and Steffen Taeger. For technical support we are grateful to Charles Waskiewicz (whose persistence in getting the NRI grant transferred to Yale was a crucial step in moving our research forward), Jen Liner, and Elizabeth Vierra. Thanks to Sarah Davidson for going over the entire document with fresh eyes and perspective.

The Rainforest Alliance provided an important arena for Newsom for practical learning, and we are grateful to Rebecca Butterfield and Richard Donovan for granting Newsom a leave of absence to help complete this project. Gary Bull, Justin Stead, and Sally Aitken provided important mentorship during Auld's year at UBC. His interactions during this year with Connie McDermott, George Hoberg, Paul Wooding, and Warren Mabee fostered his thinking and analysis.

Kafé International and its owner, Heather Whitehouse, in Cheshire, Conn., deserve special recognition for welcoming Cashore, who spent countless hours writing the final product in her welcoming café.

We were fortunate in that a number of individuals outside of our institutional affiliations provided comments either at conferences or simply out of the goodness of their hearts. They include Chris Elliott, Errol Meidinger, Charlene Zietsma, Erika Sasser, Michele Michelleti, Rudi Rüdiger Wurzel, Magnus Bostrom, John Schelhas, Mark Richenbach, Hannah Scrase, Ilan Vertinsky, Dietrich Soyez, Mark Haddock, Michael Conroy, Aseem Prakash, Kernaghan Webb, Jackie Best, Andy White, Ulli Klins, Peter Sprang, Jason McNichol, Connie McDermott, Ben Gunneburg, Justin Stead, Stuart Goodall, Tom Jorling, Tim Mealey, Nigel Sizer, Scott Wallinger, Rick Cantrell, Richard Donovan, and Heiko Leideker. We are especially grateful for detailed comments on the different chapters from Cassie Philips (chapter 1), Hannah Scrace (the UK case), Barbara Lang (the German Case), Scott Wallinger, Fran Raymond Price, John Heissenbuttel (the US case), Tage Klingberg and Magnus Boström (the Swedish case). We must note sincere appreciation to Michel Becker and Carol Grossman of Freiburg University's Institute of Forest Policy, Markets and Marketing Section, whose hosting of Newsom for a full year and Cashore for three months allowed us the time to produce a key part of this book in a most idyllic and relevant setting. Thanks go to Sandra Steinert for translating many interviews conducted in German to English.

The reader can tell from our hundreds of interviews for this book that the biggest research debt we owe is to all those in the North American, European, and International forest policy communities who took times from their busy schedules to speak with us. We promised them that we would not attribute statements to them directly without their expressed permission. However, we do list their organizations in the first appendix to this book. They know who they are and we are eternally grateful.

We are extremely grateful to Jean Thomson Black at Yale University Press, who immediately recognized this project as innovative and important. We shudder to think what might have happened without her determination, support, and perseverance. It has been an absolute pleasure working with her. We know how lucky we were. We are also grateful to Margaret Otzel, senior editor of the manuscript department, and Julie DuSablon for her outstanding editing of the original manuscript.

We now turn to our families, though we are unable to put in this short space what they mean to us. Auld is tremendously thankful and indebted first and foremost to his parents, whose encouragement of dinnertime debates helped develop his interest in questioning and exploring human behavior. This appreciation also extends to his broader family in Vancouver and to his wonderful relations in Scotland. Auld's group of Vancouver friends have been a source of great stability, companionship, and intellectually stimulating conversation. Newsom's deepest thanks go to her husband, Volker Bahn, not only for his love and willingness to let her studies and research forays take precedence during the course of this project, but also for the many discussions they have had comparing qualitative and quantitative research approaches. She would like to thank her parents and brother for their unfailing encouragement, which began long before this project did. And though he is too young to remember, the impending arrival of little Jonas provided Newsom with additional incentive to complete the manuscript, and in the years to come she looks forward to relaying his role in this book's journey. Cashore owes undying gratitude to his parents for nurturing his interest in forestry and the environment, to his mother- and father-in-law for their longstanding and generous support, and to his brother Harvey for his critical and constructive analysis during decades of discussions on policy and politics. Cashore is eternally grateful to Donna Cashore who continues to make unimaginable career (and other) sacrifices so that she could spend time with the children while Ben finished this book, the latest in a series of never-ending projects. Theresa, Joseph, and Walter's absolute zest for life and adventure provided the happiness through which he got through the tough days. It is to our families that we dedicate this book, as a small token and recognition for the support they have given us.

Abbreviations

AAC	annual allowable cut
AF&PA	American Forest and Paper Association
ATFS	American Tree Farm System
BBC	British Broadcasting Corporation
BC	British Columbia
BLM	Bureau of Land Management, United States
BMP	best management practice
CFPC	Certified Forest Products Council
COFI	Council of Forest Industries of British Columbia
CPPA	Canadian Pulp and Paper Association
CSA	Canadian Standards Association
CSFCC	Canadian Sustainable Forestry Certification Coalition
DANI	Department of Agriculture for Northern Ireland
DIY	do it yourself
EMAS	Eco-Management and Auditing System
EPA	Environmental Protection Agency, United States
ERP	expert review panel (later external review panel)
EU	European Union
FABC	Forest Alliance of British Columbia
FAO	Food and Agricultural Organization

FICGB   Forestry Industry Committee (later Council) of Great Britain
FIDC   Forestry Industry Development Council
FOE   Friends of the Earth
FSC   Forest Stewardship Council
FTA   Forestry and Timber Association
GB   Great Britain
GFTN   Global Forest and Trade Network
GMO   genetically modified organism
ITTO   International Tropical Timber Organization
MB   MacMillan Bloedel
MSC   Marine Stewardship Council
NGO   non-governmental organization
NIPF   non-industrial private forest
NRDC   Natural Resource Defense Council
NSMD   non-state market-driven
PEFC   Pan European Forest Certification
RAN   Rainforest Action Network
SBFEP   Small Business Forest Enterprise Programme
SCS   Scientific Certification Systems
SFB   Sustainable Forestry Board
SFI   Sustainable Forestry Initiative
SFM   Sustainable Forest Management
SIC   State Implementation Committees
SSNC   Swedish Society for Nature Conservation
TGA   Timber Growers Association
TIMO   Timber Investment Management Organization
TREES   Training, Research, Education, Extension, and Systems
TTF   Timber Trade Federation
UK   United Kingdom
UKWAS   United Kingdom Woodland Assurance Scheme
UNCED   United Nations Conference on Environment and Development
UNCSD   United Nations Committee on Sustainable Development
USDA   United States Department of Agriculture
VDP   Verband Deutscher Papierfabriken
VDZ   Verband Deutscher Zeitungsverleger
WFP   Western Forest Products
WHO   World Health Organization
WTO   World Trade Organization
WWF   World Wildlife Fund

PART I

The Transformation of
Global Environmental Governance

I

The Emergence of Non-State Market-Driven Authority

Hubert Kwisthout made bagpipes. He practiced his craft for over twenty years and, through hard work and perseverance, established a reputation for creating exquisite instruments capable of producing beautiful music. For Kwisthout, his bagpipes were much more than a product for sale — they represented to his customers an extension of himself and his art. It was for these reasons that Hubert Kwisthout became increasingly concerned about a moral dilemma that he could not rationalize away. Kwisthout relied on tropical hardwood imports to produce the wood components of his bagpipes, and during the 1980s, evidence accumulated indicating that the wood was coming from tropical forests that were being degraded by poor forest management practices or destroyed by conversion to other uses. The beauty he was creating with his bagpipes was contributing to the destruction of tropical forests and the endangerment of a wide range of species and habitats. The response by citizens and environmental groups in Europe and North America concerned about tropical forest destruction at this time was to promote boycotts of tropical timber. These efforts treated wood products from the tropics, regardless of the source, as contributing toward deforestation and species loss. This approach bothered Kwisthout, who, dependent on tropical wood for his bagpipes, felt there had to be a way to buy wood from companies and land-owners who practiced *sustainable forest management*. But how could this happen? Illegal logging, poor governmental institutions, and widespread corruption in many developing countries meant that relying on government might not produce the desired results.

There had to be another way. Kwisthout's first solution was to form a trading company in the United Kingdom whose purpose was to purchase timber from environmentally sound sources. Kwisthout quickly learned, however, that those claiming to sell sustainable timber had no way of verifying that the sources were actually sustainable. Another idea came to Kwisthout. Why not have environmental groups and other social interests devise a set of rules governing sustainable forest management, independently certify companies who practice these standards, and thus enable consumers to purchase wood products from companies who pass the certification process?

It was a relatively simple solution that would have enormous and complex impacts. While Kwisthout had no way of knowing it at the time, his support of this kernel of an idea would trigger a series of events that would present global and domestic environmental governance with one of the most innovative and startling institutional designs of the past 50 years: forest certification. The advent of forest certification systems ushered in a new breed of sustainable development institutions outside of traditional governmental processes that would offer fundamentally different ways of creating policy and implementing policy choices. And as we reveal in this book, something else happened that Kwisthout could never have envisioned. Thus far, the regions in the world where certification has been the most debated, supported, and institutionalized are not the tropics but rather the countries of the North, where well-organized environmental, social, and business interests, caught in longstanding struggles over temperate and boreal forest resource use, have sought to define and shape forest certification efforts.

This book is an effort to explore the emergence and support for these new institutions and understand better their effects in terms of addressing and promoting sustainable development. We refer to these new institutions as "non-state market-driven" governance systems because rule-making clout does not come from traditional Westphalian state-centered sovereign authority but rather from companies along the market's supply chain, who make their own individual evaluations as to whether to comply to the rules and procedures of these private governance systems. Environmental groups and other non-governmental organizations (NGOs) attempt to influence company evaluations through economic carrots (the promise of market access or potential price premiums) and sticks (public and market campaigns aimed at pressuring companies to support certification).

Understanding how these private governance systems emerge and gain rule-making authority is an important question for at least four reasons. First, an intense competition is currently being waged between the program originating from Kwisthout's idea, the international Forest Stewardship Council (FSC) certification program, which has strong support from most of the world's leading environmental groups, and industry and landowner initiated certification programs that have emerged in North America and Europe. We argue that the outcome of this competition is crucial to understanding whether non-state market-driven governance in the forest sector will result in privatizing "up" (Vogel 1995) existing public policy rules governing sustainable forest management, or whether a privatizing "down" or benign "sideways" incremental approach will emerge. Second, depending on how they emerge and their ultimate form and function, non-state market-driven governance systems could come to be seen by key actors as a more legitimate arena (Bernstein 2001b) in which to create environmental policy than traditional public or shared public/private forms of decision-making (Coleman and Perl 1999; Clapp 1998). Third, supporters argue that these private governance systems could improve environmental performance because perceived or potential market benefits may work to encourage commitments in ways that traditional command and compliance public policy processes have been unable (Gunningham, Grabosky, and Sinclair 1998). While such conclusions appear premature, the fact that so many groups believe that certification programs will have such an impact renders these private systems worthy of rigorous and sophisticated research. Fourth, non-state market-driven governance systems are quickly emerging in other sectors, such as food (Food Alliance 2001), coffee (Fair Trade.org 2001), tourism (Rainforest Alliance 2002), and fisheries (Simpson 2001). An analysis of the forest sector will facilitate hypothesis building for research in other sectors, thus contributing to a comparative cross-sector research agenda into the ways in which non-state market-driven governance systems emerge within different environmental policy problem arenas.

We have chosen to address these questions by comparing support for forest certification in the United States, British Columbia (Canada), Germany, the United Kingdom, and Sweden. These cases were chosen because they have two pertinent characteristics. First, they are all actively involved in the production and consumption of industrial wood and paper products. For example, in the UK and Germany, higher reliance on imports exists, while in BC, exports are the driver of local production. These differences allow us to explore how they might influence support for certification in producing versus consuming countries.

These cases also vary according to the structure of their forest sector and public policy approaches to sustainable forest management — factors that we argue are key to understanding the emergence of and support for non-state market-driven governance systems. We chose industrialized countries for two reasons. First, this choice permitted us to control for broad patterns of economic development that distinguish the developing from developed world. Second, these cases are especially significant, as they were all early participants in forest certification, giving us the opportunity to identify key factors that lead to differing paths of non-state market-driven governance development. We fully expect future research to explore developing countries, especially those tropical countries that first provided the interest in forest certification. At the same time, and as we detail below, it does appear that if non-state market-driven governance fails to institutionalize fully in developed countries, it could very well disappear as an innovative policy instrument to address forest deterioration globally. And the inverse may well be true — institutionalizing certification in the North may create the strongest and most effective way of developing certification institutions in the global South.

The Puzzle

Guided by the above questions and an analytical framework detailed in chapter 2, we noted remarkable patterns of initial convergence, then divergence, among the cases under review. In all cases under review, support for certification followed similar patterns where environmental groups and their allies supported the prescriptive and wide-ranging international FSC program, while forest companies and forest landowner associations expressed skepticism. In all the cases, the vast majority of companies and landowners initially either failed to support any form of certification or created their own rival domestic programs designed to compete with FSC.

These quite remarkable stories of convergence quickly turned into divergence (table 1.1), as FSC and FSC-competitor programs and their core supporters undertook a wide variety of market-based strategies aimed at changing forest company and forest owner support toward the FSC. The impacts of these strategies were uneven across our cases. In BC, forest companies and industry groups dropped their exclusive support for the industry-initiated program and participated in the FSC standards development process, attempting to work within the FSC to forge an agreement on sustainable forest management standards. Likewise in the UK initial reluctance on the part of most forest landowners to support the FSC turned into grudging support and eventual acceptance of the FSC standards. And in Sweden, industrial companies quickly dropped their hesitation over the FSC process and chose early on to

Table 1.1 General patterns of forest company and landowner support for the FSC

	BC (Canada)	United States	United Kingdom	Germany	Sweden
Initially	Scant	Scant	Scant	Scant	Scant
After efforts to gain support	Widespread pragmatic	Scant	Significant pragmatic	Weak	Pragmatic industry; landowner scant

support the FSC, while non-industrial private landowners eventually walked away from the FSC and created their own certification program. In the US, however, very few forest companies and forest owners have given the FSC any type of support, instead focusing massive efforts and resources on creating and modifying the industry and landowner alternative programs. Likewise in Germany the FSC has not been able to alter forest sector support significantly, with most state forestry agencies and private landowners preferring the industry program.

Why is it that forest companies and forest landowners in some countries and regions altered their support for forest certification programs to include the environmental group–conceived FSC, while in other regions, forest companies and forest landowners have remained steadfast in their support of industry or landowner programs?

The purpose of this book is to systematically explore this question. We chose to focus on "forest company and forest landowner" support for the FSC as our dependent variable because it presented the most striking divergence when we applied our framework to each case and because forest companies and landowner choices are the ultimate target of certification programs. Why some companies and forest owners would choose to support environmental group–supported certification programs when industry and landowner associations offer more flexible and less prescriptive alternatives is a curious, seemingly counterintuitive result. The answer to this question may shed light on the conditions under which companies embrace private standards that go beyond what their own associations would prefer and hence is important not just to the forestry case, but to other sectors as well where non-state market-driven governance systems are emerging. We focus on support for the FSC not because it is necessarily more important than FSC-competitor programs but because it was the observed differences in FSC support over time that emerged as a key puzzle worthy of scientific exploration. And though our question focuses on divergent levels of support for the FSC, answering this question necessarily requires that we systematically address why different interests, institutions, and actors came to support FSC-competitor certification programs, or none at all. These are inquiries that necessarily contribute to broader questions of non-state market-driven governance that we return to in the conclusion to this book.

The Argument

Our general argument is that three structural factors — the place of the country/region in the global economy; the structure of the domestic forest sector; and the history of forestry on the pubic policy agenda — strongly affect efforts by the FSC supporters to gain forest company and forest owner support. In some countries these factors combine to create a hospitable environment to such efforts, while in other countries these factors combine to limit such efforts. And the logic of this argument is important: when a hospitable environment exists the FSC and its supporters are able to "convert" forest companies and owners, without having to change significantly the FSC and its approaches. However, when a relatively inhospitable environment exists, the FSC and its supporters will have to "conform" to forest company and owner criticisms of the FSC, if they are going to have any chance of gaining some degree of support. If our argument is correct, these differences have profound implications for understanding the general "upward" and "downward" movement of rules concerning sustainable forest management, as non-state market-driven governance systems, with very different conceptions regarding both governance processes (who gets to shape the rules) and standards, compete for legitimacy and rule-making authority. This is because efforts to gain forest company and landowner support may result in the FSC certification program *reducing* the stringency of its rules as it conforms to landowner and company concerns, while FSC-competitor programs may *increase* the stringency of their rules as they attempt to conform to baseline minimum certification standards being demanded along the market's supply chain. Our book thus traces how efforts to achieve legitimacy put pressures on both the FSC and FSC competitors to alter their rules upward or downward.

This argument is not deterministic. Instead, we follow David Vogel's approach (Vogel 2001, 1995) in which he argues that increased market integration *could* lead to increased environmental and social production standards, but only when active environmental and other groups specifically pressure governments to put such wording in rules governing increased market transactions. Hence we argue that while each country or region's structural features offer unique incentives and approaches to the FSC and its supporters in their efforts to gain forest company and landowner support, we cannot predict whether the FSC and its supporters will make choices consistent with this environment. Nor can we predict the specific choices made by those strategists who are aware of the impact of these features on the ability of the FSC to gain support. This means that our analysis must carefully explore the interaction between a region or country's structural environment, the specific choices made by the FSC and its supporters, and the path dependencies[1] that are created as they engage in trial and error efforts to gain legitimacy. Similarly we must trace policy choices made by the FSC-competitor programs in their efforts to head off FSC efforts and the effects of these efforts on FSC policy choices. We review the interaction between the promotion of the FSC by environmental and social agents and the structural features that both mediate and inform these efforts. We theorize in the conclusion about the implications of these interactions for the institutionalization of forest certification as an alternative for global and domestic environmental governance.

Chapters 1 and 2 comprise part I of this book, which we use to introduce and detail our argument in five analytical steps. The next section in this chapter identifies the origins of forest certification and the two conceptions of forest certification that have emerged in the forest sector — both in terms of their decision-making structures and also in terms of their approach to sustainable forest management. We then identify key features of non-state market-driven governance, drawing on forestry and other sectors, that reveal why this phenomenon is unique from other types of shared private/public governance or voluntary industry self-regulation approaches. Chapter 2 addresses the analytical and explanatory issues that confront our study, carefully develops seven hypotheses that detail our broad argument regarding divergent responses outlined above, and explains our choice of research design.

The Emergence of Forest Certification and Its Two Conceptions of Non-State Market-Driven Governance

ORIGINS

A number of key trends have coalesced to produce increasing interest in non-state market-driven governance systems generally and within forestry specifically. The first trend can be traced to the increasing interest in market-oriented policy instruments with which to address matters of concern to global civil society. Originally research on economic globalization found that increased capital mobility, international trade, and foreign direct investment appeared to reduce or constrain domestic policy choices, sometimes leading to downward protection in environmental and social standards (Berger 1996:12). However, other scholars noted that a parallel process was taking place in which domestic policy arenas were facing increasing scrutiny by transnational actors, international rules, and norms (see Risse-Kappen 1995; Keck and Sikkink 1998), sometimes leading to a reversal of the "downward" effect of globalization, a process Bernstein and Cashore (2000) refer to as internationalization.[2] In these cases, market-based boycott campaigns were often used to force "upward" governmental and firm-level environmental protection. These internationalization efforts were often deemed easier than attempting to influence domestic and international business dominated policy networks providing important lessons to environmental NGOs about the power of using market forces to shape policy responses. This recognition increased the salience of market converting–oriented campaigns generally, but also for forestry specifically (Stanbury, Vertinsky, and Wilson 1995).

At the domestic level, these international trends mirrored increasing interest in the use of voluntary-compliance and market mechanisms (Harrison 1999; Rosenbaum 1995; Tollefson 1998). Innovative market-based solutions, including trading of pollution credits (Voigt and Cubbage 2000), are ever more popular with governments attempting to address environmental problems. Likewise, US federal agencies such as the United States Environmental Protection Agency (EPA) permit business to avoid specific regulatory requirements through such mechanisms as habitat conservation plans (HCPs), if they can devise innovative measures that address fundamental environmental goals. In the US forest sector, voluntary "best management practices" (Alabama Forestry Commission 1993) are also examples of approaches that vary from traditional "command and control" approaches.

Cashore and Vertinsky (2000) have noted that beginning in the 1980s, policy makers in Canada, the US, and elsewhere often faced the competing pressures of reduced resources available to combat environmental problems and increased demands from civil society to address environmental protection. The privatization of environmental governance appears to be an implicit way out of this conundrum, creating domestic policy climates in the late 1990s in North America and Europe hospitable toward expanding market-based environmental governance (Fletcher and Hansen 1999).

At the international level Bernstein (2001a) has noted that a norm complex of "liberal environmentalism" has come to permeate international environmental governance, where proposals based on traditional command and control "business versus environment" approaches rarely make it to the policy agenda (Esty 1998; Esty and Geradin 1998). In this context, Speth (2002) and others have noted that business-government and business-environmental

group partnerships have created the most innovative and potentially reward-
ing solutions to addressing massive global environmental problems.

A second trend can be traced within forestry generally, where efforts to
address tropical forest destruction have been pervasive. As we detail in the
chapters to follow, the late 1980s witnessed widespread concern among environmental groups and the general public about tropical forest destruction.
Boycotts were launched, and a number of forest product retailers, such as
B&Q in the UK, Ikea in Sweden, and Home Depot in the US, paid increasing
attention to understanding better the *sources* of their fiber and whether their
products were harvested in an environmentally friendly manner. These concerns led to the creation of the International Tropical Timber Organization
(ITTO) and the International Tropical Timber Agreement, through which
tropical exporting countries made commitments to promote increasing liberalization of tropical trade and *sustainable* development of the forest resource
(Gale 1998). It was in ITTO discussions that Kwisthout's idea of a product
label was given serious attention.[3] While an agreement could not be reached
that satisfied the broad array of ITTO stakeholders, labeling had now entered
the consciousness of those interested in shaping global forestry management —
a solution that fit well with the broader norms of Bernstein's (2001a) liberal
environmentalism.

The final trend favoring an interest in non-state market-driven environmental governance in the forest sector can be traced to the failure of the Earth
Summit in 1992 to sign a global forest convention (Humphreys 1996a). Indeed, participation in the forestry PrepComs for the Rio Summit led many
officials from the world's leading environmental NGOs to believe that Rio
would either produce no convention or a convention that would look more
like a "logging charter."

Two Conceptions

By 1992 these forces converged to create an arena ripe for a private
sector approach. But unlike other arenas in which business took the initiative
in creating voluntary self-regulating programs (Prakash 2000, 1999), environmental groups took the initiative in creating certification institutions. Transnational environmental and social groups, led by the World Wide Fund for
Nature (WWF), eventually embraced Hubert Kwisthout's sustainable certification idea noted earlier[4] and created the international FSC program.[5] An
initial exploratory meeting of environmental groups, social actors, retailers,
governmental officials, and a handful of forest company officials was held in
Washington, DC, in spring 1992, and a founding convention was held in

Toronto in 1993. The FSC turned to the market for influence by offering forest landowners and forest companies who practiced "sustainable forestry," according to FSC rules, an environmental stamp of approval through its certification process, thus expanding the traditional stick approach of a boycott campaign by offering carrots as well.

The FSC created nine "principles" (later expanded to ten) and more detailed "criteria" that are performance-based and broad in scope and that address tenure and resource use rights, community relations, workers' rights, environmental impact, management plans, monitoring and conservation of old growth forests, and plantation management (see Moffat 1998:44; Forest Stewardship Council 1999b). The FSC program also mandated the creation of national or regional working groups to develop specific standards for their regions based on the broad principles and criteria.

The FSC program is based on a conception of non-state market-driven governance that sees private sector certification programs forcing upward sustainable forest management standards. Perhaps more important than the rules themselves is the FSC "tripartite" conception of governance in which a three-chamber format of environmental, social, and economic actors has emerged. This structure is further distinguished by equal voting rights in the FSC general assembly for each chamber as well as equal representation for Northern and Southern stakeholders (Domask 2003). The aim of this institutional design is to ensure that no one group can dominate policy making and that the North cannot dominate at the expense of the South—two criticisms that had been made of failed efforts at the Rio Earth Summit to sign a global forest convention (Lipschutz and Fogel 2002; Domask 2003; Meidinger 1997, 2000).[6] The lumping together in one chamber those economic interests (i.e., companies and non-industrial forest owners) who must actually implement sustainable forest management rules with companies along the supply chain who might demand FSC products and with consulting companies created by environmental advocates has been the source of much controversy and criticism. Certainly industrial forest companies and non-industrial private forest (NIPF)[7] owner associations have revealed their discomfort with such an approach (Sasser 2003; Cashore, Auld, and Newsom 2003). Surveys of forest owners and companies reveal that these concerns are shared among the broader forest sector interests (Auld, Cashore, and Newsom 2003; Newsom et al. 2003; Vlosky 2000).

What is clear is that procedures are developed with a view toward eliminating business dominance and encouraging relatively stringent standards, with the goal of ensuring on-the-ground implementation (table 1.2).

The FSC certification program was quickly matched by forest industry and

Table 1.2 Conception of forest sector non-state market-driven certification governance systems

	Conception One	Conception Two
Who participates in rule making	Environmental and social interests participate with business interests	Business led
Rules — substantive	Non-discretionary	Discretionary, flexible
Rules — procedural	To facilitate implementation of substantive rules	End in itself (belief that procedural rules will result in decreased environmental impact)
Policy scope	Broad (includes rules on labor and indigenous rights and wide-ranging environmental impacts)	Narrower (forestry management rules and continual improvement)

Source: Cashore (2002)

forest landowner programs developed in Canada, the US, Europe, and most other countries in which the FSC is active (Cashore, Auld, and Newsom 2003). In the US, the FSC competitor program is the American Forest and Paper Association's (AF&PA) Sustainable Forestry Initiative (SFI) program, and in Canada, it is the Canadian Standards Association (CSA) program initiated by the Canadian Sustainable Forestry Certification Coalition, a group of twenty-two industry associations from across Canada (Lapointe 1998). In Europe, a number of national FSC competitor programs developed early in the 1990s. By the late 1990s, the Pan European Forest Certification (PEFC) system, created by landowner associations that felt especially excluded from the FSC process, emerged as the key competitor program in Europe. Efforts were also taken to create an umbrella program of these "FSC competitors" in order to obtain an international presence.

The SFI and CSA programs originally tended to emphasize organizational procedures and discretionary, flexible performance guidelines and requirements (Hansen and Juslin 1999:19). The PEFC is itself a mutual recognition program of national initiatives with the result that the substantive rules and their discretionary or non-discretionary nature vary from nation to nation. The SFI originally focused on performance requirements, such as following existing largely voluntary "best management practices," legal obligations, and regeneration requirements. Procedurally AF&PA member companies were required to file a report with the SFI regarding their forest management plans and the objectives they were addressing. Specific company data were not publicly reported but were aggregated and given to a panel of experts for review. The SFI later developed a comprehensive approach through which companies could choose to be audited by outside parties for compliance to the SFI standard. As we reveal in chapter 4, the SFI also created a "Sustainable Forestry Board" independent of the AF&PA with which to develop ongoing standards. The SFI also allows non-AF&PA members to be audited according to the SFI through licensing arrangements and has in place a mutual recognition agreement with the American Tree Farm System as the appropriate standards setting body for non-industrial private forest owners.

Similar to the SFI, the CSA focus began as "a systems based approach to sustainable forest management" (Hansen and Juslin 1999:20) where individual companies were required to establish internal "environmental management systems" (Moffat 1998:39). It allows firms to follow criteria and indicators developed by the Canadian Council of Forest Ministers, which are themselves consistent with the International Organization for Standardization (ISO) 14001 Environmental Management System Standard and include elements that correspond to the Montreal and Helsinki governmental initiatives on developing criteria and indicators for sustainable forest management (Noah and Cashore 2002). Overall, the CSA emphasis is on firm-level processes and continual improvement, though it does depart from the SFI approach in that it contains more extensive procedures regarding stakeholder participation in standards setting. The CSA program contains two standards: one explains how to develop an environmental forest management system, and the other focuses on auditing requirements (Hansen and Juslin 1999:20).

The PEFC, created in June 1999, is a framework for the development and mutual recognition of national and sub-national forest certification schemes. It is based on criteria identified at the Helsinki and Lisbon Forest Ministers Conferences in 1993 and 1998, respectively (PEFC International 2001b), though national initiatives are not bound to address the agreed-upon criteria and indicators (Ozinga 2001) as the PEFC leaves the development of certification rules and procedures to the national initiatives. A PEFC secretariat and council that tends to be dominated by landowners and industry representatives determine the acceptance of national initiatives into the PEFC recognition scheme (Hansen and Juslin 1999).

From the start, the program was explicitly designed to address forest managers' criticisms that the FSC did not adequately take private landowners' interests into account. The PEFC council's membership comprises twenty-five national governing bodies, nineteen of which are European. Authority to endorse these schemes rests with the PEFC Council, thirteen of which have been endorsed as of January 2003. The CSA of Canada and the SFI and Tree Farm of the US are also members of the council, though their programs are not (yet) endorsed under the PEFC (membership in the council is a precondition to achieving endorsement).

PEFC provides for single, group, and regional forest certification. Regular audits are conducted of forest owners participating in a group certification (Noah and Cashore 2002). Under regional forest certification, an applicant's region must be certified by a third party as meeting the requirements of the national standard. Landowners within a defined geographical area that has been granted regional certification status can apply to be recognized participants in the PEFC system only after committing to implement the national performance standards. Once the regional certification is complete and the landowner demonstrates his or her individual commitment to participating in the program (that is, he or she is committed to complying with national criteria), forest owners can apply to the PEFC council or the relevant PEFC national governing body acting on behalf of the PEFC council to obtain permission to use the PEFC logo. The PEFC offers a chain-of-custody certificate, based on physical separation of the certified product from non-certified products or based on a "percent in, percent out" type approach.

These FSC-competitor programs initially operated under a different conception of non-state market-driven governance than does the FSC: one in which business interests should strongly shape rule making, while other nongovernmental and governmental organizations act in advisory, consultative capacities. Underlying these programs is a strongly held view that there is incongruence between the quality of existing forest practices and civil society's perception of these practices. Under the SFI, CSA, and PEFC conceptions, certification is, in part, a communication tool that allows companies and landowners to better educate civil society. With this conception procedural approaches are ends in themselves, and individual firms retain greater discretion over implementation of program goals and objectives. This conception of governance draws on environmental management system approaches that have developed at the international regulatory level (Clapp 1998; Cutler, Haufler, and Porter 1999a).

These structural differences have resulted in significant differences in terms of which type of "on-the-ground" forestry practices end up being permitted under each system. Though specific "on-the-ground" rules may vary according to national and regional standards (as in the case of the FSC), table 1.3 identifies key differences between the FSC and FSC competitors with regard to their overall approaches to sustainable forest management.[8] Table 1.4 illustrates how the FSC's approach to many key forest management rules are nondiscretionary, including the use of clearcuts, harvesting in streamside riparian zones, the use of chemicals, and the role of forest plantations. Landowners and

Table 1.3 *Comparison of FSC and FSC-competitor programs*

	FSC	PEFC	SFI	CSA
Origin	Environmental groups, socially concerned retailers	Landowner (and some industry)	Industry	Industry
Types of standards: performance or systems based	Performance emphasis	Combination	Combination	Combination
Territorial focus	International	Europe origin, now international	National/bi-national	National
Third-party verification of individual ownerships	Required	Required	Optional	Required
Chain of custody	Yes	Yes	No	Emerging
Eco-label or logo	Label and logo	Label and logo	Logo, label emerging	Logo

Source: Adapted from Moffat (1998: 152); Rickenbach, Fletcher, and Hansen (2000); and http://www.pefc.org.

Terms: "Performance based" refers to programs that focus primarily on the creation of mandatory on the ground rules governing forest management, while "systems based" refers to the development of more flexible and often non-mandatory procedures to address environmental concerns. "Third party" means an outside organization verifies performance; "second party" means that a trade association or other industry group verifies performance; "first party" means that the company verifies its own record of compliance. "Chain of custody" refers to the tracking of wood from certified forests along the supply chain to the individual consumer. A "logo" is the symbol certification programs use to advertise their programs and can be used by companies when making claims about their forest practices. An "eco-label" is used along the supply chain to give institutional consumers the ability to discern whether a specific product comes from a certified source.

Note: The PEFC is included in this table for comparative reasons, but it is difficult to make universal characterizations about program content or procedures, since they vary by country or sub-region (though they must meet the minimum level set by the PEFC council).

companies *must* meet a number of specified on-the-ground requirements to become FSC-certified, while the competitor programs, in most cases, originally took the approach of offering non-mandatory indicators that could be followed. And when the FSC competitors required certain standards, they generally focused on procedural requirements, such as a written plan, rather than on mandatory on-the-ground changes. These differences underscore the importance of understanding the conditions under which competing certification programs gain or fail to gain support from forest companies and landowners, since the outcome of the competition affects which specific forestry practices are deemed acceptable under a legitimate certification program.

These different conceptions about the procedures and rules governing sustainable forest management have been reviewed here to illustrate the starting point from which the FSC and the "FSC competitors" began. As we detail below and in the chapters to follow, all programs have changed their procedures and rules in an effort to compete for rule-making authority or legitimacy. We illustrate empirically in our chapters how the FSC and the "FSC competitors" have moved closer to one another in their ongoing efforts to gain widespread support. We reflect on this phenomenon in the conclusion and whether this competition may result in the "privatizing up" of global sustainable forestry standards or whether a watered-down version of the FSC, quite distinct from its original conceptions, will emerge.

Characterizing Non-State Market-Driven Governance

In the past ten years there has been an explosion of important scholarly work on the way governmental processes either share decision-making authority with non-governmental interests or are bypassed all together, as private firms and organizations become the direct target of environmental, social, and other civil society groups pushing for change (Prakash 1999, 2001; Gunningham, Grabosky, and Sinclair 1998; Kollman and Prakash 2001; Porter and van der Linde 1995; Cutler, Haufler, and Porter 1999b; Zietsma and Vertinsky 1999–2001; Haufler 2001). Despite this scholarly interest, there have been fewer efforts devoted to conceptualizing distinct institutions and arrangements within these broad phenomena. Yet we argue that there are four features of non-state market-driven governance systems that render distinct their policy-making authority and legitimacy. We devote this section to better understanding the "non-state market-driven" phenomenon and why it is worthy of specific attention in forestry and elsewhere. Drawing on cases from the forest and other sectors, we identify in table 1.5 four related features that, taken together, comprise an ideal type of non-state market-driven governance system.

Table 1.4 *Comparison of standards influencing on-the-ground aspects of FSC and key FSC-competitor forest certification programs (as of summer 2002)*

Issue	Programs			
	FSC	CSA	SFI	PEFC[a]
Plantations	Limit establishment of new plantations; some existing plantations not eligible	Not specifically addressed	Not specifically addressed	Not specifically addressed
Chemicals	Minimize and monitor use; certain chemicals banned	Follow government regulations	Minimize use	Minimize use
Clearcuts	Size and location restricted (varies by region)	Follow government regulations	Average size must not exceed 49 ha (120 ac)	No specific policy (varies among national initiatives)
Genetically modified organisms	Prohibited	Follow government regulations[b]	Follow government regulations	Most national initiatives do not prohibit use (UK and France are exceptions)

Exotics	Permitted, but not promoted; monitor use	Follow government regulations	Minimize use	Permitted, but not promoted
Reserves	Identify significant sites and ensure protection	Identify significant sites and ensure protection	Identify significant sites; management at discretion of company/ landowner	Identify significant sites; management at discretion of company/ landowner
Streamside riparian zones	Harvesting limited or prohibited in identified areas; increased rules where harvesting is permitted	Develop plan	Develop plan, follow "best management practices"	Follow government regulations, develop plan (varies among national initiatives)

[a] The PEFC is included here for broad comparative reasons only. It is difficult to compare the PEFC with other programs because its national initiatives are the place in which most rules are developed. In some instances such as in Sweden and the UK, PEFC rules have looked quite similar to FSC rules, while in other regions such as Germany and Finland, significant differences exist.

[b] Under the draft provisions of a new CSA standard, the use of genetically modified organisms is given as an example of a forestry practice that a company should address with its public advisory group (Canadian Standards Association 2002).

Source: See Appendix 1

Table 1.5 Key conditions of non-state market-driven governance

Role of the state	State does not use its sovereign authority to directly require adherence to rules
Role of the market	Products being regulated are demanded by purchasers further down the supply chain
Role of stakeholders and broader civil society	Authority is granted through an internal evaluative process
Enforcement	Compliance must be verified

Source: Cashore (2002).

NO USE OF STATE SOVEREIGNTY TO FORCE COMPLIANCE

The Westphalian sovereign authority that governments possess to develop rules and to which society more or less adheres (whether it be for coercive Weberian reasons or more benign social contract reasons) does *not* apply. There are no popular elections under non-state market-driven governance systems, and no one can be incarcerated or fined for failing to comply. In the case of the FSC non-state market-driven governance system, for example, governments are expressly forbidden from being members or voting in decision-making processes.

This does not mean that governmental agencies and actors are unimportant. Rather, there are conditions under which the state can act as another "external audience" in accordance with non-state market-driven dynamics and other cases where the state uses its sovereign authority to force compliance, thus removing the external audience evaluations as important explanatory factors in the granting of rule-making authority (Cashore 2002).

Government Acting in Ways Consistent with Non-State
Market-Driven Governance Systems

Government can act in a number of ways that do not directly invoke its sovereign authority. The most obvious example is that existing rules and policies beyond the non-state market-driven program itself—from rules governing contract law to common law issues regarding property rights—play an important background role in non-state market-driven governance systems. Markets never operate in isolation from a broad array of governmental policies, and the same is true of a non-state market-driven governance system. (We argue in chapter 2 that public policy responses to environmental forestry conflicts in part explain divergent levels of support for non-state market-driven governance.) Second, governments can act as traditional interest groups in an attempt to influence non-state market-driven governance systems' policy-making processes by advocating that the FSC undertake a specific course of action that they desire. However, just as interest groups do not have direct policy-making authority in state-sanctioned processes, the fact that governments seek to influence and shape non-state market-driven governance rules does not mean they are the source of authority. Third, governments can act as any large organization does by initiating procurement policies and other economic actions that may influence the market-driven dynamics. Fourth, when landownership is the source of regulation by non-state market-driven governance systems, governments can become an important source of authority granting if, qua landowner, they agree to abide by these standards on their own forestlands. Likewise, in the case of certification in the fisheries sector, governments have acted as property owners to grant support. For example, the Alaska Department of Fish and Game has had the salmon fishery under its authority certified under the Marine Stewardship Council (MSC), the FSC equivalent for global commercial fisheries operations.

Fifth, governments may provide resources to groups that are attempting to become certified — an action that works to enhance the legitimacy of non-state market-driven governance systems, rather than remove it. For example, in the case of organic food certification, European governments have provided seed money for small non-organic farmers to convert to organic farming and have assisted organic farmers financially in order to persuade them to remain organic (Tovey 1997; Axelson 1996).[9]

Sixth, governments can participate in standards development by providing expertise and resources. This is an important point because while some certification programs such as the FSC forbid direct government involvement, there is no reason, in terms of constructing an ideal type understanding of non-state market-driven governance, why governments could not be involved in rule development. Again the key distinction is whether governments use their sovereign Westphalian authority to require adherence to the rules or rather work to facilitate rule development. The former role reduces the legitimacy of non-state market-driven governance, whereas the latter enhances it. Examples abound of governments facilitating non-state market-driven governance. The chapters to follow reveal governments increasingly lending support when they see certification as providing them some relief from constant environmental group scrutiny to the point that they become unofficial observers in FSC processes. The role of government is even more direct in the fisheries sector, where governments were directly involved in drafting the UN FAO's "Code of Conduct for Responsible Fisheries," upon which the MSC standards are based. Similarly the US EPA was heavily involved in creating standards governing the certification of organic food in the US.

Government Acting in Ways Inconsistent with Non-State Market-Driven Governance Systems Dynamics

When governments do use their sovereign authority to require adherence to private standards setting rules, our conception of non-state market-driven governance no longer exists, since the government, rather than the market, explains why the certified company or landowner is complying. Governments may also use their policy authority to influence only a key target audience of a non-state market-driven governance program, thus creating a hybrid effect in which non-state market-driven logics apply to some audiences but not others. This scenario is what happened in the case of the US EPA's role in creating standards governing certified organic food. The rules eventually took the form of US law, meaning that sovereign Westphalian authority was imposed on any farmer who wanted to become certified as organic.[10] However, most features of non-state market-driven governance remained, since no one was required to become organic. In such cases, it is crucial that the policy analyst understand and explore whether compliance by a key audience is the result of non-state market-driven governance dynamics or of state authority. For example, a government law requiring that all forest landowners become certified according to the FSC would negate any need to understand landowner evaluations of the FSC, as they would be doing it as a matter of law rather than individual calculations. However, if there were no similar requirement for value-added manufacturers or retailers, then non-state market-driven logic would apply to these other important audiences, rendering crucial an understanding of their evaluation process.

AUTHORITY GRANTED THROUGH INTERNAL EVALUATIONS

Recognition that governments do not use their sovereign authority to require adherence to non-state market-driven governance systems turns our attention to understanding that an array of actors and organizations make their own *evaluations* about whether to grant authority. Evaluations are key because actors and organizations cannot be fined or incarcerated for failure to comply. The range of groups that will consider granting authority is virtually the same as in traditional public policy processes. Environmental groups, businesses, professional and trade associations, as "immediate audiences," make specific choices, while the broader public is important for the value-based support it gives to organizations such as environmental groups and potentially as consumers of certified products. Governmental actors, as noted above, are also treated as an interest group (albeit a special one), trying to influence or get access to non-state market-driven governance systems.

MARKET'S SUPPLY CHAIN PROVIDES INSTITUTIONAL SETTING IN WHICH AUTHORITY IS GRANTED

The market's supply chain provides the institutional setting and incentives through which evaluations of support occur. Under non-state market-driven systems, the location of authority is grounded in market transactions occurring through the production, processing, and consumption of economic goods and services. The supply chain directs and frames political struggles of the external audience detailed above. At each stage of the economic production chain economic actors make choices as to whether they support and are willing to operate under the rules and procedures of the non-state market-driven governance system.

This is the most important feature of our conceptualization of non-state market-driven governance systems, setting it clearly apart from most other forms of shared public/private governance. Under non-state market-driven governance systems, compliance incentives (or disincentives) are created along product supply chains, as the promise of price premiums, market access, or prevention of negative boycott campaigns provide incentives to producers of the product to comply with the standards coming from non-state market-driven governance systems. In most cases non-state market-driven governance systems focus their standards on the first stage of the supply chain — namely, those who harvest the natural resource, whether it be coffee, forestry, mining, or fisheries. However, supporters of these systems will go to all parts of the supply chain in an effort to force or encourage support for their systems. Thus, environmental groups in support of the FSC will target lumber retailers, such as Home Depot, attempting to indirectly influence forest company and forest owner evaluations of the FSC. (While some assert that non-state market-driven governance is successful when individual consumers purchase certified products, all that is really needed for non-state market-driven dynamics to exist is that there is some demand along the supply chain). And those supporting alternatives to the FSC will likewise attempt to convince the range of companies along the supply chain that their programs are as appropriate, or more appropriate, than the FSC.

This is important for our study because it means that for companies along the supply chain to support non-state market-driven governance systems they must evaluate them as providing some kind of economic benefit, either direct or indirect. If they evaluate these programs as hurting a company's profit maximizing goals, they will necessarily have to reject them or risk going out of business (of course the more perfect competition that exists in a market the more this statement holds). Recognition of this is important. As our story to follow reveals, non-state market-driven governance systems tend to be supported by "core audiences" who grant support for value-based reasons, rather than economic ones. Hence, most environmental groups support the FSC because its approach resonates with their concerns about global forest deterioration, while many non-industrial private forest owners and forest company officials view alternatives to the FSC as morally more appropriate than the FSC. This means that supporters of these programs have to walk a tightrope between maintaining support from their "core audience," while attempting to use the market's supply chain to increase economic incentives needed to attract forest companies and forest owners to supporting their system.

Understanding what motivates the firms that must actually implement the non-state market-driven governance rules versus those who simply demand their products down the supply chain is thus an important question with respect to non-state market-driven governance. A non-state market-driven governance system is supported by different interests for very different reasons, with fundamental implications for the nature of non-state market-driven governance. For example, in the case of FSC certification, forest companies and landowners operate under different constraints than demand-side audiences, such as forest product purchasing firms (homebuilders, lumber dealers, publishers, retailers) who do not actually have to implement the certification standards themselves. In fisheries, for example, pressures from buyers down the supply chain (in addition to competition from the already-certified Alaskan salmon industry) led the BC Salmon Marketing Board to pursue MSC certification (BC Salmon Marketing Council 2002).[11] Much of the FSC efforts to promote and encourage on the part of forest companies and forest owners are focused further down the supply and demand chain toward those value-added industries that demand the raw products and ultimately, to the retailer and its customers (Bruce 1998, chap. 2; Moffat 1998:42–43). To satisfy this demand, the FSC grants not only "forestland management" certification but also "chain-of-custody" certification for those companies wishing to purchase and sell FSC products.[12] This was an important development in the case of forestry because, as our empirical cases reveal, the FSC competitors, which initially relied on "logos" rather than "eco-labels," learned that in order to gain support from companies further down the supply chain, they needed to guarantee that their fiber was coming from sustainably managed sources.

Similar supply chain issues are occurring in other emerging non-state market-driven systems. In the case of certified fair trade coffee, efforts to bring it to the marketplace have included activist campaigns and threats of demonstrations. In the US for example, the California-based NGO Global Exchange succeeded in getting a commitment from Starbucks to offer fair trade certified coffee in 2000 using this strategy. This demand from retailers was translated through brokers and large roasters to the coffee farmer, who, under a "fair trade" coffee certification program[13], for example, may receive two-, three-, or even fourfold the price of non-certified coffee (Sasser 2002).

Likewise, support for the MSC was tied directly to harnessing "consumer purchasing power to generate change and promote environmentally responsible stewardship of the world's most important renewable food source."[14] MSC and its supporters have managed to gain commitments from large natural foods chains, such as Whole Foods Market, to promote MSC-certified products. And Unilever, which supplies 25 percent of the frozen fish in Europe, is a strong supporter of the MSC.[15]

Identification of the supply chain as the ultimate source of authority and the different conceptions of certification noted above leads to the recognition that a law of "monopoly interests" exists for the FSC and FSC competitors along different stages of the supply chain. The supply chain is key to our story because it creates *inverse* monopoly interests among competing conception one (FSC) and conception two (FSC-competitor) programs at different points in the supply chain. At the beginning of the supply chain (where forest management occurs and where the actual focus of policy is directed), the supporters of the FSC have no interest in having a monopoly (being the only certification program accepted by firms). It does not matter to those wishing to see the FSC emerge as a dominant non-state market-driven governance system whether other certification programs are also accepted by firms, since their desire is to influence forest management according to their standards and policy processes. However, FSC-competitor programs do have a monopoly interest at this stage of the supply chain, since they were created by company and landowners associations in order to offer governance systems and rules that they felt were more appropriate and more flexible than the FSC approach. Thus, these FSC-competitor programs largely lose their raison d'être if the FSC is widely accepted at the forest company and private forest landowner level. *Recognition of this situation is important. A choice by a forest company or landowner to support both the FSC and the FSC competitor is not neutral with respect to the competition for legitimacy. In this case dual support for the FSC and FSC competitors does much more for FSC efforts to gain support than for the competitor program. This is important in the story to follow and explains why we treat forest companies who support both the FSC and FSC competitors as distinct from those who support only the competitors.*

The exact opposite logic is true moving down the supply chain, which is illustrated best in the following chapters by strategies targeted at lumber retailers, such as Home Depot, Lowe's, and Ikea. Since support from these retailers is meant to influence forest companies to support a particular certification program, the FSC necessarily requires that its program be the only acceptable one at this stage of the supply chain. FSC-competitor programs require no such monopoly of support; they only need their programs to also be considered an acceptable source of certified products. The reason this is so is that if retailers accept a number of programs, then there will exist no economic incentive for forest companies to accept the less discretionary, wider-ranging FSC standards. The result of this logic is evident in the cases to follow: FSC competitor programs are calling for "mutual recognition" of different certification programs at the retailer and consumer level, often arguing to these audiences that their programs are largely similar. Calls for "competition of choice" and an end to "monopolies" at the retail and consumer end of the supply chain by proponents of the competitor programs are evident in all our cases. However, at the landowner and forest company end of the supply chain, FSC competitors are working hard to limit support for the FSC by informing companies and landowners about important differences between the FSC and their own systems, which they assert are more friendly to landowners and forest companies.

ON-THE-GROUND COMPLIANCE IS VERIFIED

Non-state market-driven governance systems must involve a verification procedure to ensure that the regulated entity meets "on-the-ground" "performance-based" standards. In the case of the FSC and CSA, mandatory auditing processes are conducted by external auditors. The SFI originally developed looser verification procedures, but voluntary independent third-party auditing is now the method of choice for most companies operating under SFI. Verification is important because it provides the validation necessary for products to be distinguished within the marketplace.[16] This last feature is distinct from procedural verification of environmental management systems such as the ISO 14001 approach (Prakash 1999) or Eco-Management and Auditing System in Europe (Kollman and Prakash 2001) because they must verify on the ground performance, rather than changes only in procedures or systems.

We can use the four characteristics (see table 1.5) to explore whether other sectors' voluntary or "shared governance" programs might be emerging as non-state market-driven systems. For example, the certified sustainable tourism sector has experienced rapid growth in the past ten years, resulting in the emergence of over 100 ecotourism certification programs — with no interna-

Table 1.6 Comparison of emerging non-state market-driven governance systems across sectors

	Conditions of non-state market-driven governance				
	State does not require direct adherence to rules	State does not control standard-setting process	Products are demanded by purchasers further down the supply chain	Authority is granted through evaluative process	Compliance is verified
Forestry	✓	✓	✓	✓	✓
Fisheries	✓	✓	✓	✓	✓
Coffee	✓	✓	✓	✓	✓
Organic foods[a]	✓		✓	✓	✓
Ecotourism	✓	✓		✓	Usually

[a] Refers to organic food certification in the US.

tionally recognized accreditor. Because these programs have different levels of verification, ranging from no verification to rigorous yearly audits, and address different aspects of tourism, which might include the ecological sustainability of hotels and transportation or the safety and wages of tour operators, for example, concerns have emerged about inconsistencies among certifiers and the proliferation of labels. As a result, the Rainforest Alliance, which certifies sustainable tourism through its SmartVoyager program, is currently conducting a feasibility study to determine the potential demand for and optimal structure of an international accreditor for certified ecotourism, an effort that could lead to a full-fledged non-state market-driven system in this sector. Table 1.6 reviews certification programs in key sectors, identifying, as of 2002, how close they are to our "ideal type" non-state market-driven governance systems.

By carefully delineating an "ideal type" of non-state market-driven systems, we can identify examples of private governance that clearly fall outside our classification system. For example, the chemical industry's voluntary Responsible Care program, designed to reduce its impact on the environment, is not a case of non-state market-driven governance because there is no attempt to recognize its program along the chain of production as being environmentally beneficial. It rests instead in the category of codes of conduct to which the industry adheres in order to be viewed as socially responsible (Prakash 2000); Gunningham, Grabosky, and Sinclair 1998, chap. 4). There is no direct market

Table 1.7 Comparison of non-state market-driven governance sources of authority with other forms of governance

Feature	Non-state market-driven governance	Shared private/public governance	Traditional government
Location of authority	Market transactions	Government gives ultimate authority (explicit or implicit)	Government
Source of authority	Evaluations by external audiences, including those it seeks to regulate	Government's monopoly on legitimate use of force, social contract	Government's monopoly on legitimate use of force, social contract
Role of government	Acts as one interest group, landowner (indirect potential facilitator or debilitator)	Shares policy-making authority	Has policy-making authority

Source: Cashore (2002)

mechanism at play but rather the more abstract desire to obtain a "social license to operate." Importantly, our classification of non-state market-driven governance systems purposely avoids a discussion of specific rule content, which we argue is more the result of the competition for rule-making authority.

Failure to accurately conceptualize non-state market-driven governance has led existing political science research to collapse this phenomenon under the broad rubric of "private" or shared governance. Yet such thinking ignores the fundamental basis on which these programs gain rule-making authority (table 1.7). For instance, international relations literature focusing on the role of "private authority" (see Cutler, Haufler, and Porter 1999a, b; Clapp 1998; Lipschutz and Fogel 2002; Lipschutz 2001; Haufler 2001) has taken as given that traditional Westphalian sovereign authority is always present. For example, Cutler, Haufler, and Porter (1999a) examine why and how "the framework of governance for international economic transactions increasingly is created and maintained by the private sector and not by state or interstate organizations." Their projects seek to understand better why private governance emerges and the ways in which it operates. Their project's strength lies in identifying a broad range of private organizations with growing influence both in interstate rule-making procedures and in the development of firm-level collaborative relationships that undertake functions historically accomplished by state actors.

For the most part, however, their approach excludes the case of non-state market-driven governance. Cutler, Haufler, and Porter (1999a) argue that private international authority only exists when "private sector actors" are *"empowered either explicitly or implicitly by governments and international organizations with the right to make decisions for others"* (19, italics added). But this characterization does not capture the non-state market-driven cases in forestry, fisheries, tourism, coffee, and food production, where state author-ity is neither direct nor indirect, and where programs initiated by environmen-tal groups attempt to woo businesses to support them.[17] John Ruggie has argued that "the interplay between [civil society organizations and transna-tional corporations] . . . is creating, for the first time ever, a truly global social domain, however inchoate it still may be — a space that allows for the direct expression of human interests and values in global governance, *not simply those mediated by states*" (Ruggie 2002:10–11, italics added).

Like the work of most of their colleagues in international relations, the highly sophisticated work by students of policy instrument choices do, for the most part, treat market initiatives as one kind of instrument among many available to governmental decision makers (Howlett 1999). Much of this liter-ature is preoccupied with the complex range of substantive policy instruments that policy makers have at their disposal (Hood 1986; Howlett 2000) and whether and when markets can be relied upon to provide public goods (Wolfe 1988). In a sophisticated wide-ranging review of this policy instrument litera-ture, Howlett (1999) makes the same assumption that the state retains ulti-mate authority. While acknowledging that "truly voluntary instruments are totally devoid of state involvement" (8) he sees them as the product of "nega-tive" public policy decisions in which governments consciously decide to "rely" on these measures, a choice largely influenced by the nature of the subsystem and state capacity. Similarly, Gunningham, Grabosky, and Sinclair (1998) see policy instruments that are totally devoid of state involvement as a strategy available to government and tend to view them as "voluntary," more flexible programs associated with industry self-regulation as opposed to the more prescriptive and regulatory conception of non-state market-driven gov-ernance envisioned by many environmental groups and the FSC. As a result of this approach, the public policy literature also limits its discussion on whether and how *governments* maintain, achieve, or lose legitimacy. Indeed, the lack of attention by political scientists to the political aspects of consumerism issues has led Micheletti (1999) to develop a similar theoretical perspective to ours, noting that "the difference between [existing political science] theoretical frameworks and the one suggested here is that *the state is neither necessarily the moderator of nor the institution that initiates cooperative endeavors. Nor does it need to sanction actions for citizen well-being*" (3, italics added). But Micheletti and other literature on political consumerism also collapse an array of processes that are not united in their private governance attributes but instead are united around the issues of consumer labeling and auditing (Power 1992).

Recognition that existing literatures have not explicitly conceptualized the unique nature of non-state market-driven governance poses significant challenges to the construction of an appropriate research design. Given the lack of conceptual and empirical attention to non-state market-driven governance, we must first develop an analytical framework that will permit a better understanding of how non-state market-driven governance is emerging and facilitate our comparative research into understanding differing levels of support for non-state market-driven governance across countries/jurisdictions. Chapter 2 is devoted to addressing these challenges. For those readers more interested in understanding the descriptive account of forest certification politics in each of our cases and less interested in our explanatory analysis, most of chapter 2 can be skipped. We would, however, urge all readers to familiarize themselves with the explanatory hypotheses in chapter 2 since these hypotheses frame our accounts of certification politics in each region.

The remainder of this book proceeds as follows. Part II addresses the case of forest certification in BC and the US. Part III explores the European cases of Germany, the UK, and Sweden. Each chapter follows the same format: first a detailed description of the explanatory factors, followed by a historical account of the way these factors mediated efforts on the part of the FSC to gain support from forest companies and non-industrial forestland owners.

The concluding chapter to this book contains three sections. The first section summarizes the cases and reflects on the ability of our argument to explain cross-national and regional divergence and whether other explanations or hypotheses might contribute to a more complete understanding. The second section reflects on what our cases say about non-state market-driven governance in general, its durability as a new and emerging kind of environmental governance system, and the ways in which it might emerge in the future in forestry and other sectors. The third section addresses what our cases tell us about the impacts of forest certification on addressing sustainable forest management, focusing particularly on whether environmental standards in the forest sector will be "privatized up" and whether certification can be relied upon to address the serious issues influencing the deterioration of forests across the biosphere.

2

The Research Design: Toward an Analytical and Explanatory Framework

The presentation in chapter 1 of forest certification as an emerging form of global and domestic non-state market-driven governance immediately presents a comparative policy analyst with important analytical and methodological decisions. We must first develop an analytical framework, appropriate and applicable *across* cases, that permits us to uncover and classify the phenomenon we seek to study. This analytical framework must also facilitate broader reflections on the nature of the new phenomenon and facilitate the development of explanations for noted differences.

This chapter proceeds in three steps. First, we present an analytical framework designed to both classify and highlight differences in the emergence and support for non-state market-driven forest certification governance in our cases. Second, we develop explanatory hypotheses to explain divergence in levels of support. These hypotheses draw on our analytical framework in order to categorize and explain causal relationships. The hypotheses themselves are drawn both inductively from preliminary research on our cases and from existing relevant literature that indirectly pertains to our cases. Third, we outline and justify the methodological choices designed to illustrate and test our argument with reference to literature on research methods for political science and social science in general.

Analytical Framework

In order to develop a heuristic and theoretical framework for conducting our comparative research, we reach beyond the boundaries of political science to empirical work done on legitimacy within organizational sociology. This work provides a broader approach — one that looks at the way different organizations, particularly firms, may gain legitimacy (Jennings and Zandbergen 1995; Oliver 1991). Drawing on Cashore (2002), we focus on Suchman's (1995:574) seminal 1995 review essay, which defined legitimacy as "a generalized perception or assumption that the actions of an entity are desirable, proper, or appropriate within some socially constructed system of norms, values, beliefs and definitions."

Suchman's article is important to our study for two reasons. First, it addressed how organizations actively seek "legitimacy." This approach fit well with the activities we observed being undertaken by the Forest Stewardship Council (FSC) and its competitors, where frantic efforts to obtain support have led them to either change their own programs (conforming) or to alter by market campaigns and other sources of leverage the evaluations of external audiences (converting). Second, Suchman conceptualized very different types of legitimacy that are granted to organizations, which we also found applicable to the case of emerging certification programs where environmental groups tend to grant "moral" support to FSC-style certification, while the FSC's survival rests on gaining economic "pragmatic" support from forest companies and owners. Suchman also linked the issues of legitimacy and durability, a connection we explore in the conclusion of this book.

STRATEGIES FOR GAINING LEGITIMACY

Suchman noted that organizations seeking legitimacy are rarely passive but rather actively pursue strategies that conform to the external audience, convert the external audience, or inform unaware audience members of the organization's activities[1] (table 2.1). This finding fits our research on forest certification, where the battle for legitimacy has been an active one, fought on different fronts using a variety of strategies on the part of the FSC and competitor programs and their respective core supporters. Suchman identified three types of strategies that are important for our consideration.

Converting strategies refer to those cases in which an organization actively seeks to change the preferences of those groups or individuals from whom it is

Table 2.1 Types of strategies undertaken by certification programs and their supporters

Strategy	Approach	Effects on certification program processes and standards	Effects on external audiences
Converting	Create incentives for external audiences to change evaluations	None	Change evaluations to gain increased support for program
	Create new external audiences	None	New organizations support program
Conforming	Change certification program to address external audience's interests/values	Become stricter or laxer	None
Informing	Identify like-minded external audiences	None	None

seeking legitimacy.[2] Thus, the FSC and its core supporters, such as the World Wildlife Fund (WWF), have been actively involved in the creation of new interests through their assistance in developing "buyers groups" (Rametsteiner et al. 1998; Hansen 1998) now operating in Europe (Mirbach 1997; World Wildlife Fund United Kingdom 2001b), North America (Certified Forest Products Council 1999) and globally with the development of a global forest and trade network (Global Forest and Trade Network 2002). Converting can also occur through negative boycott or direct action campaigns as well as through the use of "moral suasion" techniques. Converting strategies are important because, if successful, they do not result in any changes to the forest certification program seeking legitimacy, thus allowing the program to stay closer to the core conception of certification around which it was initially created.

In addition, forest certification programs can attempt to achieve legitimacy by *conforming* to external audiences. For example, the cases to follow show that the FSC has changed rules governing harvesting in old growth forests, added a new rule on forest plantations, and developed small landowner initiatives all in an attempt to gain pragmatic legitimacy from these "supply side" interests. Likewise, FSC-competitor programs are constantly adding new rules and including new stakeholders in an effort to appeal to forest product processors and retailers who are currently demanding only FSC wood. Conforming strategies are seen as less desirable by certification programs and stand in contrast to converting strategies because they alter certification programs—often moving the program slightly away from its original conception. Indeed, a key question is to what extent FSC and FSC-competitor programs can conform to gain legitimacy from non-core interests without risking disapproval from their core audiences. Conforming strategies are often undertaken when converting strategies have failed to significantly alter levels of support.

Finally, certification programs may attempt to achieve legitimacy by neither conforming to external audiences nor converting them but rather by invoking active *informing* strategies that focus on those audiences likely to grant legitimacy if they were simply aware of the program. Informing strategies might include conducting advertising campaigns to the general public or targeting information to like-minded organized interests. Our cases to follow show that informing was most certainly used as a strategy by the certification programs; however, the programs used this strategy less frequently than the converting and conforming strategies. We found that informing strategies were usually subservient to conforming or converting strategies because efforts to gain legitimacy were focused on those actors along the supply chain who, for the most part, were quite aware of certification but were, at first, nonplussed.

Identifying whether a program is able to use conforming, converting, or informing strategies is important because it in large part determines whether certification programs remain close to their original conceptions or whether they must change their approach to forest sustainability in an effort to become accepted in the marketplace. And because conforming strategies are less preferred than converting ones (because the program has to change from its original conception), identifying conforming strategies helps us to establish when a certification program is having difficulty gaining support.

Different Types of Legitimacy Granting

Suchman noted that in addition to different strategies for achieving legitimacy, different types of legitimacy are granted: an interest-based pragmatic legitimacy, a value-oriented moral legitimacy, and a culturally focused cognitive legitimacy (table 2.2).

PRAGMATIC LEGITIMACY

Suchman argues that *pragmatic legitimacy* rests on the "self-interested calculations of an organization's most immediate audiences" (1995:578). Under this process of legitimation, "Audiences are likely to become constituencies, scrutinizing organization behavior to determine the practical conse-

Table 2.2 Types of legitimacy external audiences may give to certification programs (non-state market-driven governance)

Type	Source
Pragmatic	Narrow self-interest
Moral	Guiding values about the "right thing" to do
Cognitive	From a cognitive evaluation that something is "understandable" or that "to do otherwise is unthinkable"

Source: Cashore (2002)

quences, for them, of any given line of activity." For Suchman, the key here is that the granting of legitimacy rests on some type of an exchange between the grantor and the grantee, in which the organization being granted legitimacy affects the audience's "well being" giving the grantor a direct benefit (589).[3] We use the pragmatic legitimacy category to capture choices made by profit maximizing firms along the supply chain to support a forest certification system.[4]

Examples of pragmatic legitimacy abound in cases where non-state market-driven governance systems are emerging because firms choosing to comply want to know "what is in it for them." Indeed, the story told by the chapters that follow reveals that, thus far, when firms grant the FSC support it is at the level of pragmatic legitimacy. Firms that have opted to operate under FSC governance have almost always done so by evaluating whether their participation can improve market access or reduce market decline. For example, the BC chapter reveals how the commitment of MacMillan Bloedel (now Weyerhaeuser) in BC to follow FSC certification came only after FSC's environmental group supporters launched intense market campaigns against forestry practices in BC (Zietsma and Vertinsky 1999–2001; Cashore, Vertinsky, and Raizada 2001). Likewise, Swedish industrial forest companies expressed interest in FSC-style certification only after their purchasers in the UK and Germany gave indications they would give preference to FSC products.

In the chapters that follow, we show that the granting of pragmatic legitimacy occurs all along the supply chain. For instance, Home Depot committed to purchase wood from FSC sources (Home Depot 1999) only after a two-year campaign on the part of the Rainforest Action Network (RAN) (Carlton 2000a; Rainforest Action Network 1999; Sasser 2001). This campaign set the stage for RAN to obtain commitments for purchasing FSC-certified wood from a number of homebuilders and retailers even before they were the targets of any direct action (Rainforest Action Network 2000; Forestry Source 2000; Hannigan 2000).

MORAL LEGITIMACY

In contrast to short-term incentives associated with pragmatic legitimacy, *moral legitimacy* reflects a "positive normative evaluation of the organization and its activities. It rests not on judgments about whether a given activity promotes the goals of the evaluator, but rather on judgments about whether the activity is 'the right thing to do'" (Suchman 1995:579). Suchman asserts that "at its core, moral legitimacy reflects a prosocial logic that differs fundamentally from narrow self interest" (579). It is therefore difficult for an organization to achieve moral legitimacy through "false statements" or lip service without either being denied such support in the end or "buying into their own initially strategic pronouncements" (579).

In the case of forest certification, the granting of moral legitimacy plays two different roles, depending on which non-state market-driven conception is being evaluated. With regard to the FSC's core audience of most major environmental groups in Canada, the US, Europe, and elsewhere, moral legitimacy is not only given, it appears to be a prerequisite for participation by these groups. Since these groups exist to promote environmental protection, they will only grant moral legitimacy to a non-state market-driven governance system that reflects their overall values. Second, moral legitimacy acts as a brake on how far non-state market-driven governance systems can conform, or not conform, to achieve pragmatic legitimacy from non-core economic audiences. It is thus important to understand how one type of legitimacy granted by one set of external actors and interests may influence the choices of other actors and interests to give, or not give, a different type of legitimacy.

Moral legitimacy finds its roots in values permeating civil society but is often expressed or articulated through values or actions that environmental groups deem morally acceptable. For instance, environmental organizations, such as the RAN, will use media campaigns to appeal to popular support in their efforts to force those in the supply chain, such as homebuilders and lumber retailers, to adopt environmentally friendly purchasing policies. As Suchman says, the judgments about "the right thing to do" are made by "immediate" audiences but "usually reflect beliefs about whether the activity effectively promotes societal welfare, as defined by the audience's socially constructed value system" (579). Likewise, small landowners may believe that FSC-style certification is destroying the livelihoods of family forestry operations — a concept that finds its roots in values about the contribution of small-scale rural life to social well-being.

COGNITIVE LEGITIMACY

Cognitive legitimacy is based on neither interests nor moral motivations but rather on "comprehensibility" or "taken for grantedness." In the former case, legitimacy is given because the actions of an organization are understandable; in the latter case, legitimacy is given because "for things to be otherwise is literally unthinkable" (583). While absent from the development of certification to date, exploring any future achievement of this form of legitimacy will be fundamental to understanding the potential durability of certification in forestry and other industrial sectors.

Implicit in Suchman's theory is a "durability" pendulum, wherein pragmatic legitimacy is the least durable owing to its emphasis on short-term material incentives, and moral legitimacy is "more resistant to self-interested converting." Cognitive legitimacy is asserted to be the most durable: "If alternatives become unthinkable, challenges become [virtually] impossible, and the legitimated entity becomes unassailable by construction" (583). While these concepts are useful, we detail in the conclusion that the link between durability and legitimation granting is more complex than that envisioned by Suchman.

Two clarifications are in order with respect to our application of Suchman. First, Suchman's category of "pragmatic legitimacy" has been critiqued for not being legitimacy at all but rather a narrow self-interest category that generally falls outside the traditional concept of legitimacy, which rests on morally and culturally engrained senses of "appropriateness."[5] There are two ways to accommodate this critique that do not take away from our explanatory and analytical framework. The first is to say that not every "self-interest" action would fit under Suchman's "pragmatic legitimacy" category. Suchman would insist that regardless of the motives, the evaluator must believe that the organization (in this case, FSC-style certification) is "appropriate" in some sense, even if it is an economic self-interested one. While it is possible that a forest company or owner could be forced to accept FSC certification "kicking and screaming" (i.e., did not consider the FSC as "appropriate" in any pragmatic, moral, or cognitive sense), this appears to be rare in our cases to follow. Indeed, we document in our cases that those who did not consider the FSC appropriate in any sense ended up strongly supporting the FSC-competitor programs.

The second clarification is that the approach we develop in this book is not dependent in resolving ongoing debates about whether or not "pragmatic" and "legitimacy" are mutually exclusive. Our main empirical puzzle is why forest companies and forest landowners either granted pragmatic legitimacy to the FSC or gave it no support at all. As the narratives to follow reveal, we did not witness any significant patterns of support that could be characterized as moral or cognitive from forest companies and forest landowners — though we did find such support when it came to environmental non-governmental interests attempting to influence supply chain support. Whether or not one agrees with our use of the pragmatic legitimacy label, few can disagree with our operationalization of the concept and the support we seek to explain. In the end, we chose to stay as close as we could to Suchman's terms to facilitate cumulative research and for the strong definitional reasons noted above. Indeed, it became very clear early on in our research that Suchman's "converting" and "conforming" categories captured well the way the FSC and its competitors attempted to gain and maintain support along the supply chain.

Explanatory Framework:
Explaining Differences Through Hypothesis Development

Guided by the above concepts of "conforming" and "converting" strategies, we took care to classify and operationalize our dependent variable (presence or absence of pragmatic legitimacy granted to the FSC by forest companies and forest landowners) and then developed hypotheses designed to account for the factors that shaped differences in (pragmatic) support from the FSC across the regions.

CLASSIFYING AND OPERATIONALIZING THE DEPENDENT VARIABLE

We were confronted with two tasks in articulating our dependent variable: the first was to identify clearly just what population was covered by our dependent variable, and the second was to develop a way to measure "pragmatic" legitimacy, since we could not actually get inside the heads of the company officials and non-industrial forest owners making the evaluations.

Classifying

The ensuing cases reveal complex and distinct landownership patterns that have emerged in the world of forestry. The role of state lands, for example, varies widely among cases. In some cases, most land is owned by the government (such as in BC), but private timber leases for the public lands have placed many management responsibilities, including choices over forest certification, in the hands of large industrial forest companies. In other cases, government-owned forestland has come under such pressures for conservation that its role in commercial production is minimal (such as the US national forestlands) or is driven primarily by non-market pressures because other environmental or conservation objectives have taken precedence (such as forestlands owned by US state governments). And yet, in other cases, government-owned forestland provides a commercially viable timber harvest (such as in Germany and the UK). The role of non-industrial private forestland also varies. In some cases, it dominates commercial production (such as Germany and the US); in others, industrial and non-industrial production are roughly equal (such as Sweden and the UK); and in BC non-industrial private forestland provides minimal fiber to that region's forest sector.

These differences posed significant challenges in operationalizing our dependent variable in a way that would facilitate comparisons and contribute to broader understandings. We ultimately decided to operationalize and focus our dependent variable on industrial forest companies (who managed and/or owned their own forestland) and non-industrial forest owners. This choice allowed us to include in our analysis both non-industrial private forest owners as well as governments who owned significant amounts of commercial forestland. We decided to concentrate on those landownerships that were used to produce a commercial harvest of forest products, since certification was ultimately designed to address these lands. The extent to which certification was applied to other forestlands that do not focus primarily on commercial harvesting is noted in our cases, but data on these lands do not figure in our explanatory analysis. Likewise, we note other kinds of cases, such as that of US national forests lands, where certification has been rejected as a policy instrument both by environmental groups, who fear certification might open up commercial harvesting on these lands, and by the US Forest Service, who remains uncertain about the applicability of certification to already highly regulated lands.

This operationalization is meant to capture key landownership patterns and at the same time permit us to reflect on differences between industrial forest companies and non-industrial forest owners and identify areas of future research to better understand these differences.[6]

Measuring Pragmatic Legitimacy

Given that we could not actually go inside the heads of those company officials and non-industrial forest owners who offered, or did not offer, support to the FSC, we had to determine what pragmatic support might actually look like in practice. With reference to our discussion above on pragmatic legitimacy, we decided that a number of different actions would qualify as giving pragmatic legitimacy. These actions include participation in FSC standards development, monetary or other in-kind donations to the FSC, public statements that a company was either considering the FSC or was pursuing the FSC, or, of course, actual certification under the FSC system. Since all of these could have indicated potential moral or cognitive support as well, we analyzed the specific details and context of statements and support to see if there was any indication of "moral" or "cognitive" support. For example, if a company accepted the FSC after a boycott campaign and cited this campaign as influencing its decision, we inferred from this that the acceptance of the FSC was based on pragmatic evaluations. If statements referred to the decision to accept the FSC as owing primarily to potential increases in market share, we again placed this support in the pragmatic category. In each of our cases we identify the support, or lack of support, of the leading industrial forest companies, as well as leading landowner associations, and where appropriate, the government as a forest landowner. We also refer to surveys of forest companies and landowners where data were available. Because we seek to generalize across cases, and since there were no generalizable patterns of moral legitimacy granted to the FSC, our analysis does not explain the limited outliers that *did* grant the FSC moral legitimacy. Nor do we present detailed analyses of every company's and every landowner's decision-making process. We consciously chose a broad definition of our dependent variable in order to present and analyze overall trends. We recommend to future researchers that more nuanced firm-level analysis also be undertaken to augment and enhance our approach.

EXPLAINING DIVERGENCE

This section specifies how three structural features — place in the global economy, structure of the forest sector, and the history of forestry on the public policy agenda — work to facilitate or hinder efforts to have forest companies and non-industrial forest owners support the FSC. These factors, through both their independent and intersecting effects, help us understand why the FSC has gained pragmatic support from forest companies and forest landowners in some countries or regions, but little or no pragmatic support from forest companies and landowners in other countries or regions. Because our research focused on understanding how supply chain dynamics create or do not create economic incentives for forest companies, we necessarily focused our empirical inquiry and hypothesis development to the issue of whether or not forest landowners and forest companies grant the FSC pragmatic legitimacy. In the conclusion, we reflect on whether identifiable broad patterns might exist of forest companies and forest landowners in also granting moral and cognitive legitimacy to the FSC, and if so, what the effects of such support might be for non-state market-driven governance.

Important implications stem for our hypotheses about how the FSC and its supporters will attempt to obtain support from forest companies. Where favorable conditions exist, the FSC and its supporters will generally be able to use *converting* strategies — that is, to use actions that allow the FSC to gain forest company and landowner support without having to modify the processes or standards that are central to its program. When converting strategies fail to work (i.e., when the factors we describe below are not present), then FSC supporters will be forced to take less desirable "conforming" strategies, in which the FSC program alters its procedures or rules (usually "downward" from its original conception). Thus, our hypotheses contribute to the broader literature on "converting" and "conforming" approaches as well as provide two important conceptual terms that help us frame the cases to follow. Because the non-state market-driven phenomenon is new and relatively poorly studied within social science, we develop our hypotheses inductively from our cases and deductively from existing literature on economic globalization, internationalization, and public policy.

Place in the Global Economy

Much of the research on globalization and policy convergence has pointed to the susceptibility of a region to global and transnational pressures when it relies on foreign markets for sale of its products (Berger and Dore 1996) or when producers face a high level of competition from producers in another country or region. Through an inductive analysis of our cases, we draw on these broad findings to argue that a high proportion of forest product *imports* from, or *exports* to, foreign markets appears to have been an important factor in creating an environment conducive to FSC supporters' efforts to achieve forest company and non-industrial forest owner support for the FSC.

Hypothesis 1: Forest companies and non-industrial forest owners in a country or region that sells a high proportion of its forest products to foreign markets are more likely to be convinced to support the FSC than those who sell primarily in a domestic-centered market.

Hypothesis 1 helps account for divergence among our cases because it addresses those heavily export-oriented regions in which environmental efforts by non-governmental organizations (NGOs) to force change have been directed toward non-domestic purchasers of a region's forest products. The perspective of the environmental NGOs, supported by existing research, is that it is easier to wage internationally focused boycott campaigns in countries that consume the products than in the countries where those products are manufactured (Barker and Soyez 1994; Bernstein and Cashore 2000). Certainly even before certification emerged on the policy agenda, the case of BC (i.e., a region highly reliant on export markets) is revealing of these dynamics.

Environmental groups, such as Greenpeace (Greenpeace UK 1998; Greenpeace International 1993), targeted environmentally conscious consumers in the UK and German markets. They also placed pressure on consuming companies and retailers in the UK and Germany to cancel their contracts with companies operating in BC. Such efforts had effective results in producing large-scale concern about market loss on the part of forest companies and governmental officials (Stanbury 2000; Stanbury and Vertinsky 1997).

But what explains the greater effectiveness of boycotts and market campaigns launched outside the domestic market? After all, it seems intuitive that market demands along the supply chain should work regardless of the country in which that pressure happens to reside. There are two reasons why this is not so.

The first reason lies in understanding the effects of "popular sovereignty," what Bernstein and Cashore define as the "idea that government authority ultimately derives from the people being governed."[7] Bernstein and Cashore argue that such a notion works to limit the efforts of international groups to "infiltrate" the domestic policy-making process because they are often deemed inappropriate by those organizations and individuals who reside *within* the domestic system being targeted. And while this phenomenon applies to governmental processes, we believe it has *even greater explanatory capacity* in our cases because of the importance of *evaluations* of actors and organizations along the market's supply chain in granting legitimacy to non-state market-driven systems.

While FSC certification bypasses governmental decision-making processes, it is still open to popular cries that it represents ruling from "outside" the political system, since it is international in scope and originates outside any one country's domestic processes. We would expect those organizations initially opposed to the FSC to argue that it breaches popular sovereignty, but these ideas would only have an effect on the evaluations of those retailers and companies along the supply chain who reside in the same political system. Cries that the FSC breached popular sovereignty simply would not matter to producing companies and retailers outside of the political system. When customers of forest products make demands that companies outside of their own political system support forest certification, they are relatively immune to cries within the producing country that FSC certification breaches that region's popular sovereignty *since, they, too, are outside* that political system. The logic is clear: FSC supporters have an easier time pursuing converting strategies in regions that export to another political system because charges of breaching popular sovereignty (i.e., of the region where production takes place) have little effect on choices made further down the supply chain in other markets. The reverse also appears to be true: when the supply chain is largely contained in one domestic market, charges that the FSC breaches popular sovereignty, or "ruling from outside," are more easily made and may tip the scales against FSC converting strategies.

The second reason that converting strategies work best when economic pressure originates outside the domestic market is because there tend to be a greater number of domestic policy tools with which to fend off these pressures (Esty 1994, 2000). For example, Frontiers of Freedom Institute attempted to reduce market pressure from environmental activists by asking the US Internal Revenue Service to revoke the non-profit status of the Rainforest Action Network — one of the leading direct action, market campaign groups in the US (Kuipers 2001; Martin 2001). The ability to use these domestic levers is hence important both for the evaluations of forest companies and landowners of the market pressure and for potentially limiting the power of domestic-centered market pressure.

At the same time we must note that openness to the global economy mediates environmental activism to have companies support the FSC, but it does not determine this pressure. The outside market must also be receptive to supporting "green" products. The cases to follow reveal that it is the combination of demand for green products, and the location of this demand outside of the country or region, that facilitates FSC efforts to gain forest company and non-industrial forest owner support.

Hypothesis 2: Forest companies and non-industrial forest owners selling wood to a domestic market in a country or region that imports a large proportion of all the forest products it consumes are more likely to be convinced to support the FSC than those in a country or region that imports a small proportion of all the forest products it consumes.

Hypothesis 2 reveals our effort to push beyond existing research on the effects of market boycotts to hypothesize that these issues are also important for countries that import significant quantities of forest products. Importing large amounts of forest products can influence the susceptibility of forest companies and landowners to FSC converting strategies in two distinct ways. First, forest companies and producers in a region that imports a large proportion of its forest products will be especially susceptible to competition from FSC-certified producers outside its borders if their own domestic market is demanding FSC-certified products. Fear of losing market share to foreign imports makes these domestic producers more susceptible to FSC converting strategies. Second, forest companies and landowners in a region that imports a large proportion of its forest products will be more susceptible to moral suasion to practice the same sustainability requirements that their foreign producers are being required to do. Otherwise they risk facing accusations of promoting a double standard.

It is for these two reasons that, as with the exporting regions, those regions or countries dependent on a high degree of foreign imports for domestic consumption are likewise less influenced by charges that the FSC contravenes "popular sovereignty." Forest companies and forest landowners likely to make such charges tend to be less important to the economic well-being of the region or country, and hence the manufacturers and retailers pursuing FSC are less likely to be moved by claims relating to "popular sovereignty."[8] Second, if the civil society of a country or region is demanding FSC-style certification elsewhere, it is hard to argue that the FSC is challenging popular sovereignty in the domestic market when the same citizens are promoting contravening popular sovereignty in other countries (i.e., it is hard to argue that what is good for the goose is not good for the gander). It is for these reasons we hypothesize that a heavily import-dependent country is more susceptible to FSC converting strategies, everything else being equal, than in regions that are not highly dependent on foreign imports.

Structure of the Forest Sector

The structure of a region's or country's forest sector is highly complex and context specific. Forestry scholars have expended considerable research effort to describe and explain the unique characteristics of the forest sector within each of the regions or countries. While it is beyond the scope of this book to thoroughly review these findings, we identify three important forest sector "structural" features that appear to significantly influence FSC efforts to gain support from forest companies and non-industrial forest owners.

Hypothesis 3: Large and concentrated industrial forest companies are more likely to be convinced to support the FSC than relatively small and less concentrated industrial forest companies.

Large, concentrated industrial forest companies are characterized by two features. First, they are "vertically integrated," a term which refers to the extent to which a forest company occupies different positions along a supply chain. Within forestry, the term has been applied to industrial companies that either own or manage their own forestland as well as to primary processing facilities (whether they manufacture solid wood, composite, or pulp and paper products). Second, they are "horizontally integrated," a term that refers to the extent to which firms dominate production at the same place of the supply chain (for our analysis, the landownership "place" is arguably the most important measure of horizontal integration since it directly affects choices regarding forest certification).

A forest sector that is marked by large, concentrated industrial forest companies (i.e., has large vertically and horizontally integrated forest companies) is more susceptible to converting strategies by FSC supporters for two key reasons. The first is that these enterprises are big and easily identifiable. These characteristics make them more desirable targets for environmental campaigns than smaller, less recognizable companies (Sasser 2002). Since they own forestland and produce forest products, the large, concentrated companies are logical targets for market campaigners wishing to bring attention to their cause. Hence, when international environmental groups went to the UK and Germany in the early 1990s to protest harvesting of old growth forests in BC, they were able to target key companies, such as MacMillan Bloedel, when making their demands. And this permitted German and UK customers of key products to cancel, or threaten to cancel, contracts with these large companies — a much easier feat than offering a blanket refusal to purchase any products from BC, which would have been required if BC had been characterized by hundreds or thousands of small companies.

The second reason is that their "bigness" makes it easier to adopt FSC-style certification owing to reduced transaction costs associated with economies of scale. In terms of FSC certification, these large, concentrated companies enjoy benefits from not having to convince primary processing companies along the supply chain to support "chain of custody," since they *are* their own primary processors (vertical integration), and from being able to ensure consistent implementation of FSC forest management standards (horizontal integration). Similarly, the more horizontally integrated, the less forest companies interested in tracking certified products to the marketplace need to expend efforts to convince non-industrial suppliers to become FSC certified.[9] This hypothesis is key for our study because if accurate, it follows that the more a region is characterized by large, concentrated forest companies, the greater likelihood the FSC will gain widespread support.

Hypothesis 4: Unfragmented non-industrial forest ownerships are more likely to be convinced to support the FSC than fragmented non-industrial forest ownerships.

There are three rationales behind this hypothesis, which draws from traditional economic theory and sociological research on the attitudes of small landowners. First, regions with a large number of small landholdings will show a low acceptance of FSC certification owing to the diseconomies of scale associated with certifying small forest tracts. Both costs of forest certification and access to capital have been identified as a key concern by many small non-industrial forestland owners who may only harvest their lands once every generation, or even less frequently. Second, just as large forest companies are easily targeted by FSC supporters pursuing converting strategies, the more non-industrial forest owners there are, and the smaller the size of their hold-ings, the less impact FSC campaigners will have in their targeting efforts. Hence, FSC converting strategies become more difficult and costly.

Third, existing research indicates that small, non-industrial, and private forest owners tend to be philosophically opposed to an environmental-group initiated program creating rules for their forestlands and also opposed to a program in which non-industrial private forest owners do not have a lead role in decision-making processes (Newsom et al. 2003). These factors mean that the more a region is characterized by fragmented small non-industrial private forest ownerships, the less susceptible its forest sector will be to FSC convert-ing strategies. If this hypothesis is accurate, it follows that the more a region is characterized by fragmented non-industrial forest ownership, the less likeli-hood there is that FSC will gain widespread support.

Hypothesis 5: Forest companies and non-industrial forest owners in a country or region with diffuse or non-existent associational systems are more likely to be convinced to support the FSC than those in a country or region with rela-tively well-coordinated, unified associational systems.

The roots of this hypothesis can be traced back to literature on associational systems (Schmitter and Streeck 1981; Coleman 1988) that asserts that the structure and integrative capacity of business organizations is a strong deter-minant behind the ability of businesses to influence policy making. This litera-ture forms the basis for existing work on policy networks that reviews how integrative associational systems have greater influence in sectoral-level policy networks than do poorly integrated systems (Coleman and Skogstad 1990). The policy networks research found that integrative associational systems have more resources, professional expertise, and strategic abilities to accom-plish the coordination of their memberships. We use this logic to argue that the better organized an associational system is, the better able it is to "fend off" pressures from the FSC by undertaking well-coordinated and strategic re-sponses (Oliver 1991). For example, inhibiting the production of FSC-certified products in a region and, therefore, blocking FSC supporters' market strat-egies by keeping supply of FSC certified wood low would seem to be an effective way to prevent the FSC from gaining headway.

Further, such an association is better poised to limit the ability of individual members to defect or break ranks, such as in the case of a company or landowner who wishes to take advantage of relatively high demand for FSC certified products. And we expect the ability of a region's forest sector to keep supply-side members from defecting will be heightened in a region where forest industry associations represent a large part of the industrial forest sector. Well-represented and unified industries appear not only to be less fertile ground for FSC market campaigns but also able to create a cultural environment in which forest companies are not receptive to certification market pressures.

The History of Forestry on the Public Policy Agenda

In the traditional public "policy cycle," a problem is brought to the government's attention, policy options are formulated, and a particular course of action is put into effect and then monitored (Howlett and Ramesh 1995). ~policy cycle~ When controversial forest management issues have reached a region's institutional policy agenda,[10] forest companies and non-industrial private forest owners may choose to rely exclusively on the government's solutions or may choose to take independent action to address the issue. We would expect to observe a large degree of company responsiveness in an internationalized sector such as forestry, where transnational environmental groups have the resources and expertise to publicize controversial issues quickly. We propose the following hypothesis related to the history of forestry on the public policy agenda:

Hypothesis 6: Forest companies and non-industrial forest owners in a country or region with sustained and extensive environmental group and public dissatisfaction with forestry practices are more likely to be convinced to support the FSC than those in a country or region with less dissatisfaction.

In each of the regions under review, the appropriateness of certain forest management practices were, to some extent, an issue on the political agenda. To the extent that the perceived problem is a threat to profits or "social license," we expect forest companies and non-industrial forest owners to be more likely to grant pragmatic legitimacy to the FSC. Unlike the traditional state-centered policy responses to ecological crises, the FSC offers a set of standards endorsed by both domestic and international environmental NGOs. This factor affects the receptiveness of forest companies and landowners to see the FSC as an attractive solution, since if they supported the FSC, they would be agreeing to a set of standards that were supported by the very groups critical of the public policy regulations. We expect this effect to be more pronounced in cases where the conflict is sustained over longer periods of time (because companies in these cases would more likely be interested in finding alternative solutions) and in cases where non-government environmental and social organizations have acted in a concerted effort. The effect is most likely less pronounced in cases where a single organization has acted alone.

Hypothesis 7: Forest companies and non-industrial forest owners in a country or region where access to state forestry agencies is shared with non-business interests are more likely to be convinced to support the FSC than those in a country or region where forest companies and non-industrial forest owners enjoy relatively close relations with state forestry agencies vis-à-vis non-business interests.

The logic of this hypothesis is that when the forest industry and/or non-industrial private forest owners enjoy close relations with governmental agencies (i.e., the subsystem is categorized as "clientelist" or "agency captured"), forest companies and landowners are less likely to support an FSC-style certification program because it represents a fundamentally different approach in which business cannot dominate forest policy development (Wilson 1990). Supporting the FSC would mean giving up a comfortable policy-making arena in exchange for one in which business, by institutional design, can no longer dominate or even lead the policy process (Hoberg 1993b; Wilson 1998a; Yaffee 1994). On the other hand, if the subsystem had already opened up to include an array of interests groups in which business is one of many, then, everything else being equal, business is more likely to support FSC-style certification.

Since certification is relatively new, there is little empirical evidence beyond our cases to support our argument. However, in the US the multi-stakeholder forest dispute resolution processes that were convened in the late 1980s and early 1990s to advise Washington State over the development of private forest regulations, illustrates the logic of this argument. Washington State governmental official regions experimented with multi-stakeholder bodies (Washington State 1987; Halbert and Lee 1990) as a means to resolve longstanding conflict over use of forest resources on private forestland. However, the dominance of business interests and their supporters on the state forest practices board meant that business was reluctant to veer too far from existing approaches, and environmental groups eventually withdrew from these processes (Washington Environmental Council 1999). If business had not dominated forest practices networks, they might have been more likely to attempt a consensus with the environmental groups, since a failure at the multi-stakeholder level would have led to uncertain policy responses elsewhere. But the close relationship that business has with governmental agencies in traditional policy networks meant that it could feel confident that these traditional networks would develop a business-friendly policy response if the multi-stakeholder experiment failed.

Indeed, the story of public forest policy networks in BC illustrates how business will embrace multi-stakeholder processes when they no longer dominate traditional public policy networks. In BC, a government with a strong environmental agenda was elected in 1991 and established multi-stakeholder bodies designed to develop forestland management plans requiring a doubling of the protected areas in the province from 6 to 12 percent. And here, the forest industry — recognizing that the government was intent on opening up the closed policy networks and would act on its own, potentially in an unfavorable direction, if there was no solution — worked hard to gain consensus with environmental groups and other stakeholders (Wilson 1998a). There thus appears to be indirect support for our hypothesis that industry calculations to support a multi-stakeholder FSC in part depends on whether they enjoy more access in the public policy arena.

INTERSECTING EFFECTS

For clarity, we have limited our hypotheses to address the independent effects of the specific factors identified above in influencing the ability of the FSC and its supporters to gain support from forest companies and non-industrial forestland owners. We illustrate in the cases to follow that when these facilitating conditions exist, the FSC is able to rely largely on market-based converting strategies (where the FSC does not have to change its own program to gain support) rather than on conforming to forest companies and non-industrial forestland owners demands by changing aspects of the FSC (table 2.3). The cases to follow show that once active efforts are underway to gain forest company and non-industrial forest owner support, the relationships identified by our hypotheses have both independent and *intersecting* effects. In some cases, the independent effects of one variable are *reversed* when intersected with another. A good example concerns our hypothesized effects of cohesive associational structures, which we assert make it more difficult for FSC supporters to gain support because forest companies are better able to resist FSC pressures through concerted action. At the same time, however, if other factors, such as a region's place in the global economy, eventually lead forest companies to pursue the FSC, then a strong associational system actually helps build support because defection is rare. Hence, strong associations make it difficult to gain support, everything else being equal, but once support is granted, strong associations make the support more widespread than it would have been otherwise.

Owing to what would have been an inordinate number of hypotheses had

Table 2.3 Factors that facilitate FSC efforts to obtain forest company and non-industrial forest land owner support

---

Place in global economy

  Dependence on foreign markets

Structure of forest sector

  Large, concentrated industrial forest companies

  Unfragmented non-industrial forest land ownership

  Diffuse or non-existent associational systems

History of forestry on public policy agenda

  Sustained and extensive public dissatisfaction with forestry practices

  Forest companies and non-industrial forest land owners share access to state
    forestry agencies with other societal interests

---

we explored every possible permutation of intersecting effects, we instead have chosen to carefully investigate intersecting effects in our cases where they seem to reverse the direction of our hypothesized direct effects. We turn to the role and influence of intersecting effects in the conclusion.

Methodological Choices

What is the most appropriate research design to follow, given the complex, just emerging, and multifaceted nature of non-state market-driven governance and the divergent patterns of support we seek to explain? The discipline of political science yields no easy answers in this regard where debates about the relative merits of quantitative versus historically qualitative approaches[11] have created deep philosophical chasms, as supporters of each approach attempt to define the debate to favor their preferred method. These debates are illustrated in King, Keohane, and Verba's seminal work (1994) on research design, in which they argue that the scientific logic of inference is exactly the same whether conducting qualitative or quantitative research. This argument has been met with praise by many scholars (see, for example, Van Evera 1997) but has also been directly (McKeown 1999) and indirectly (Bernstein et al. 2000) criticized by others as severely constraining the quest for broad understandings about social phenomena in its assertion that everything scientific has a "statistical" logic.[12] We make no effort to resolve this debate. Rather, we seek to elaborate our methodological choice and, in so doing, acquaint those unfamiliar with qualitative methods and small-*n* case studies with the "value added" of such an approach in addressing complex social phenomena.

We chose early on in this study to invoke what is known as a "comparative qualitative" case study approach to exploring forest certification as an emerging non-state market-driven governance system. This approach is common in the fields of comparative politics (Dogan and Pelassy 1990; Lijphart 1975) and comparative public policy (Heidenheimer, Heclo, and Adams 1990)[13] and is often invoked when there is "causal complexity," contextual influence (Ragin 1987), and a limited number of cases (King, Keohane, and Verba 1994). As with most comparative case study approaches, the distinction between quantitative and qualitative is somewhat artificial. Our research methods include in-person interviews,[14] which we used to develop an accurate description of what transpired in our cases, supplemented by primary and secondary document analysis (e.g., media and press releases and policy statements from organizations) and by quantitative data elaborating landownership patterns and the role of a domestic forest sector in the global economy. Where relevant and feasible, we also draw on our own related but independent projects, as well as on surveys conducted by others, that explore the values and opinions of forest company and landowner officials.

CHOOSING OUR CASES

Perhaps the most important decision in our approach is the choice of cases, where, as King, Keohane, and Verba (1994) and Geddes (1990) assert, many scholars run into difficulty by choosing only one value of the "dependent variable" or phenomenon to be explained (in our case this would be the level of forest company and non-industrial forestland owner support for the FSC). Geddes details three pitfalls to avoid. First, she explains that if one only chooses cases where the event has occurred (support for FSC certification), it is impossible to explore what makes these cases different from those in which the event has not occurred (lack of support for FSC certification).[15] Along these lines, Geddes argues that a wide variation of the dependent variable (in our case support for FSC certification, lack of support, and mixed support) is needed in order to understand and rule out different causal hypotheses. Choosing only one measure of the dependent variable, whatever its value, will not permit broad conclusions about how the dependent variable is, actually, *dependent,* on identified independent (explanatory) variables. Third, Geddes argues that there are pitfalls in choosing end points in time series studies because the dependent variable could itself change, even when the independent variables remain constant. While this is always a pitfall when attempting to explain change or stability, our research project, as we elaborate below, partly addresses this because our model is designed to address the dynamic changes that took place over time in support for the FSC. Hence, our model focuses on the process through which change (or stability) occurs, rather than incorrectly assuming a static value of the dependent variable for a particular case.

With this direction in mind, our cases conform to a three-pronged approach. First, as explained in chapter 1, our cases focus only on developed countries, which allowed us to control for overall patterns of economic development (all of our cases are industrialized countries). Second, our cases are selected from countries or regions[16] where certification had been promoted to forest companies and non-industrial forest owners. While such an approach at first glance appears to go against advice to select cases taking on a range of values of the dependent variable, a closer look at our choices reveals that we do not ignore this guidance. This is because our central question is to understand choices made by forest companies and non-industrial forest owners over forest certification, and how responses changed over time. It would be an entirely different study to explore why certification was not promoted in some countries compared to other ones (though we would note that in virtually all European and North American countries that have a domestic forest sector, forest certification has been promoted, and domestic forest companies and non-industrial forest owners have made responses). Third, our cases represent a range of the dependent variable. As table 1.1 reveals, BC had obtained widespread pragmatic support, the UK eventually gave significant support, Sweden was split, Germany's level of support was weak, and support in the US was extremely low. By choosing a range of the dependent variable we were well positioned to explore the influence of our independent variables in shaping different patterns of support.

HOW MANY CASES TO CHOOSE?

The number of cases to choose in a comparative qualitative case study analysis is the subject of much debate. In general, the smaller the number of cases, the more in-depth "thick description" that can take place. But the drawback with this approach is that it is harder to select a range of values of the dependent variable. Thus one's ability to generalize beyond the specific case, or cases, is uncertain, although such an approach is well-designed for theory development. On the other hand, the larger the number of cases chosen, the greater the likelihood that contextually specific explanations could be missed or glossed over, as researchers attempt to uncover generalizations across cases (i.e., we may miss an important explanatory variable and thus potentially introduce positive or negative bias into our results). There is no perfect answer or solution, and over the long run, large, small, and medium *n* studies should all be pursued in an inductive or deductive effort to develop theory that can then be rigorously tested and revised.

We chose a "happy medium" approach for our study: exploring five key cases in depth. Each case could easily have been the topic of an entire book; as a result, the drawback to our approach is that we abstract events to a degree that might frustrate some historians or participants who have detailed knowledge of the certification developments in their particular region. But the benefit, as our empirical case studies reveal, is that we have been able to explore significant themes and interactions that our "independent" variables play across cases, allowing us to both offer and test wide-ranging applicable hypotheses that would not have been possible had we limited our cases to one or two. We also assert strongly that the subject matter we identify—with its complex and emerging nature—is best suited to our "happy medium" approach. Our conclusions can be further tested by those wishing to utilize large *n* analyses or can be examined by those who wish to further develop our theory by exploring a single case in greater depth.

CERTAINTY OF FINDINGS, LOGIC OF INFERENCE

Given the drawbacks and limitation of any approach, we are obliged to identify and elaborate the certainty of our findings and method. There are essentially two ways in which we "test" our hypotheses (outlined above). The first test is to see if the story of support for forest certification is *consistent* with our hypotheses.[17] Did what transpired fit within the parameters of our hypotheses, or were our hypotheses contradicted by the story?[18] The second test, or illustration of the validity of our hypotheses, is to "process trace" (King, Keohane, and Verba 1994; Van Evera 1997; Ragin 1987), a procedure in which our historical account traces the degree to which our independent variables affect other variables which in turn affect our dependent variable. Our case is somewhat special in that we trace the efforts by business and environmental NGOs to gain support for their respective certification programs, and we hypothesize about the way in which each region or country's structural features mediate these struggles and produce divergent levels of support across countries. Hence, our process tracing involves exploring over time the way in which these structural features influenced active agent-based efforts to obtain support. [19]

How confident can we be that this approach will yield accurate results? The answer to this question partly depends on where one sits in the quantitative or qualitative divide, but we offer some general observations. First, given the complexity of our cases and the seven hypotheses that we argue influence FSC efforts to achieve support, our approach will not yield the kind of confidence levels that would have come with a study examining only one of our explanatory variables and controlling for the rest. In this sense our study both raises hypotheses (which are important for future research) and also rigorously assesses them through our qualitative comparative approach, in which we constantly shifted between inductive and deductive reasoning in order to build an appropriate model.[20] We expect future quantitative and qualitative work to focus more specifically on one or more of the hypotheses and to apply our argument to other cases for refinement and testing. But these qualifications are not meant to downplay the significance of our findings, nor are they meant to imply that we only focus on descriptive inference, rather than causal inference (King, Keohane, and Verba 1994). In the spirit of Bernstein and others (2000), our choice of research methodology and cases was chosen intentionally to focus more on *accuracy* than *precision* (McKeown 1999). This is an important point. A historical narrative approach is not used simply when large-*n* approaches are not possible (and hence when they are assumed to be second best), but because they are often the best when addressing complex historical processes. Such an approach may help in identifying factors that are important to the model but that the model did not specifically capture—a benefit that large *n*-analyses are less able to provide. As Buthe (2002) explains, "A preference for narratives then is due not to the unavailability of analytical techniques that lead to other forms of presenting our results, but to particular *strengths* of the narrative form. The most important of these strengths is that narratives, in addition to presenting information about correlations at every step of the causal process, can contextualize these steps in ways that make the entire process visible rather than leaving it fragmented into analytical stages. Moreover, narratives allow for the *incorporation of nuanced detail and sensitivity to unique events, which may be necessary to understand the particular manifestation of an element of the model but which are beyond the model*" (486, italics added).

For these reasons our approach is arguably the most appropriate research design for those wishing to yield broader understandings of the new and important non-state market-driven phenomenon and its implications for addressing forestry practices in diverse regions. In fact, so sensitive are we to "unique" events outside the model that, in our conclusion and in the interests of transparency and supporting future cumulative research, we raise new hypotheses that emerge from our cases that we had not originally incorporated.

The cases to follow reveal both the story we seek to explain and our "historical narrative" approach in doing so. Our argument is laid out in a transparent fashion so that future researchers and social scientists can choose to build upon, explore in further depth, or cast their line in new directions. What does arise from our comparative narrative is a powerful account of the way these new non-state market-driven governance systems emerge in different countries, their ultimate form and function, and their ability to address the complex problems facing the earth's biosphere.

PART II

North America

3

British Columbia, Canada

Arguably no greater interest has been shown in forest certification as a policy instrument for addressing sustainable forestry than in British Columbia, Canada, where longstanding environmental forestry conflicts have garnered the attention of activists far beyond the province's borders (Bernstein and Cashore 2000). The combination of domestic and international pressures on BC to preserve more of its forestland base and to develop environmentally sensitive harvesting practices resulted in a series of policy changes in the 1990s (van Kooten, Wilson, and Vertinsky 1999; Stanbury 2000; Wilson 1998). The New Democratic Party government doubled the amount of protected areas, introduced a Forest Practices Code that regulated an array of harvesting practices, and initiated a host of related processes from reviews of the annual allowable cut to timber pricing policy (Cashore et al. 2001).

While these efforts were aimed at striking a balance between environmental concerns and maintaining a healthy forest industry, they did little to stop international and domestic criticism. The rollback of the Forest Practices Code's biodiversity guidelines, legislated efforts to reduce impacts of these policies on the annual harvest level, criticism of the implementation of the Forest Practices Code on streamside riparian zones, and initiatives that relaxed forest management procedures all combined to increase the perception that BC forest management was in crisis. As a result, frustration levels were high, not just on the part of environmental groups and their supporters, but on the part of the ailing forest industry, which was suffering from the collapse of the Asian market in the late 1990s.[1] And after a decade of introducing an array of forestry initiatives, many governmental officials felt beleaguered.

As a result, BC was ripe for the introduction of Forest Stewardship Council–style forest certification. Not only were environmental groups looking for another route with which to address their concerns, but there was also a logic in supporting the FSC that many companies came to understand: accepting the FSC would make it difficult for environmental groups to criticize BC forestry internationally, since environmental groups themselves would have played a central role in fashioning the rules to which industry was now agreeing to adhere.

This chapter traces the development of support for the FSC in BC, along with that of the FSC-competitor program, which was created for the Canadian Pulp and Paper Association (CPPA) under the auspices of the Canadian Standards Association (CSA). The BC case illustrates that a region's place in the global economy, the history of forestry on the public policy agenda, and the structure of the forest sector intersected to create conditions conducive for FSC strategists to convert forest companies into supporting the FSC. The combination of BC's dependence on foreign markets and continued perception of forestry as a problem on the public policy agenda were large factors in creating this environment. The FSC supporters were able to use the same direct-action protest and campaigning tactics they used previously in the mid-1990s to obtain new commitments from UK and German publishers and paper producers to purchase FSC wood, which were accompanied later by similar commitments from US retailers and home builders.

These commitments were aimed directly at altering the preferences of BC forest companies, which were at this point reluctant to support the FSC. The same conditions that led the FSC to pursue converting strategies resulted in the CSA often pursuing conforming strategies, attempting to fashion the CSA approach in such a way as to conform to the demands of these foreign companies further down the supply chain, while attempting to maintain the discretionary spirit of the CSA approach. The increasing fragmentation of the BC forest sector's associational system further facilitated FSC efforts to convert forest companies, as the sector was unable to offer a systematic or unified approach that might have helped in promoting the CSA.

At the same time, we show below that some factors did push the FSC in the opposite direction — forcing the program to expand beyond converting strategies. The BC industry's reliance on old growth forests and the entrenched system of "forest tenures" meant that the FSC had to pursue conforming strategies as well — but these were of a more limited nature and were largely used in cases where companies could not, by themselves, alter these constraints. We also show that both the FSC and CSA used informing techniques as secondary efforts to shore up traditional support, but these techniques, for the most part, played a background role to the converting and conforming strategies. *Importantly, the chapter also reveals that even when conditions favor FSC efforts, support is never unconditional.* When the FSC regional standards process resulted in significantly stricter standards in spring 2002, industry became hesitant, asserting that the economic costs of these rules would outweigh the anticipated benefits. We explain below that this is an important finding: an imperative of non-state market driven systems is that they maintain economic incentives for profit-maximizing firms. No amount of factors facilitating FSC efforts can change this logic.

This chapter reviews this story in three analytical steps. First, it reviews the three factors that together largely shaped FSC efforts to gain support from BC forest companies. Second, it describes the early patterns of support for certification in BC that closely resembled those present in the other cases: forest companies were virtually unanimously opposed to the FSC and spent considerable time and effort supporting the creation of its rival program, the CSA. Third, it traces both the efforts of FSC supporters to target the preferences of those further down the supply chain whose procurement policies would influence the decisions of BC forest companies and the counteractions of the CSA and its supporters. This section reveals how the factors above provided different avenues and fundamentally shaped whether market-based converting, program-changing conforming, or informing strategies were pursued as well as the relative success of these strategies. The conclusion of this chapter reviews the way in which the case illustrates the specific hypotheses identified in chapter 2.

Factor 1: Place in the Global Economy

DEPENDENCE ON FOREIGN MARKETS

The BC forest sector has become increasingly dependent on foreign markets. From the mid-1970s to the mid-1990s, BC's share of the world's softwood lumber market rose from about 20 percent to 30 percent (BC Stats 1996; Council of Forest Industries of British Columbia 1999). During this same time, BC increased its share of the US market to over 30 percent, more than double its total of the mid-1970s. Reflecting these trends, BC exported over 85 percent of its provincial lumber production in 1997 (Council of Forest

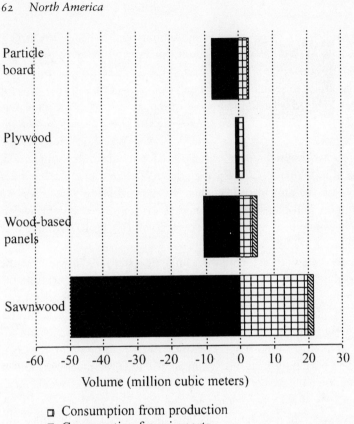

Figure 3.1. Volume (in cubic meters) of Canadian apparent consumption (production plus imports less exports) for various aggregate wood products in 1999. Trends in Canadian forest products trade are used as proxies for BC's trade patterns because comparative data at the subnational level are not readily available. BC consistently represents approximately 50 percent of Canada's total production. In 1999, Canada sent 86 percent of its total sawnwood exports (by volume) to the US market, 97 percent of its particleboard exports, and 66 percent of its plywood exports. Source: FAO 2001.

Industries of British Columbia 1999). BC's softwood lumber exports account for almost half of the Canadian total and over one quarter of the *world's* total (Council of Forest Industries 2000). Figures 3.1 and 3.2 present Canadian trade data.

During the 1990s, the BC supply side sent an increasing share of its exports to two markets: Japan and the US. Exports to Japan, particularly from coastal companies, were on a steady rise during the early 1990s and then fell dramatically in 1997 and 1998 during the Asian economic downturn.[2] However,

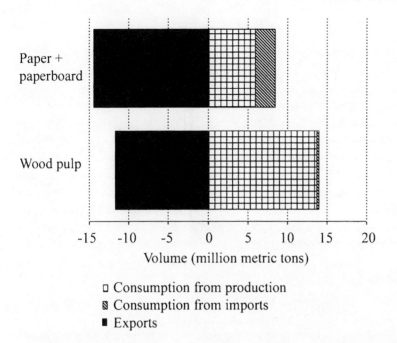

Figure 3.2. Volume (in metric tons) of Canadian apparent consumption (production plus imports less exports) for wood pulp and paper plus paperboard products in 1999. Trends in Canadian forest products trade are used as proxies for BC's trade patterns because comparative data at the sub-national level are not readily available. BC consistently represents approximately 50 percent of Canada's total production. In 1999, Canada sent 43 percent of its total wood pulp exports (by volume) to the US market and 82 percent of its paper plus paperboard exports. Exports to Germany, Italy, France, and the UK represented 22 percent of its total wood pulp exports and 5 percent of its total paper plus paperboard exports. Source: FAO 2001.

redirecting shipments to the US market was restricted at this time, owing to export restrictions imposed by the Canada–US Softwood Lumber Agreement. As a result, the European market and environmental pressures there were even more important than they otherwise would have been.

The logic of this export dependence is that it created an ideal environment in which the FSC could pursue more aggressive converting efforts directed at foreign purchasers of BC timber, thus rendering BC companies vulnerable to international pressures (Stanbury 2000; Bernstein and Cashore 2000). As we show below, customers who had previously been exposed to boycott campaigns were especially quick to meet environmental group demands, particularly since the object of their pressure — BC forest companies — was far removed from their own domestic political concerns.

Factor 2: *Structure of the Forest Sector*

Two structural features of the BC forest sector strongly influenced efforts by the FSC to gain forest company support: high levels of industrial forest company concentration and the relatively diffuse associational system (Coleman 1987, 1988). While in other cases patterns of non-industrial forest ownership exerted independent effects on FSC efforts to gain support, here virtually all forestlands — approximately 95 percent of the province's 95 million hectares — are owned by the provincial government (Council of Forest Industries 2000). Rights to harvest timber from these lands are allocated through tenure arrangements that vary from short-term cutting rights to more secure long-term licenses that better approximate fee simple ownership (Zhang and Pearse 1996). The tenure system is important for understanding support for forest certification because industrial control of the large majority of the provincial annual harvest creates significant degrees of horizontal forestland integration.

INDUSTRIAL FOREST COMPANY CONCENTRATION

Companies operating in BC are the most concentrated of those examined in this book. Not only are they highly vertically integrated (Marchak, Aycock, and Herbert 1999) — a single company will often control all phases of timber production, from forest management to manufacturing to marketing — but they also enjoy extensive horizontal integration at the landownership level.

The rights to harvest BC forestlands are distributed to companies through a system of tenures, which are predominantly volume-based forest licenses and area-based tree farm licenses (figure 3.3). The "tenure system" and accompanying sustained yield policies existing in BC limit changes companies can make to the amount of timber they harvest and where they can harvest it from (Haley and Luckert 1998). The tenure system is designed to create a long-run sustained yield of fiber in the province, and it is based on a "liquidation conversion" (Wilson 1990, 1998) project in which old growth forests are converted into faster growing second growth forests cut on rotation ages (varying by forest type and climatic zone) when average annual growth falls near to its maximum. Absent changes to the tenure system and annual allowable cut calculations, old growth forests will necessarily have to be part of most forest harvesting[3] (Cashore et al. 2001, chap. 4 and 6).

Responsibilities for management of provincial forests are shared between the Ministry of Forests and companies. The largest forest companies in BC manage hundreds of thousands of hectares (Marchak, Aycock, and Herbert

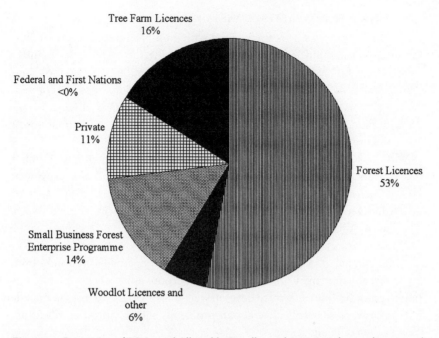

Figure 3.3. Proportion of BC Annual Allowable Cut allocated to various forms of tenure and land ownership. Source: British Columbia, Ministry of Forests (2002).

1999). This characteristic of the land base has meant that, in contrast to their counterparts in the US who often rely on hundreds of private non-industrial landowners for raw materials, BC companies are spared many of the logistical difficulties in tracking their fiber sources, as required under the FSC, and thus would be able to maintain a consistent supply of certified wood.

While policy analysts in BC have pointed out the potential negative effects the industry concentration has on public interests through, for instance, inefficient allocation of timber resources (Haley and Luckert 1998), this characteristic facilitates converting strategies used by the FSC and its supporters. A high level of concentration combined with the old growth logging controversy (described below) means that many BC companies are easily targeted in market campaigns, and those companies that do pursue certification are more visible to their competitors. High concentration also means that companies are relatively well prepared to implement FSC requirements if they make the choice to do so. Given these vertically integrated firms already had their own internal systems in place for tracking their forest products, companies faced relatively minor financial costs and logistical hurdles in modifying their existing systems to comply with FSC chain-of-custody tracking.

now
Forest Product Assoc.
of Canada

ASSOCIATIONAL SYSTEM COHESION

The BC supply side is fragmented with multiple associations represent-ing its provincial and federal interests. Nationally there are such organiza-tions as the CPPA and the Canadian Plywood Association. In 1994, the Cana-dian Sustainable Forestry Certification Coalition was developed to represent twenty-two supply-side organizations from across Canada (Elliott 1999). None of these associations has been the sole representative of Canadian forest companies on sustainable forestry issues.

Within the province, regional bodies, such as the Coast Forest Lumber Association and the Northern Forest Products Association, and provincial bodies, such as the Forest Alliance of BC (created in 1991) and the Council of Forest Industries (COFI), represent key forest companies on different issues. The Forest Alliance acts as the forest sector's public relations wing, while COFI is an association of associations (Wilson 1998), with membership in-cluding regional associations as well as individual companies. Because both the Forest Alliance and COFI contain diverse groups as members and differen-tiated responsibilities, efforts to develop well-coordinated responses to outside issues that represent collective action problems, such as forest certification, have proven difficult.

Factor 3: History of Forestry on the Public Policy Agenda

PUBLIC DISSATISFACTION WITH FORESTRY PRACTICES

In BC, widespread criticism of forest policies began in the 1980s, as domestic environmental groups and other critics of BC forest practices high-lighted examples of poor management. The battle by domestic environmental groups at this time was generally conducted on a valley-by-valley basis on the BC coast, especially on Vancouver Island (Wilson 1998). Especially conten-tious issues included BC's "sustained yield" policies and specific forest prac-tices such as clearcutting.

In the early 1990s, environmental groups from abroad, particularly Europe, joined domestic environmental groups in criticizing BC forest practices. This shift greatly increased the environmental movement's leverage; foreign en-vironmental groups such as Greenpeace in the UK worked alongside domestic groups, including the Western Canada Wilderness Committee and the Sierra Club. Together, these and other groups laid the groundwork for the future market campaigns that would lead to forest certification by publicizing BC's controversial practices to decision makers and the general public in Europe.

In 1991, European environmental groups began pressuring buyers of BC pulp to demand chlorine-free products, with the implicit threat that the environmental groups would call an end-consumer boycott of the buyers' products if they did not comply. These efforts were effective, and companies such as *Time Magazine* announced that they would stop buying paper from chlorine-bleached pulp (Stanbury 2000). By 1994, it became clear that the central environmental group tactic was boycott threats; now, however, the main issue shifted from the use of chlorine bleach to the clearcut logging of old growth forests (Stanbury 2000). Under pressure from Greenpeace UK, Kimberly-Clark and Scott Ltd. both committed to "reviewing their purchasing policies and [pledged] not to take wood pulp from clearcut temperate rainforests" (Greenpeace UK 1994). Sainsbury's, a British grocery chain, was also reported to be revising its policies on purchasing "pulp from virgin temperate rainforest" (Greenpeace UK 1994). The most important market pressures on BC forest managers at this time came from German publishers.[4]

This controversy revolving around BC forest practices threatened the entire industry with a serious loss of both profits and "social license," making the market access and green image benefits of FSC certification attractive. And, because the FSC was created and promoted by the very environmental groups that had criticized BC forest practices and orchestrated boycott campaigns, it was thought that participation in the FSC would be a de facto acknowledgment by those groups that a company was performing forestry to a high standard.

FOREST-COMPANY AND LANDOWNER RELATIONS WITH
STATE AGENCIES RELATIVE TO OTHER SOCIETAL INTERESTS

The array of sustainable forestry policy initiatives on the part of the BC government in the late 1980s and early 1990s, including the development of a wide-ranging but much maligned Forest Practices Code[5] did little to abate the perceived environmental crisis in BC forests. Environmental groups, motivated by the ecological significance of British Columbian forest ecosystems (Bryant, Nielsen, and Tangley 1997) produced an array of reports indicating that the initiatives were failing to protect endangered species and ecosystems (Greenpeace Canada 1994; BC Wild 1994; Sierra Legal Defence Fund 1998, 1999) and that the government was backtracking in its effort to focus instead on industry profitability (Lush 1998; Hamilton 1998a). These criticisms are taken seriously in a province where the forest sector is relied upon for significant governmental revenue and employment.[6] Moreover, wide-ranging consultative processes initiated in the early 1990s by Premier Harcourt were used in a more limited and narrow fashion by the mid-1990s, with critics arguing that closed industry–Ministry of Forests processes were returning as dominant methods of policy making.

The persistent inability of the provincial government[7] to pacify the environmental community meant that industry was open to new ideas and solutions. A key illustration of receptiveness to new approaches is illustrated by Mac-Millan Bloedel's (MB; purchased by Weyerhaeuser in 1999) conversion in addressing international boycott campaigns and outside environmental pressures. MB's initial response to such pressures in the 1990s was one of intransigence, arguing that it had long practiced sustainable forestry. However, controversial protests of its intention to log the intact temperate rainforests of Clayoquot Sound on BC's Vancouver Island tarnished its reputation, threatened its European and US markets, and led to strong doubts internally about the wisdom of its strategic choice (Cashore, Vertinsky, and Raizada 2001; Zietsma and Vertinsky 1999–2001; Cashore and Vertinsky 2000). Following the hiring of a new chief executive officer and creation of a new vice president for the environment, MB altered course and began efforts to find common ground with environmental campaigners. The result was MB's forest project, in which the company promised to phase out clearcutting on BC's central coast, which environmental groups had dubbed the "Great Bear Rainforest." This commitment was followed by a market-based agreement on the part of prominent environmental groups and forest companies, including MB, about forest management planning in the central coast (Hamilton 2000b; Coastal Rainforest Coalition 2000; Coady and Hackman 2000; Coady 2000). While other actors, including governmental officials, would become important in finalizing the deal, the importance of this experience for our analysis here is that companies and environmental officials learned that by bypassing the government they could address international environmental pressures much faster, and with greater satisfaction to both industry and environmental group interests. As a result, BC forest sector was receptive to new market-based initiatives that bypassed governmental processes, creating a relatively amicable environment for FSC efforts to gain forest company support.

The factors discussed above worked to create the most favorable conditions for FSC converting strategies of all the cases explored in this book (table 3.1). Even still, the BC story clearly shows that even in highly fertile ground, the FSC must find ways to ensure that some form of economic incentive exists. We elaborate on this point in the section that follows as we trace the early success of FSC market-based converting strategies that were followed by slow progress in translating company commitments into certified lands.

Table 3.1 Influence of explanatory variables on FSC efforts to gain support from BC forest companies

Factor	Existence
Place in global economy	✓
Dependence on foreign markets	
Structure of forest sector	
Large, concentrated industrial forest companies	✓
Unfragmented non-industrial forestland ownership[a]	✓
Diffuse or non-existent associational systems	✓
History of forestry on public policy agenda	
Sustained and extensive public dissatisfaction with forestry practices	✓
Forest companies and non-industrial forestland owners share access to state forestry agencies with other societal interests	✓

Note: A check mark indicates strong presence.

[a] The lack of non-industrial forestlands equates with unfragmented non-industrial forestland ownerships as market pressure only has to be directed at one audience, that is, forest companies.

Introduction of the FSC and Initial Efforts to Gain Forest Company Support (1993–96)

The FSC's conception of forest certification first entered the BC forest policy community as an idea in the mid-1990s, following the FSC's founding meeting in Toronto in 1993. A national FSC office and FSC-BC were both officially launched in 1996. Supporters of the FSC immediately seized upon an opportunity to expand its core audience by altering FSC institutions to court aboriginal support. The existing FSC approach to divide stakeholders among environment, economic, and social distinctions was altered by adding a fourth aboriginal arena for national board deliberations and for regional standards setting processes (Forest Stewardship Council 1999a). Aboriginal and forestry issues had become intertwined in BC in the 1990s (Hoberg and Morawski 1997; Cashore et al. 2001), so much so that aboriginal support became a key strategy in "tipping the scales" in the environment versus industry struggles. This preemptive effort to court aboriginal interests into certification paved the way for FSC converting strategies below.

The BC initiative was poorly funded initially, receiving little in-kind support from domestic and international environmental groups nor support from US philanthropic foundations that had been the lifeblood of the province's environmental movement. Most environmental groups at this time were focusing mainly on public forest policy in BC. The effectiveness of the provincial Forest Practices Code, which came into effect on June 15, 1995 (British Columbia 1995), became the focus of their interest (Sierra Legal Defence Fund 1996). However, high-profile environmental organizations, such as Greenpeace, recognized the value of the FSC, offering the program their support in order to apply further pressure on the BC government and forest companies to reform their forest management practices and policies (Greenpeace n.d.). But at this time FSC was raised more as a strategic idea, acting as a cover for continued boycott and protest campaigns. The idea that BC companies might actually be able to meet FSC's high standards was not deemed likely.

The initial response of the forest industry to the FSC came from the national CPPA (now called the Forest Products Association of Canada) — an association that had been on the front line of the European boycott battles of the 1990s, which were largely targeted toward BC forest practices. Under the CPPA's leadership, Canadian forest managers joined ranks to form the Canadian Sustainable Forestry Certification Coalition.[8] The coalition quickly agreed that its mandate was the creation of an international, third-party certification system, and approached the Canadian Standards Association, which operates under the rules and discipline of the National Standards System, about creating a Canadian forest certification standard. The CSA based its performance requirements on the Canadian Council of Forest Ministers' criteria for Sustainable Forest Management (Abusow 1997), linking its program to the intergovernmental Montreal Process's criteria and indicators (Moffat 1998; Elliott 1999). Like many FSC-competitor programs, the CSA approach began by conforming to ISO 14001 protocols, an internationally accepted program recognized as consistent with World Trade Organization international trade rules. The CPPA explained that it was involved in CSA forest certification for marketing reasons, as a benchmark for achieving good business practices, and for achieving transparency and public trust (Forest Products Association of Canada 2000).

The BC forest industry's initial strategy was to address pressures for forest certification using the CSA process (Paget and Morton 1999), hoping that the CSA program would meet the requirements of its buyers, avert boycott threats, and ensure international customers that "Canada is working towards sustainability in its forests" (Forest Alliance 1996). The CPPA committed three years of funding for CSA certification standards (Elliott 1999). The Canadian Council of Forest Ministers, the Canadian Forest Service, and Industry Canada also gave early support to the CSA certification process as an important way to secure market access (Elliott 1999). As a result most BC forest companies gave their early and quick support to the CSA, which they viewed as much less intrusive and more appropriate than the FSC program. And for the same reasons, most major environmental groups, along with other social organizations,[9] ended up boycotting the CSA process (Gale and Burda 1997), arguing that the CSA was an effort to reduce the stricter environmental regulations offered by the FSC (Mirbach 1997), with BC activists arguing that the CSA would permit continued clearcutting in the province (Curtis 1995). Even before CSA standards were complete, groups such as Greenpeace stated in 1995 that CSA standards "would allow products derived by large-scale clearcutting and chemical pesticide use to be called ecologically responsible," noting also that "all major environmental groups have come forward to condemn it" (cited in Greenpeace Canada, Greenpeace International, and Greenpeace San Francisco 1997:25).

By 1996 the CSA program standards were completed. While more flexible and discretionary than FSC on environmental performance requirements, the CSA was viewed by many as rigorous on rules for community consultation and a multi-stakeholder standards development process, with some industry officials believing that in this area the CSA rules were potentially more onerous than the FSC's requirements.[10]

Many BC companies latched onto the CSA initiative and explicitly criticized the FSC. Many felt that the FSC was something that had to be fended off.[11] For instance, as the chair of the Forest Alliance of BC, Jack Munro stated: "Even the FSC's broad principles have been generated by people with no experience in developing international standards, have been developed without any input from those involved in sustainable forest management and have had no input from government. Companies in Canada are lining up to be certified with an independent and more objective model developed by the Canadian Standards Association. And those standards will be very high indeed. What's required, and what Greenpeace Canada cannot provide, is credibility and independence" (Forest Alliance of British Columbia 1997).

The Effects of FSC and CSA Efforts to Gain Support

The introduction of the FSC and CSA into BC forestry debates quickly spiraled into a high-stakes competition over which program would be seen as having the legitimate authority to create certification rules. The forest industry was, for the most part, content to support and portray the CSA as the credible national certification program, while it viewed the prescriptive, wide-ranging

Table 3.2 *Strategic efforts to either gain or maintain support between 1994 and 2002*

| Year | FSC competitor program supporters | | FSC supporters | |
	Action	Strategy	Action	Strategy
1994	CSFCC approaches CSA about creating a national SFM certification program	Conforming		
1995			Environmental groups back out of CSA standards development, condemning the program	Converting
1996	CSA releases specification and guidance documents for its Canadian SFM certification program	Conforming	Founding of FSC National initiative and creation of a fourth "aboriginal" chamber	Informing
	Industry launches domestic public relations campaign to fight international market pressure	Informing	Marketing campaigns in Europe begin to make direct demands for FSC certification	Converting
	CPPA and government establish international public relations campaign to counter "misinformation" given to foreign customers by environmental groups	Converting/ Informing	WWF 95 group renamed the 95+ group; group membership continued to expand and demands for FSC became clearer	Converting
1997		Conforming	WWF "buyer groups" established in other European countries and in the US	Converting
			European customers begin to cancel contracts with BC forest companies or threaten to do so in the future	Converting

Year	Conforming	Event	Converting
1998	Conforming	Creation of the CFPC in the US	Converting
		RAN and Coastal Rainforest Coalition (now ForestEthics) begin campaign against Home Depot	Converting
1999	Conforming	Home Depot releases wood procurement policy that states preference for FSC certified wood (August)	Converting
		Wickes (no. 2 DIY in the US) announces procurement policy preferring FSC	Converting
		FSC changes Principle Nine, emphasizes maintaining and enhancing high conservation value forests	Conforming
2000		Further US companies (e.g., Centex Corp., Kaufman and Board Home Corp., Lowes, and Anderson Windows) announce procurement policies preferring the FSC	Converting
2001	Conforming	CSA participates in the PEFC, while not yet seeking endorsement for its program from the PEFC Council	
2002	Conforming	CSA releases "Forest Products Marking Program"	
	Conforming	CSA releases draft revised standards with increased emphasis on consultation procedures and enhancements to its standards (Feb)	

rules of the FSC as threatening. This program's supporters recognized that if other programs gained support in the marketplace, the very vision of the FSC would be compromised, and their reasons for creating the FSC in the first place, all but lost.

Supporters of the FSC and CSA undertook an array of strategic actions as a result (table 3.2). Amid this seeming chaos, significant patterns of successful efforts resulted, with the three factors described above facilitating, for the most part, FSC converting strategies. The CSA was on a more defensive and conforming mode not only in its efforts to gain acceptance further down the supply chain (among wholesalers and retailers), but also in its efforts to remain the *sole* choice of certification program among BC companies.[12]

The FSC supporters began their efforts with aggressive market-based converting strategies aimed at generating FSC demand further down the supply chain — demand that was most easily created outside the Canadian political arena. Drawing on successful boycott campaigns, groups were now returning to the same companies to offer them a carrot (public recognition that they were supporting sustainable forest management) alongside their usual stick (that they would also be subject to a boycott if they did not comply).

By pinpointing BC's heavy reliance on export markets, these environmental groups directed their energies toward convincing international buyers to avoid BC products (Stanbury 2000). By changing the demands made in a normal customer-supplier exchange, the FSC and its supporters hoped BC forest companies would pursue FSC certification. Most of the efforts were focused on German and UK purchasers of forest products, from the British Broadcasting Corporation's magazine publishing division, the British home retailer B&Q, and key German companies such as publisher Springer-Verlag and paper producer Haindl.[13]

In addition to the targeting of individual companies by groups including Greenpeace, Coastal Rainforest Coalition (now ForestEthics) and Friends of the Earth, the World Wildlife Fund (WWF), a core supporter of the FSC, undertook a comprehensive converting strategy that would significantly affect how BC companies viewed the FSC: the creation of buyer's groups whereby new "environmentally and socially aware" organizations were created, and where member companies would be recognized by the WWF as supporting environmentally sensitive harvesting practices. The first example was the creation of the WWF 95 Group, later changed to the 95+ Group, established in anticipation of the FSC in 1991. Originally it brought together fifteen UK-based retail companies willing to commit to purchasing wood from "well managed" sources by the end of 1995 (World Wildlife Fund United Kingdom 2001b; Hansen and Juslin 1999).[14] By 1997, FSC buyers groups existed in

Germany, Belgium, Austria, Switzerland, and the Netherlands (Hansen and Juslin 1999).

The efforts of campaigning environmental groups such as Friends of the Earth, Greenpeace, and Rainforest Action Network played a role by threatening to boycott companies who did not enlist with the WWF (Paget and Morton 1999) or make similar independent purchasing policies (Hansen and Juslin 1999). These environmental groups used media campaigns showing large clearcuts in BC as an effort to focus international attention on management practices, and while these demands were at first not directly linked to FSC certification, they provided companies greater incentives than seen elsewhere to consider seriously participation in the program as a way to show they were doing the right thing (Baldrey 1994).[15]

The results of these efforts were mixed at first. The FSC was now seen as something that could not be ignored, yet there was a growing recognition that the CSA was viewed in Europe as an industry effort to avoid the teeth behind FSC certification. Companies, however, were still far from ready to commit to the FSC. Strong concerns about specific FSC principles buttressed this position. A key issue, for instance was the FSC's Principle Nine (Jordan 1997), whose original wording stated that "primary, natural and semi-natural forests . . . shall be maintained, conserved and/or restored" (Moffat 1998:44). This posed a significant barrier for BC forest companies since the vast majority of their long-term harvesting plans relied on continued harvesting of primary or old growth forest stands (Ministry of Forests British Columbia 2001).[16] Principle Nine, as worded at the time, was interpreted by industry and labor organizations to mean that the FSC would not permit certification of the vast majority of forestland in the province. Jack Munro, the Forest Alliance chairman, argued in 1997 that the result of this was that "[the FSC] is designing regulations that make it impossible for us to operate" (Stanbury 2000:293).

This initial commitment to the CSA and concerns that Principle Nine would destroy the BC industry resulted in the CPPA waging its own European market efforts to counter the FSC. Yet the campaign was of a very different nature compared to FSC campaigns. The CPPA began its efforts domestically by emphasizing, through informing strategies, the negative impacts of market campaigns, a critique that resonated with communities, the general public and labor unions (IWA Canada 1996). BC forest companies and the CPPA asked federal and provincial governments to become involved, and both levels of government were generally receptive. They offered funding and technical support to the international converting and informing strategies that were designed to correct the "misinformation" being distributed by environmental groups to customers of BC's forest products (Greenpeace Canada, Greenpeace

International, and Greenpeace San Francisco 1997, 20–25; British Columbia 1994). Reflecting their trade-oriented mission, Canadian embassy officials in Europe played a key role, setting up meetings where local buyers were invited to presentations made by BC forest-company and government officials joined by provincial social interest groups (e.g., First Nations, labor unions, and community representatives).[17] Action in the media included articles published in European and US newspapers (Stanbury 2000). [18]

In other instances, the CPPA, with its office in Brussels, took a lead role with media relations and information dissemination. Its staff was instrumental in coordinating converting and informing strategies directed at European customers of BC products (Barclay 1993). It had a history of involvement in personal networking, distributing printed information material, public communication, and responding to specific crisis events (Stanbury 2000).[19] Overall, these efforts were ineffective; the CPPA was on the defensive, attempting to make the case that the CSA program did conform to international environmental concerns. Specific arguments notwithstanding, most of the UK and German publishers were interested in supporting a program that had environmental group support, since it was this condition that gave them cover from being targeted themselves.

By the mid-1990s the mood was one of continued conflict. Environmental groups remained unsatisfied with the BC government's forest policy initiatives and were unwilling to offer support to the CSA program. They continued pressing international buyers of wood from BC's large vertically integrated firms in the hope that BC companies would modify their forest management practices. However, the FSC Principle Nine—which addresses the management of high conservation value forests—continued to pose problems to BC forest companies who might otherwise have been willing to consider the FSC. While FSC strategists now recognized that their converting efforts in Europe would have greater success in BC if the FSC permitted some degree of harvesting of old growth forests, many of the FSC core audience supporters, who had long fought battles to preserve these forests, were reluctant to allow changes to the rules that might see environmental groups actually supporting logging in regions of the province that they were still fighting to protect.[20] At this point a stalemate existed: BC forest companies were unwilling to accept the FSC and maintained sole support for the CSA, while the environmental community and foreign purchasers of BC products supported the FSC for BC forests.

With no side willing to back down, market pressure was ratcheted up another notch. Market pressure occurred at a vulnerable time for the BC industry, which was suffering the double effects of the Asian economic collapse and its restricted access to the US market given import duties set by the Canada-US

Softwood Lumber Agreement. Market efforts expanded as the Global Forest and Trade Network was created in September 1998, which was designed to coordinate the activities of the national buyers groups around the world (World Wildlife Fund for Nature 1999). The Certified Forest Products Council was launched officially in 1998, merging the former US buyers group with the Good Wood Alliance (World Wildlife Fund for Nature 1999). Its members had less specific policies than many of their European counterparts, yet the threat that this development posed was significant for the BC forest companies as together they sent approximately 73 percent of their softwood products to the US market (Council of Forest Industries 2000). Market campaigners led by the Rainforest Action Network, based in San Francisco, decided at this time to target much of their efforts on the US do-it-yourself-giant, Home Depot. They launched demonstrations against the company across the United States and Canada and purchased advertisements informing readers that Home Depot sold products from endangered forests.

These developments occurred alongside the parallel market campaign led by Greenpeace to force logging companies to stop harvesting in BC's central coast region, which, as noted above, was important because it illustrated how environmental groups and forest companies might work together, offering a way out of the continued polarization of the public policy debates. The central coast campaign focused on MacMillan Bloedel (now Weyerhaeuser), Western Forest Products, and Interfor, all of whom were suffering economically owing to their reliance on the collapsed Asian markets and their inability to move into the US market due to the Softwood Lumber Agreement quota system. This situation rendered FSC European market converting strategies even more effective on BC companies than they might otherwise have been (Taylor and van Leeuwen 2000).

The market-based campaigns were further facilitated by the structure of the BC forest industry, whose area-based tenure rights described above places forest harvesting rights in the hands of a few large vertically integrated firms. As a result, environmental groups could focus their efforts on a small number of large companies (Stanbury 2000). Two illustrations provide evidence of these dynamics. First, the *British Broadcasting Corporation Magazine*, another member of the WWF 95+ Group, placed specific pressure on Western Forest Products operating on the central coast. Having been informed by Greenpeace UK that some of its products bought from German suppliers (publishers) might be coming from the "Great Bear Rainforest," the BBC queried these German suppliers for verification, who subsequently decided they would suspend their contract with Western Forest Products.[21] Second, B&Q chairman Jim Hodkinson announced in a meeting with the World Bank in Washington on January 9, 1998, that by the end of 1999 his stores would only carry wood products certified by the FSC (DIY 1998; National Home Center News 1998), which was directly connected to the large companies on the BC coast, as a media report noted: "B&Q is phasing out hemlock stair-parts sourced from British Columbia, where there is a reluctance to go for FSC certification" (DIY 1998).

The persistence of these market pressures and the large vertical integration of the BC industry paved the way for BC companies to reevaluate their opposition to the FSC. As would be expected, companies under the most direct pressure were the first to reconsider their position. In mid-1997, Western Forest Products responded to customer demands by conducting an internal assessment of its ability to achieve FSC certification (Western Forest Products Limited 2000).[22] By June 1998 Western Forest Products became the first BC company to announce its application for FSC certification (Hayward 1998; Hogben 1998). Not more than a week later, MacMillan Bloedel also announced intentions to pursue FSC certification (Alden 1998; Tice 1998) with Chief Executive Officer Tom Stephens, explaining that the decision "was in response to market demand. Nothing else" (Hamilton 1998c).

Yet while market demand was the driving force for change, the centralized and vertically integrated structure of BC forest companies facilitated these choices. And, at the same time, there was no single industry associational structure well positioned to develop more proactive industry responses, thus limiting other options available to BC companies. Choices to support or not support the FSC were now viewed clearly as company-specific decisions, in contrast to the view in the US that strategies concerning certification were to be developed at the national associational level.

Indeed, the FSC's success in targeting specific companies weakened the already fragmented associational system, with companies like MacMillan Bloedel terminating their membership in the Forest Alliance (Hamilton 1998b) and an informal industry group on certification quickly and quietly dissolving. The announcements by Western Forest Products and MacMillan Bloedel to pursue FSC certification threw the entire sector into turmoil, with industry officials viewing it as a breaking of ranks of previous industry support for only the CSA.[23]

What is striking about these early commitments was that they occurred before any changes had been made to the FSC's Principle Nine, illustrating the independent effects of the market-based campaign, facilitated by many structural characteristics of BC's forest sector. However, companies operating in old growth forests felt that the regional standards, still to be developed, could be worded in such a way as to continue harvesting in these forests and still meet Principle Nine. As Western Forest Products' Chief Forester, Bill Dumont, was quoted as saying, "We do not expect in any way to have to make significant changes in our operations" (Hogben 1998). This statement stood in contrast to the previous position of the Forest Alliance of BC and illustrates the change in approach that was occurring among key forest companies in their bid to achieve FSC certification.

Standards-Setting Process and Forest Company Support

These initial firm-level decisions sparked a series of strategic decisions within the BC forest industry to participate in FSC processes in order to change the program from within, rather than fighting it from the outside. At the provincial level forest companies now joined the FSC standards-setting process, rather than boycotting it, making a decision that stands in stark contrast to most US forest company decisions to not participate in FSC regional standards-setting processes. Individuals, companies, and associations began to apply for membership (Hamilton 1999; Jordan 1999a), taking elected positions on the FSC-BC steering committee and nominating and having their members posted to the BC Standards Team (Forest Stewardship Council 1999). And in a striking move, the Forest Alliance of BC, soon to be joined by BC's Industrial Wood and Allied Workers Union, decided to apply for FSC membership, attending its meetings, and influencing policy debates. Importantly, this increased support was occurring as the Home Depot announced its pro-FSC purchasing policy in August 1999 (Carlton 2000a). While movement had already started in the BC case, this announcement certainly served to shore up support, with industry officials now recognizing that BC's largest market, not just Europe, was becoming an increasingly important factor. And while the US chapter reveals that US forest companies reacted to the Home Depot announcement by altering the Sustainable Forestry Initiative, BC companies took this announcement as another indication that their steps toward the FSC were going to prove productive.

The result of these moves was that the CSA was being marginalized as a player in certification debates in BC, since companies were focusing on changing the FSC, rather than attempting to make the CSA more palatable. (This stood in stark contrast to certification debates south of the border, where industry was frantically readjusting its program to conform to retailer certification requirements, while focusing on ensuring member companies remained unsupportive of the FSC.)

The BC industry strategy was effective in responding to the fact that companies felt the original draft standards poorly addressed their concerns. As one industry official noted, "In BC . . . [the first FSC standards development process] turned out to be a complete mess, so they wiped the slate clean and they're starting over again. The industry is making damned sure that they're [at the standards development process] this time, so they get something out of it, if they have to do it."[24]

The response of many provincial governmental officials toward the FSC mirrored industry changes. Governmental officials in the Ministry of Forests and trade agencies were at first highly skeptical, laying out conditions under which certification must work in the province (British Columbia 2000). Though industry was clearly the target, support from the government was key, since, as both the regulators and owners of 95 percent of the forestland base, their support would be important for removing any obstacles that might exist. Despite significant opening up of the BC forest policy-making process in the 1990s, the BC Ministry of Forests historically has had the closest ties of any agency to the forest industry (Wilson 1998b), and as a result industry changes in FSC certification may have facilitated the forest ministry's changes as well. The FSC and its supporters noted the changes, with one FSC supporter explaining that the BC government "has now embraced the FSC as one of the certification schemes, and even goes so far as to insinuate that they were integral in having it come to BC."[25] The Ministry of Forests has recently chosen to officially take a "cooperative" role toward certification (BC Ministry of Forests 2001). The new Liberal government has indicated it will work to address conflicts between provincial legislation and the FSC standard (Haddock 2000). An example is the efforts by Timfor Contractors to obtain FSC certification for its temporary five-year non-replaceable forest license in Knight and Call Inlets, located on the mainland coast opposite the northern tip of Vancouver Island. After the FSC auditor indicated that it needed a letter of commitment that the license would be managed in line with FSC standards after the forest license ran its course, the Ministry of Forests obliged (SmartWood Program 2000).[26]

Far from its hesitant position of a few years earlier, BC forest ministry officials also participated in the post-industry FSC-BC standards-setting process by offering its expertise to the Standards Team. It gained two non-voting ex-officio positions in which its role was to comment on redundancies and conflicts with existing public regulations. And in its own bid to become certified, the government had the Small Business Forest Enterprise programme (SBFEP) assessed to determine the changes that would be required to achieve certification on SBFEP lands, including FSC-style certification (PricewaterhouseCoopers 1999).

As a result of these dynamics, six of the ten largest forest companies in BC

Table 3.3 Actions taken by ten largest BC forest companies[a] to support the FSC and CSA certification programs

Company	Announcement of intention to pursue certification[b]		Other support for FSC[c]
	CSA/ISO	FSC	
Slocan Forest Products	✓		
Weyerhaeuser/MacMillan Bloedel	✓	✓	✓
Canfor	✓	✓	
West Fraser Timber Co.	✓		✓
Domain Industries (Western Forest Products)	✓	✓	✓
International Forest Products	✓	✓	✓
Skeena Cellulose	✓		
Riverside Forest Products	✓		
Weldwood	✓		✓

[a] Based on percentage of provincial AAC in 1998 (from Marchak, Aycock, and Herbert 1999) and adapted slightly to account for recent company mergers.
[b] Values in this table are from status report at Sustainable Forest Management Certification Coalition Web site (http://www.sfms.com/decade.htm) and company Web sites.
[c] "Other show of support" includes membership in the FSC, participation in FSC-BC standards or steering committees, and/or "major gifts or in-kind resources and services" to FSC (as listed on FSC-BC Web site http://www.fscbc.org).

had either made an announcement of their intention to pursue FSC certification or had made other proactive overtures toward the FSC (table 3.3). This support was clearly pragmatic in character, with all of these companies maintaining support for the CSA program at the same time.[27]

The struggle in BC over certification and its accompanying rules had clearly shifted, by 2000, from an "FSC versus CSA" competition, to an internal struggle within the FSC. However, two important caveats are in order to describe this period. First, companies in BC were clearly hedging their bets — they had not given up on the CSA approach and could easily turn to only support the CSA if the market pressure ended. Second, CPPA efforts to support the CSA in European markets had not in any way abated. Still, European buyers continued to view the CSA as unable to satisfy their own certification requirements, including the lack of an international profile.[28] The CSA has responded to the latter criticism by joining the Pan European Forest Certification (PEFC) program, although it has not yet gone further to seek endorsement from the program's council (PEFC International 2001b). It also addressed its credibility issue by launching a new "Forest Products Marking Program" which introduces a chain-of-custody system and a product label (Canadian Standards Association 2001).

While leaving options open with the CSA, BC forest companies and their allies were able to use their decision to support the FSC to target what had long been considered a key obstacle: the fear that, if not clarified or changed, Principle Nine on old growth forests would make successfully pursuing FSC certification difficult. As a result, forest companies were able to use their access to the FSC, along with their continued pursuit and support for the CSA, to pressure the FSC to make compromises on Principle Nine. The BC government echoed these concerns, arguing in a press release: "We urge European buyers to support certification processes which are compatible with the sustainable forest management practiced here, but we are opposed to approaches that inherently discriminate against jurisdictions like BC which retain and protect significant amounts of primary forests while continuing to harvest in them" (British Columbia 1998).

In part recognizing that the BC case could lead to significant gains for the FSC if the Principle Nine obstacle could be removed, the FSC made an important conforming move, altering Principle Nine to focus not on preserving old growth forests, but in maintaining or enhancing high conservation value forests.[29] Previous interpretations of the old wording that it forbade logging in old growth forests were now negated, arguably paving the way for FSC certification of at least some harvesting in BC old growth forests.[30]

Changes to Principle Nine and increasing industry roles in the FSC have created a tension among some environmental groups, illustrating the difficulty the FSC program sometimes has in maintaining moral support from its core audience while simultaneously achieving pragmatic support from forest companies. Partly in an effort to limit the number of conforming actions, the Good Wood Watch was created by Greenpeace, Sierra Club of BC, the Friends of Clayoquot Sound, West Coast Environmental Law, the David Suzuki Foundation, and the Rainforest Conservation Society to specifically "[ensure] that the FSC-BC Regional Standards develop into a credible standard that upholds ecological integrity and social responsibility" (Good Wood Watch 2001). Still, BC's place in the global economy, history of sustained and extensive dissatisfaction with public forest policy, and structure of its forest sector meant that conforming strategies were relatively minimal here, compared to other cases in the book, such as Germany and the United States.

By the end of 2001 the FSC in BC was in the rather enviable position (compared to most other cases in this book) of working to maintain forest company support rather than still striving to achieve it. Forest companies were working within the FSC to make it more hospitable to their profit-maximizing goals, while the environmental groups pressed to keep the standard as high as possible.

The 2002 Regional Standards Decision and an Industry U-Turn

Despite the rosy picture painted for FSC supporters in BC, the year 2002 would witness what some observers had predicted — increased acrimony between industry and environmental groups over the final draft of the regional standards and a signal from industry actors that their support of the FSC might be short lived. The conflict can be traced to the process that led to revisions of the second draft standards. Produced in Summer 2001, the final draft standard was crafted by an eight-person technical standards team and was then revised by the working group's steering committee after having been subject to widespread public comment. By a 7 to 1 margin with Bill Bourgeois, the sole industrial forestry representative, voicing his opposition, the committee voted to send the standards to FSC-Canada for approval. While different actors have different interpretations of what transpired, the overall story is not in dispute. At some point during versions two and three, when discussions over very specific but important forest practices regulations were taking place, Bill Bourgeois was becoming increasingly concerned that the emerging standard was going to place the FSC as a "boutique" standard that would be unacceptable to the major industrial forest companies in BC: "If it is the stated intent of FSC Canada to have a regional standard for British Columbia that will be applied in a limited number of unique circumstances, I would say that Draft 3 should be endorsed. On the other hand, if FSC Canada's intention is to have a standard that will be applied across a spectrum of sizes and types of forest operations, then Draft 3 should not be endorsed. In which case further work is required to develop a certification standard that measures the achievement of good forest management, and which has broad applicability in British Columbia" (Bourgeois 2002).

Of specific concern were emerging standards on riparian zone harvesting, stand level retention, the setting of "threshold indicators," and forest reserves, which would have placed BC's already comparatively high forestry standards[31] even higher vis-à-vis their North American competitors (Bourgeois 2002).[32] Mirroring industry responses to the Harcourt government's Forest Practices Act (Hoberg 2002), Bourgeois asserted that he could not support such standards without an impact assessment of the effects of these standards on industry economic health and its annual allowable cut. The announcement took other participants by surprise — they asserted this was too late in the day to perform such an assessment, while Bourgeois felt that an assessment was not possible until the final standards were known.[33] Assessments were conducted on the economic viability and costs of the standards (Spalding 2002) and impact on allowable cut (Bancroft and Zielke 2002), and both reports predicted significant cost increases to the BC forest industry and impacts on the annual allowable cut from 10 to 30 percent of existing allocations. Upon receiving the findings, Bourgeois wrote FSC-BC and FSC-Canada, asserting that: "It is the opinion of forest company managers that significant cost increases without any visible means of recovering them and severe limitations on management flexibility will be incurred if the Draft 3 standard is implemented. I encourage you to seriously take these comments into consideration in determining whether to recommend the present draft to FSC-International for approval" (Bourgeois 2002).

With the environmental participants frustrated at this turn of events, and with their belief that the standards were appropriate to certify BC forest products, the standards were passed on to FSC-Canada despite Bourgeois's objections.[34] And the FSC-Canada board likewise voted to send the standards to Oaxaca for approval (Forest Stewardship Council Canada 2002). However, this time one economic representative on the FSC-Canada board, Tembec, voted against sending the standards to Oaxaca, while the other economic member abstained. Strategic choices made by actors within the standards development process had led to an outcome in which large vertically integrated industrial companies — the very companies FSC strategists had worked so hard to woo — were now sidelined and indicating that their support was now far less certain than it had appeared just six months before.

At first glance, the story of 2002 would seem to contradict the argument presented in this chapter by implying that the BC case was not that hospitable to converting strategies after all. However, this interpretation would be incorrect; instead, what transpired in BC in 2002 reveals two key themes that pervade forest certification as non-state market-driven governance, regardless of how conducive conditions are for the use of converting strategies. First, forest company support along the supply chain is not unconditional — support from profit-maximizing forest companies for non-state market-driven governance necessarily requires some kind of evaluation that it is in the company's economic self-interest. Hence market campaigns matter and influence company evaluations but the demands placed on the companies must be such that the perceived benefits of supporting FSC-style forest certification outweigh perceived costs. This means that even in a region where all the factors facilitate converting strategies, FSC-style certification must still fit within some kind of economic cost/benefit analysis — and in this case the standards were perceived to be so high, and the impacts so costly, that it could not be supported as a province-wide industry standard. Second, BC's experience reveals in 2002 that the FSC itself is not a unified body and that strategic decisions are not often made at the same level within the organization, nor with the same goals in mind.

The FSC leadership in Oaxaca and officials from leading FSC-accredited certification bodies were clearly concerned that the stringent BC standards would not only hamper what they felt was one of the FSC's best success stories and that what happened in BC might send signals to companies far beyond the province's borders. But many environmental group participants on the BC standards committee had long focused on BC forestry and had, through years of frustration, come to see the FSC standards-setting process as a way of gaining the increased standards that they were unable to achieve at the public policy level. They strongly believed that if increased rules were not put in place, the old growth–dependent forest ecosystems would be destroyed and lost forever.[35]

During fall 2002 and winter 2003, the FSC secretariat wrestled with how it would respond to industry's protest, finding no easy way out of this difficult situation. If they moved to strike the standards, they risked losing their most solid supporters in BC and perhaps their legitimacy among environmentalists there. If they accepted the standards, they risked losing support from industry in one of the places in the North that had been most hospitable to FSC-style certification — and risked sending signals to other potential industry supporters to be very careful before offering support to the FSC.

This conundrum came at a time when the CSA had been given new life. The CPPA recreated itself as the Forest Products Association of Canada in 2001, hired a new director, moved to the national's capital, and immediately began to set a path of "approachment" with the WWF and the Global Forest and Trade Network. While efforts to have the CSA formally interact with the FSC have proven difficult, the future path in BC, and in Canada as a whole, does now seem to rest on the ability of strategic actors, both within the FSC and the CSA, to recognize what kind of strategies are most likely to be effective given the environment within which they operate and the broader constraints imposed by market-based governance.

FSC international officials recognized these constraints. They supported changes made at their General Assembly that would require broader support from national initiatives *before* standards were sent to Oaxaca for approval. And in January 2003 they proposed a compromise solution for the BC case in which standards would be approved, but subject to revisiting a number of the most controversial rules, and to involving forest companies directly in such revisions. Indeed, their report went out of its way to note that a number of the BC standards went "significantly beyond the requirements of the FSC P&C" (principles and criteria) (Forest Stewardship Council 2003:5). And in a direct rebuke to the BC regional standards setting process for moving ahead without industry support, the report asserted that such high standards would require a "higher than normal degree of agreement" (5).

Conclusions

The events of the BC case offer a rich array of insights into the way strategic efforts employed by the FSC and its supporters are mediated by a region's place in the global forest economy, the history of forestry on the public policy agenda, the structure of the forest sector, and finally, strategic choices made by FSC-competitor programs. The BC case supports hypothesis 1 (*Forest companies and non-industrial forest owners in a country or region that sells a high proportion of its forest products to foreign markets are more likely to be convinced to support the FSC than those who sell primarily in a domestic-centered market*), hypothesis 6 (*Forest companies and non-industrial forest owners in a country or region with sustained and extensive environmental group and public dissatisfaction with forestry practices are more likely to be convinced to support the FSC than those in a country or region with less dissatisfaction*), and hypothesis 5 (*Forest companies and non-industrial forest owners in a country or region with diffuse or non-existent associational systems are more likely to be convinced to support the FSC than those in a country or region with relatively well-coordinated, unified associational systems*) as well as partial support for hypothesis 7 (*Forest companies and non-industrial forest owners in a country or region where access to state forestry agencies is shared with non-business interests are more likely to be convinced to support the FSC than those in a country or region where forest companies and non-industrial forest owners enjoy relatively close relations with state forestry agencies vis-à-vis non-business interests*).

International market-based converting strategies carried out by the FSC and its supporters were clearly effective in altering forest company support. BC's reliance on foreign export markets created a unique situation in which foreign companies could relatively quickly respond to pressures to demand FSC products, while not having to worry about the political fallout they would have experienced had they been operating in Canada or BC.[36] Interestingly, the Asian collapse and the BC forest sector's economic downturn made companies more susceptible to FSC non-state market-driven governance demands. In contrast to forest companies' traditional *increase* in influence in the public policy-making process during economically depressed times (Cashore et al. 2001; Hoberg 2000), the BC case indicates that the reverse seemed to be true of their influence in shaping FSC-style private sector forest certification systems.

The poorly developed forest company associations in BC also appeared to play a key role in facilitating FSC converting strategies and frustrating CPPA attempts to respond strategically with changes to their own CSA program. In addition, the existence of large, concentrated forest companies in BC aided the success of FSC market converting strategies, since the FSC could target a small number of large companies. Such targeting alone did not facilitate success (as FSC experiences in the US reveal), but it did interact with other factors noted above to create key inroads for the FSC. Ironically, the very same large economies of scale and tenure security that economists have argued promoted industry efficiency and investments (Zhang and Pearse 1996) also facilitated BC companies' responses to FSC converting strategies, where after one company broke ranks, others quickly followed. Further, industry support for FSC was highest in a region highly dependent on foreign trade, which provokes interesting questions about the negative relationship between globalization and environmental practices posited by some groups who ardently oppose globalization.

The BC case also reveals how forest companies may participate in FSC standards processes as a new means for strategic gain. By becoming involved in the standards process and bringing traditionally close governmental allies to the table, forest industry officials were able to conform "from the inside" FSC rules that would have undoubtedly been more strict if not for their active participation. Some BC environmental groups have suggested that the forest industry became involved in the development of BC regional standards in order to "sabotage" the process and have threatened to withdraw support for the program should the program veer too widely from its original standards and goals. These dynamics reveal a tension within all our cases: the need for the FSC to maintain moral legitimacy from its core, largely environmental group supported audience. The BC case as of 2002 illustrates this tension—especially when the core audience believes that, for the environment's sake, the standards need to be *increased* compared to what other FSC-certified companies do elsewhere. This position clearly impacts FSC efforts to gain legitimacy—especially in those regions or countries that are otherwise the most hospitable to FSC-style certification. Just how far the FSC can go in conforming without losing its core support is a question we return to in other chapters and in the conclusion to this book.

4

The United States

Widespread interest in the use of forest certification as a key instrument to promote sustainable forest management arrived relatively late in the US, where, until the mid-1990s, those championing the Forest Stewardship Council (FSC) had most of their successes in niche markets whose supply came from a relatively limited number of forest management companies and non-industrial forestland owners. However, by the late 1990s, forest certification had caught the attention of most forest companies, environmental groups, professional foresters, and many non-industrial private forest landowners. The FSC, which was supported by almost all mainstream environmental groups in the US as well as a handful of US philanthropic foundations, found its ranks expanding. Following direct action campaigns by some of its supporters, the FSC gained endorsement from key retailers such as Home Depot as well as from homebuilders and lumber dealers. Yet, major industrial forest companies were, with few exceptions, opposed to FSC-style certification on their forest-lands. Through their American Forest and Paper Association (AF&PA), they vigorously promoted and defended their own program, the Sustainable Forestry Initiative (SFI), as a legitimate alternative to the FSC. Environmental groups and their supporters immediately saw the SFI certification as a threat to their efforts to increase the stringency of rules governing sustainable forestry practices, while private non-industrial forestland owners were, for the most part, opposed to both the FSC and SFI certification programs, preferring either no program at all or landowner-initiated programs, such as the American Tree Farm system or Green Tag, both of which contain more flexible and discretionary forest management standards that were deemed by many non-industrial private forest owners and forest companies to be more appropriate for smaller forest properties.

These divisions created the most polarized climate of all cases reviewed in this book, as both sides pursued diverse efforts to gain support in hopes of emerging as the dominant certification program. For the most part, however, the FSC and SFI efforts to gain support for their respective programs simply worked to entrench instead of alter the initial patterns of support. The FSC did pursue an array of program-changing conforming strategies, including program changes that attempted to address criticism that the program was too expensive and prescriptive and that it limited the influence of business interests. Yet these efforts failed to alter the initial opposition of industrial forest companies and small non-industrial forest owners to the FSC (although the FSC did gain broader support from some state-owned government lands and a minority of mid-sized companies and private forest landowners (Hayward and Vertinsky 1999).

Similarly, SFI efforts to gain acceptance from environmental groups and retailers met only partial success. In part as a result of conforming strategies aimed at making the program more independent, performance focused, and international in orientation — characteristics that pro-FSC groups claimed set the FSC apart — by the early part of 2002 the SFI had secured the participation of two leading conservation groups on its arms-length Sustainable Forestry Board. However, the majority of environmental groups interested in certification maintained steadfast support for the FSC. Likewise, SFI efforts to gain recognition as a legitimate certification program from those retailers and lumber dealers who had originally made commitments to FSC-style forest certification failed to significantly affect these companies' purchasing policies as of 2002.

This chapter illustrates how the place of the US forest sector in the global economy, the structure of its forest sector, and history of forestry on the public policy agenda created an environment where FSC and its supporters' efforts to gain greater forest company and non-industrial private forest owner support for the FSC were significantly restricted. It created a polarized atmosphere in the US, as it provided the right conditions for the FSC-competitor program, the SFI, to obtain and maintain widespread support among industrial forest landowners, while at the same time limiting SFI efforts further down the supply chain.

In the analysis to follow, an important feature of the US forest sector's place in the global economy is that the vast majority of US wood products are sold within the domestic market, with excess consumer demand being met primarily by Canadian wood imports. This meant that the market campaigns waged against European forest product buyers that clearly influenced choices made by companies in BC (chapter 3) did not have the same degree of influence over the certification policy decisions made by US companies. The most important customers for many forest companies and landowners were based in the US; therefore environmental-group pressure to promote the FSC in the market's supply chain in the US would have to be a domestic affair. And while pro-FSC market campaigns by US environmental groups met with some notable successes at the retail end of the supply chain, the size and dominance of the US forest sector in the global economy meant that such campaigns had more influence on the choices of the US industry's competitors in, for example, BC (who saw certification as a potential route for maintaining or improving market access) than on US firms themselves. Such features meant that the FSC and its supporters would be required to focus market-oriented converting campaigns on US domestic markets, rather than international ones.

Three structural features of the forest sector worked to mediate and limit FSC efforts to gain support from US forest companies and non-industrial forestland owners: (1) the existence of a well-coordinated national forest industry associational system, which provided the organizational resources for the FSC-competitor program to make proactive and effective strategic choices; (2) a production system characterized by forest companies that own a key share of their forestlands but are also reliant on hundreds of thousands of non-industrial private forestland owners and timber investment management organizations (TIMOs) for much of their fiber supply; and (3) a high degree of vertical integration within the industrial forest sector. While these conditions made the targeting of large companies by environmental groups easier, the country's millions of non-industrial private forest owners were less exposed. Further, these conditions also made FSC-style certification more costly and cumbersome, since fragmented ownerships meant that the tracking of FSC-certified wood along the supply chain (a key feature of FSC-style certification) would be more difficult to implement. Indeed, the existence of a well-coordinated industry association would greatly enhance the ability of the forest industry to promote its own program, which it felt addressed its certification issues more satisfactorily.

The history of forestry on the public policy agenda was also crucial. The longstanding debates that existed over the use of National Forest lands in the 1980s and 1990s and environmentalists' victories in significantly reducing the timber harvest from these lands meant that certification was largely focused on private and state-owned forestlands, where forestry regulations were either less strict than those on National Forest lands (the US Pacific Coast, Northeast), or nonexistent (US South). This meant that industrial and non-industrial private forest owners, who had long fought against increased regulation of private forestlands and championed private property rights, saw FSC certification as another means through which non-forest owners were trying to increase the regulatory burden on forest owners — much to their dissatisfaction. At the same time, new forestry rules proposed in 2000 by the federal US Environmental Protection Agency arguably gave companies an additional incentive to support SFI-style certification. The program acted as both an alternative to the FSC and a mechanism to fend off efforts by the federal government to increase regulations.

We argue that by the time FSC and SFI supporters had begun to engage in an all-out effort for legitimacy in the late 1990s, the three explanatory factors (place in global economy, forest sector structure, and history of forestry on the public policy agenda) intersected to create an environment where FSC faced an uphill struggle in its efforts to alter initial certification choices made by industrial forest companies and non-industrial private forest owners. Choices made by certification bodies in the US existing before the FSC's official launch created policy legacies that further entrenched the restrictions on FSC support-seeking activities. In four analytical steps, the remainder of this chapter illustrates how these conditions shaped strategic choices made by the FSC and FSC competitors (choices that themselves are important in creating policy legacies that impact subsequent strategic choices) and the development of the current polarized atmosphere. First, it details the position of the US forest sector in the global economy, the structure of its forest sector, and the history of forestry on its public policy agenda. Second, it reviews the emergence of certification as an issue in the US, beginning with its "pre-FSC roots" and the initial industry and landowner response. Third, it carefully traces the historical development of active support-seeking efforts undertaken by the SFI and FSC that increasingly preoccupied environmental and industry groups. The conclusion identifies how the US case pushes forward a theoretical understanding of the way non-state market-driven governance systems emerge in large, domestic-oriented economies, and it identifies where US legitimacy struggles might be headed.

Factor 1: Place in the Global Economy

DEPENDENCE ON FOREIGN MARKETS

The US is the world's largest producer of industrial roundwood[1] and primary and value-added forest products. US consumption of wood products

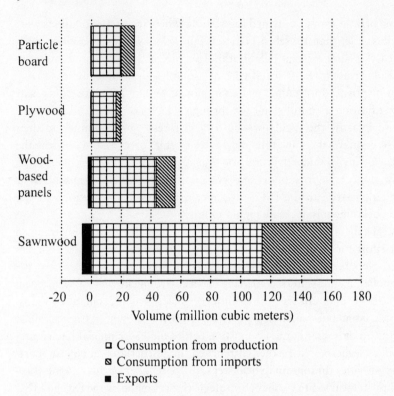

Figure 4.1. Volume (in cubic meters) of US apparent consumption (production plus imports less exports) for various aggregate wood products in 1999. In this year the US purchased nearly all its sawnwood and particleboard imports from Canada (see figure 3.1). Brazil, Indonesia, and Malaysia were the most significant sources of plywood: they combined to represent 66 percent of all US plywood imports (by volume). The US sent 23 percent of its total sawnwood exports, 33 percent of its plywood exports, and 55 percent of its particleboard exports to Canada. Exports to Spain, Italy, the UK, France, the Netherlands, and Germany represented 16 percent of its total sawnwood exports, 15 percent of its plywood exports, and 4 percent of its particleboard exports. Source: FAO 2001.

outpaces its own production,[2] resulting in the US being the world's largest importer of forest products, most of which come from Canada (figures 4.1 and 4.2).[3]

These features meant that the domestic sector was largely shielded from FSC international market pressures—be they of the "dependence on foreign markets" type characterized most strongly by the BC case, where BC companies felt pressure for FSC certification from distant European buyers, or the "moral suasion" type characterized by net importer countries such as the UK and Germany, where those retailers that had previously demanded FSC certification of their tropical suppliers were pressured to make the same demands

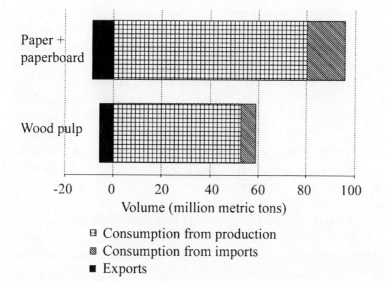

Figure 4.2. Volume (in metric tons) of US apparent consumption (production plus imports less exports) for wood pulp and paper plus paperboard products in 1999. In this year, the US purchased nearly all its wood pulp and paper plus paperboard products from Canada (see figure 3.2). Brazil was the next most significant source of wood pulp imports (14 percent). The US only sent 4 percent of its total wood pulp and 30 percent of its paper plus paperboard production to Canada. Exports to Spain, Italy, the UK, France, the Netherlands, and Germany represented 16 percent of its total sawnwood exports, 15 percent of its plywood exports, and 4 percent of its particleboard exports. Source: FAO 2001.

of domestic suppliers. Rather, US market campaigns would need to be focused solely within the domestic market, which, all else being equal made campaigns more susceptible to "backlash" by powerful domestic actors such as industry, landowners, and their respective associations.

Factor 2: Structure of the Forest Sector

Three structural features of the US forest sector are key to understanding certification legitimacy dynamics: concentration of industrial forest company interests; landownership patterns; and the existence of the strongest and most strategic associational system of any reviewed in this book.

INDUSTRIAL FOREST COMPANY CONCENTRATION

A handful of large vertically integrated forest companies dominate industrial forestry in the US. High levels of vertical integration generally mean that the FSC's chain-of-custody requirements are easier to meet than they would be if a different company performed every stage of production. The large size of many of these companies also means that the sector contains many publicly identifiable, well-known forest companies that are therefore more susceptible to pro-FSC and other direct action campaigns.

However, the fact that individual companies in the US produce substantial quantities of specific forest products[4] meant that individual companies held a degree of market power that was absent in other cases we studied, shielding them somewhat from FSC converting strategies. This power, of course, was amplified when they acted together under the auspices of the AF&PA, but it also exerted independent impacts on the legitimacy struggles. It meant that the relative market power held by large retailers was not as great, given they were not the most concentrated part in the supply chain.

Further, high levels of vertical integration did not allow these companies to skirt the issue of obtaining fiber from external sources. Industrial forest company production exceeds its own timber supply, so much so that most companies rely on non-industrial private forest lands, characterized by increasing fragmentation and small ownership size, for more than half their fiber needs.[5] The lack of horizontal integration at the crucial level of forest ownership meant that these companies were far more difficult to convert to the FSC than their counterparts in regions where companies owned enough forests to produce a significant portion of their total fiber requirements.

NON-INDUSTRIAL FOREST OWNERSHIP FRAGMENTATION

One-third of the US' 917 million hectares is forested, with ownership split among the following groups: federal government (Forest Service and Bureau of Land Management), state government, county and municipal governments, forest industry (forest companies and timber investment management organizations), and non-industrial private landowners (figure 4.3). Classified as commercially productive[6] are 204 million hectares of forestland, of which 58 percent is controlled by non-industrial private owners (118 million ha), 13 percent by industrial interests (27 million ha), 22 percent by the federal government (44 million ha; 88 percent of which is National Forests), and 7 percent by state and municipal government (15 million ha) (United States Forest Service 2000).

The US contains nearly 10 million non-industrial private forest owners, 85 percent of which own tracts 20 hectares in size or smaller (National Research Council 1998), which created a situation that was not conducive to FSC converting strategies. Not only did the logistics of such diffuse ownership present a challenge to companies interested in tracking the sources of US wood products right back to the stump, but the direct targeting of such a large number of

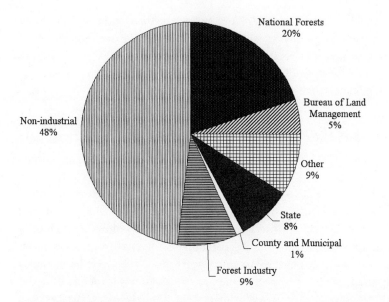

Figure 4.3. Ownership of forested lands in the United States. Source: United States Forest Service 2000.

individual landowners by FSC strategists was not feasible. And, even if targeted, the fact that non-industrial private forest landowners placed a high value on their independence and, thus, were likely not to respond favorably to a program that, they felt, did not give them a strong enough voice, created a much greater need for conforming strategies than observed elsewhere.

ASSOCIATIONAL SYSTEM COHESION

The creation of the AF&PA in 1991 through a merger of three national associations[7] resulted in a well-coordinated associational system that facilitated industry in undertaking swift choices over an array of forest policy issues. It was established in large part as a response to acrimonious debates over old growth preservation in the US Pacific Northwest in the 1980s. The original mission was to bolster the industry image among the US public and give the industry a greater political lobbying presence in Washington DC (Wallinger 1995; Meridian Institute 2001c). Its initial membership of 425 companies, associations, and organizations was reported to control 95 percent of US paper production and 65 percent of solid wood production and own approximately 90 percent of the industrial forestland base (Wallinger 1995). The AF&PA is important for our story because it provided an immediate organizational setting in which the industry could undertake proactive and strategic choices in response to the increasing pressure for forest certification.

In fact, the existence of the AF&PA permitted the forest industry to reorient its SFI program — originally envisioned as a corporate code of conduct — into a certification program (AF&PA 2002a). This meant the FSC was unable to enjoy an "early innovator" advantage that proved to be important in its success in regions such as the UK.

Factor 3: History of Forestry on the Public Policy Agenda

PUBLIC DISSATISFACTION WITH FORESTRY PRACTICES

The history of forestry on the public policy agenda is key to understanding how FSC efforts to gain support were received because it directly affected the arena in which debates over forest certification would occur. The key distinction for this analysis is the divergent approaches to forestland management that have emerged in the past thirty years in the US. On federal lands a relatively strict litigious regime has emerged, which largely (though not exclusively) affects National Forest land management (United States Department of Agriculture Committee of Scientists 1999). Conversely, on private lands a more discretionary and industry-friendly approach has developed through state-level regulations.

It has been during the past thirty years that these divergent approaches have developed (see, e.g., Cubbage, O'Laughlin, and Bullock 1993; Cashore 1999, 1997; Hoberg 1993a and b, 1997). Forest management practices on National Forest lands are directly addressed through such legislation as the National Forest Management Act and the Resources Planning Act. Other broader legislation, such as the National Environmental Policy Act, the Clean Water Act, and the Endangered Species Act, also affect forestland management. Although these address all landownership types, their impacts on National Forest lands have been the most prescriptive and wide ranging (Cashore 2001a, 1997). [8] This approach to forestland management has resulted in highly successful litigation efforts by environmental groups, especially concerning preservation of Northern Spotted Owl habitat in the Pacific Northwest and the subsequent adoption of ecosystem management by the US Forest Service (Kohm and Franklin 1997). Together these outcomes placed most federal forests off limits to commercial logging. As of 2002, timber harvested on National Forest lands amounted to only 5 percent of the national total. Not only did these changes mean that, de facto, certification of commercial logging would address mostly private forestlands (and state-owned forestlands on which logging occurs), but also that national environmental groups that promoted zero cut policies for National Forest lands in the mid-1990s, such as the Sierra Club, made a precondition of their support for the FSC that no fiber coming from these lands be certified.

FOREST COMPANY AND LANDOWNER RELATIONS WITH STATE AGENCIES RELATIVE TO OTHER SOCIETAL INTERESTS

Forest management regulations governing private forestland have been left to the state agencies to develop, where more flexible and discretionary forest practices rules have emerged (Ellefson, Cheng, and Moulton 1995). Though even at the state level, regulations differ between Pacific states such as Oregon, Washington, and California that have developed forest practices acts and most states in the US South that have no forest practices act at all (Ellefson, Cheng, and Moulton 1997, 1995). The southern states have preferred voluntary and cooperative approaches to addressing forest management (Cashore, Vertinsky, and Raizada 2001; Cashore and Vertinsky 2000). These divergent approaches to forest management are both the result, and cause, of different business-government relationships that have emerged in the US forest sector at the state and national levels.

The close relationship between state forestry agencies and the forest industry facilitated the initial creation of state-level forest practices acts in Oregon and Washington, which were designed to head off potential federal encroachment. In the US South, similar close relations exist; however, they instead led to voluntary policy instruments, as state level agencies generally work with industry and landowners in a cooperative and partnership-like fashion (Cashore and Vertinsky 2000). As a result of this state-industry relationship, environmental groups have had comparatively limited access to state-centered policy networks, though access does vary according to region (limited in the US South, highest in Pacific Northwest). Much of this limited access is reinforced by a strong ethic of private property rights.

The combination of significant success by environmental groups in influencing public policy governing forestland management on National Forest lands and the relatively limited success in influencing private forest management regulations explains why most environmental groups were interested in certification as a way to strengthen the more flexible and less stringent rules governing private forestlands (Natural Resources Defense Council 1997). Further, we show below it was recognition of this dynamic that explains why many private forest landowners and their associations came to fear FSC-style certification. As we document in the next section, this fear would ultimately increase forest company and non-industrial forest owner support in the US for FSC-competitor programs and polarize industry and environmental groups.

The factors discussed above have influenced the ability of the FSC to gain

Table 4.1 Influence of explanatory variables on FSC efforts to gain support from forest companies and forest owners in the US

Factor	Existence
Place in global economy	
Dependence on foreign markets	
Structure of forest sector	1/2
Large, concentrated industrial forest companies	
Unfragmented non-industrial forestland ownership	
Diffuse or non-existent associational systems	
History of forestry on public policy agenda	
Sustained and extensive public dissatisfaction with forestry practices[a]	
Forest companies and non-industrial forest land owners share access to state forestry agencies with other societal interests	

Note: A check mark indicates strong presence. 1/2 indicates mixed presence.
[a] For the US case, this variable refers to public policy approaches on state and private lands only.

legitimacy from the forest sector and, alone and in cooperation, create the most inhospitable environment for converting style strategies of the cases this book explores (table 4.1). As the story below illustrates, FSC efforts to use market-based converting efforts, and even program changing conforming strategies, to gain greater support from companies and non-industrial forest owners face tremendous obstacles.

Setting the Stage: Establishment of Forest Certification Factions in the US

The early experience of forest certification in the US commenced before the formal creation of FSC, and it is important for our analysis because the choices made at these early stages influenced later FSC attempts to seek support from companies and non-industrial private forests. As with Germany and the UK, most of the groups that supported sustainable forestry certification programs in the early 1990s were focused on tropical deforestation (Natural Resources Defense Council 1993; Smith 1995), and paid less attention to US forestry practices. And as in the UK and Germany, initial efforts to curb tropical rainforest destruction focused on boycott campaigns of tropical wood, which, as Kolk (1996:245) noted, did not significantly affect "production and consumption patterns in the North." The limited success of such campaigns, and recognition of their unintended effects in increasing forest destruction, led organizations in the US to focus on developing certification as a carrot for forestry operations, to complement the traditional boycott stick.

In 1989 the Rainforest Alliance's SmartWood Program, now based in Richmond, Vermont, became the world's first forest certification program. Two years later it was followed by the California-based Scientific Certification Systems (SCS) Forest Conservation Program. Both organizations developed standards for domestic and tropical forest practices, creating a system that would allow forest landowners in any part of the world to be audited for sustainable forestry practices and, if certified, sell "green" forest products in the marketplace, ideally for a premium. Early on, the SmartWood Program focused efforts on developing a regional or ecosystem approach for developing specific standards.

The development of these certification programs was taking place alongside increasing international governmental attention to tropical deforestation, which became a significant issue leading up to the Rio Earth Summit in 1992. As we detailed in chapter 1, the failure of intergovernmental processes to achieve a legally binding global forest convention at Rio led environmental groups to focus increasingly on certification — an arena they believed to be more capable of overcoming what they perceived to be slow and limited efforts to address global forest degradation.

Many of the groups that helped create the Forest Stewardship Council and provided the resources needed to do so came from the US. The creators included environmental groups such as the Rainforest Alliance, the World Wildlife Fund (WWF) and the Woodworkers Alliance for Rainforest Protection (later the Good Wood Alliance) as well as retailers such as Home Depot (Viana et al. 1996; McAlexander and Hansen 1998). From the beginning the FSC enjoyed high levels of support from several leading US environmental groups and from a small contingent of forest companies and non-industrial forest owners who saw a niche market opportunity (Hayward and Vertinsky 1999; Island Press 1998).

The FSC, however, did not appeal to the US forest sector for two main reasons. First, few companies saw potential market benefits from obtaining FSC certification. Early mobilization of support for the FSC at the demand end of the production chain was limited (Crossley 1996; Lyke 1996) and took place in the absence of large retailer-focused campaigns such as those that existed in the UK. Some local governments showed an interest in adopting wood policies banning tropical timber products, and organizations such as the Good Wood Alliance took on the role of promoting the purchase of responsibly managed wood products (Viana et al. 1996). Yet, there was no national buyers group equivalent to the UK's WWF 95 Group (chapter 5). FSC certification conducted by SmartWood and SCS was still generally viewed as a niche-market process offering little incentive for the larger industrial forest companies (Lyke 1996). As one industry observer noted: "This early interest [in certification] with certain small companies was not typical of the broader forest sector. Most large industrial forest companies were relatively unaware and not interested in the initial development of the FSC. The national focus on the federal lands and general public discontent with certain forestry practices occupied the minds of regional and national industry associations, as well as their individual member companies" (Wallinger 1995).

Second, forest companies were particularly concerned about the FSC's wide-ranging performance-based approach to forest management and its chain-of-custody requirements, which they felt were unrealistic given the structure of the US industry, where large industrial forest companies relied on thousands of private forest landowners as a supply of fiber. Under the FSC chain-of-custody rules at the time, a large industrial forest company would only be able to sell its products as "FSC certified" if it could also show that its suppliers of raw materials — regardless of size — were also FSC certified. As one company executive explained: "That is one of the real problems . . . to think that someone in the short-term could really logistically get the kind of certification that FSC requires on chain of custody is unrealistic . . . take for example, a small example, a plywood mill that [our company] has . . . About 98 percent of wood comes from outside sources and that is from dozens and dozens of small, independent landowners somewhere . . . Logistically, it is extremely difficult to figure out where all of that wood comes from and then go out and try to see that all of these dozens if not hundreds of small private landowners are practicing FSC standard of forestry. Even if you really wanted to do it and thought that there was a reason for it — and I don't — logistically it would be almost impossible."[9] Many company officials also felt that the logistical difficulties of tracking chain of custody would serve to entrench opposition by non-industrial private landowners, as they became aware of the FSC and its approach to sustainable forest management.

Even though most large industrial forest companies did not pay much attention to the FSC at this time, AF&PA officials monitoring the FSC certainly did. Learning from their lessons in the US Pacific Northwest Spotted Owl debates, AF&PA officials took a proactive response to certification. In the event that certification might later gain more interest, the association would be ready with what they believed to be a more industry-friendly alternative to the FSC.

SFI Promoted as Alternative for Forest Industry and Small Landowners

In 1993, the AF&PA began to take strategic actions to introduce a comprehensive forest management program for its member companies. Following the merger of the AF&PA and the American Paper Institute early in 1993, members of the combined association were required to adhere to the institute's Forest Management Principles (Meridian Institute 2001c). Soon after, AF&PA officials pushed to increase the scope of these principles. By early in 1994, the association's Board of Directors had approved the SFI's Principles and Implementation Guidelines. Even though the SFI was originally conceived as and reported to be a voluntary industry program designed to improve forest practices on the ground and its related public image (Wallinger 1995; Cashore 1997), industry officials soon realized that the program could also serve as a full-fledged forest certification program and thus act as an alternative to the FSC. Indeed, leading officials in the AF&PA would note that at this time the chief executives of many AF&PA companies had made a major public commitment to the SFI, and they followed through with what they asserted to be strong firm-level implementation "from top to bottom." They were intent on ensuring that their commitment to the SFI not resemble "just one more big PR program that didn't mean anything."[10] We show below that FSC supporters do not appear to have taken into account the depth and strength of this commitment to the SFI.

Indeed this early recognition meant that, unlike the industry's relatively uncoordinated response to the public policy issues that arose in the 1980s and early 1990s, the AF&PA was proactive in monitoring the FSC and portrayed the forest certification issue as a challenge to the entire US industry. It framed certification early on as a common industry issue, rather than as a challenge to be left to individual companies. As an official from one forest company explained: "If [certification] were simply a business issue for the company, you could have just gone away and done your own thing. But, I think people believe that it was one for the industry. What generally happens is you probably don't gain a lot by going away and doing your own thing because we all tend to live by regulations. For example, they don't pass one set of regulations for one company and another set for the rest of the industry. Most things happen that way, so even though there was a lot of discussion about doing it that way, I think the consensus was that we would go forward trying to do our part with a few other companies that really thought SFI was the right way to go."[11]

Environmental group response to the SFI program, once recreated as a certification system, was, not surprisingly, a negative one. At this point these groups began to openly criticize the program as lacking performance standards, transparency, a chain-of-custody system, and mandatory third-party audits (Ozinga 2001; American Lands Alliance 2000). According to the Rainforest Action Network, "The SFI is the logging industry's attempt to self-regulate. We call it a case of the fox guarding the henhouse. Members of the SFI have a weak set of standards and are free to monitor their own compliance" (Rainforest Action Network n.d.). Another environmental group official said, "90 percent of industrial timberlands is represented [in the AF&PA], so it by definition includes the status quo . . . Let's put it this way — SFI is a start, and it's necessary, but it really doesn't mean anything yet."[12]

FSC and SFI Become Institutionalized in the US

Between 1994 and 1995 heightened activism on the part of US groups in support of the FSC remained focused primarily on tropical and temperate forest degradation in other parts of the world. Support for certification in the US marketplace developed in a limited number of specialty markets, and regional standards development processes were generally ad hoc and poorly integrated at the national level. In 1994, efforts to break these patterns began to take hold, when SmartWood took steps to collaborate with other certification supporters to build an extensive network of regional groups that would work to develop standards and processes for certification in their region (Ozanne and Vlosky 1996). Yet for the most part during this period it was uncertain whether certification was going to gain the momentum necessary for it to affect the management practices of more than a few proactive forest companies and landowners. The FSC and SFI supporters were waiting to see if the US marketplace was going to take certification seriously, and if so, what the other program would do.

Indeed, the actions of both programs bore limited relation to market-access issues; rather incremental steps were undertaken to slowly build the respective programs' capacity and credibility. For instance, the SFI established an Expert Review Panel (now External Review Panel) in 1995, which was touted as an independent body and was charged with the task of reporting on the yearly progress made by the program (Meridian Institute 2001c). The FSC for its part built capacity by establishing a national contact person in 1996 and by beginning the first formal efforts to develop nine regional working groups with which to create region-specific FSC standards governing sustainable forest management (Meridian Institute 2001b). The decision to use regional working groups instead of the usual practice of creating one national standard can be traced, in part, to SmartWood's pre-FSC choice to develop a regional or ecosystem approach to rule development (Ozanne and Vlosky 1996).[13] Whereas most countries had focused on developing national standards, the US approach would create a more administrative and multi-level decision-making structure, thus developing a more cumbersome system for rule making and dispute resolution than the SFI's streamlined national approach. Significant latitude was granted to each region. Regional working groups devised their own voting structures, decided who would participate in standard setting, and had power over the scope and detail of rules developed. Of course, these processes were, in principle, guided by FSC requirements for standards development (Forest Stewardship Council 1999d); however only the US national office was formally recognized by the FSC, eliminating any formal accountability at the regional working group level. On the other hand, the granting of considerable autonomy to the regional standards development processes was successful in earning much needed support from local environmental and other groups critical of regional private forest management practices, who felt disenfranchised with their role in influencing private forest management public policy networks (Auld 2001; Newsom 2001).

This multi-layered governance approach for standards development, however, presented the program with a problem. The reality that the FSC regional standards were not going to be ready fast enough to certify forest companies and landowners already interested in the program led the FSC to permit the use of provisional standards developed by accredited auditors until regional standards were formally endorsed. This decision along with the historical legacy of SmartWood and SCS in FSC development meant that these auditors would have a significant role in FSC policy development while maintaining considerable autonomy for implementation and development of their own certification initiatives.[14]

The regional standards setting approach taken by the FSC raised two seemingly contradictory concerns on the part of forest companies and AF&PA officials. On the one hand company officials feared that sustainable forest management standards would differ significantly between regions, leaving companies in different parts of the country having to practice different forest management. On the other hand, company officials in regions that faced relatively limited public policy forest practices regulations were concerned that the FSC might increase their rules "upward" (more stringent) to be more like rules in more regulated regions. As one industry official noted: "[The] fear would be that we don't want Oregon forest practices applied in Alabama or Georgia. I can understand that — do you want to voluntarily take on significantly more regulatory load? You can see where that would cause . . . concern amongst brethren at the table that are in the same business because 'I operate here and that's all I know and [now I] have to operate by their high standards.' "[15]

The AF&PA carefully monitored these developments and undertook swift conforming changes within its own program in its effort to increase support — requiring in 1995 that all of its members adhere to the SFI Principles and Implementation Guidelines (Meridian Institute 2001c). Of the association's 250 members, 16 withdrew following this decision (AF&PA 2000b).

The Competition for Legitimacy 1996 to 2002

The remainder of this chapter traces the story of how the FSC and the SFI have struggled to gain the upper hand in the US (table 4.2). We describe how our three factors — place in the global economy, structure of the forest sector, and the history of forestry on the public policy agenda — mediated the strategic efforts of the FSC and SFI directed at securing support from key actors along the supply chain, and we look intently at why the FSC has struggled to gain the kind of support from large forest companies that it has enjoyed in regions such as Sweden and BC.

Arguably the most important development in the post-1996 period was the decision by the WWF and its US office to invest considerable resources in an effort to promote the FSC in the marketplace as the legitimate certification program for US forest companies and forest landowners. This choice was made for a number of key strategic reasons. The WWF recognized that because the US market was the world's largest forest products consumer and US companies were dominant global producers, the choices made in the US were not just important for improving forest management domestically, but also for ongoing efforts to institutionalize FSC-style certification in the tropics and elsewhere. Supporters of the FSC in the US also recognized that large-scale impacts on US biodiversity conservation were not possible if the program remained a niche player.[16]

Taking the lead from successful market efforts in the UK and recognizing that the US forest sector's place in the global economy meant that procurement choices made by foreign companies would have limited impact on certification policy choices made by most US producers, FSC supporters began efforts to build domestic demand for FSC products. Environmental Advantage, a small firm based in New York, formed the first North American buyers group, amalgamating the existing network of ad-hoc, uncoordinated retail and other consumer-end company support for the FSC (Lyke 1996; Crossley 1996).

Still, demand was weak. While some mail surveys indicated some consumers might be willing to pay more for certified products (Forsyth, Haley, and Kozak 1999; Ozanne and Vlosky 1996),[17] consumers were, for the most part, not demanding FSC-certified products. The FSC was becoming entrenched as a niche-market program (Vlosky, Humphries, and Douglas 2001) that many came to believe could easily co-exist with the more flexible and industry friendly SFI program.

Recognizing these trends, and not content to have the FSC relegated to niche market status, in 1997 the WWF stepped up its efforts to institutionalize demand for FSC products by helping to form the Certified Forest Products Council (CFPC),[18] which created a more efficient structure with which to identify and support demand for FSC products.

The creation of the CFPC greatly aided the domestic-focused market campaigns of environmental groups such as the Rainforest Action Network and Coastal Rainforest Coalition (now ForestEthics) that were starting to directly target retailers, such as Home Depot, and demanding that they commit to purchasing FSC wood (Cashore 2002; Sasser 2003; Gereffi, Garcia-Johnson, and Sasser 2001). These groups became aware that the relatively weak consumer demand for certified wood (Hansen, Forsyth, and Juslin 1999) meant that building US demand for the FSC would require targeting the retailers who arguably would be more likely to see certification as a means to increase their companies' environmental credentials.

During this period, the FSC regional standards setting processes continued. Yet, owing to the now overt competition between the FSC and SFI, almost all AF&PA members refused to participate in any FSC standards setting processes. This was in spite of some notable efforts on the part of some regional working groups to secure broad participation; for example, the Southeast US working group was able to attract individuals from all seven states and most major stakeholder groups, except the forest industry.[19] Despite multiple invitations, no industry members joined the standards working group (Humphries 1999),[20] and, in the end, the working group was forced to rely on consulting foresters as industry proxies. The Pacific Coast working group also failed to gain participation of most industrial companies (with the important exception of companies such as Collins Pine and Big Creek Lumber). Even in the Northeast working group, where industry maverick Seven Islands had already undergone FSC certification, the majority of industrial forest companies were vocally opposed to the FSC. The lack of industrial participation in FSC standards setting was undoubtedly in large part due to the ability of the AF&PA to keep its members true to an anti-FSC stance, which stemmed back to the association's experiences dealing with past public relations issues. Such experiences led supporters to believe that there were in fact a number of companies that were "actively sort of wanting to go down the FSC pathway but [were] being

Table 4.2 *Strategic efforts to either gain or maintain support between 1995 and 2002*

Year	SFI supporters Action	Strategy	FSC supporters Action	Strategy
1995	AF&PA creates Expert Review Panel to oversee publication of the SFI annual report	Conforming	SmartWood creates Resource Manager certification	Conforming
1996	AF&PA makes participation in SFI program a requirement of membership (Jan)	Conforming		
1997			Certified Forest Products Council created	Converting
			Rainforest Action Network and Coastal Rainforest Coalition begin directly targeting retailers with pro-FSC market campaigns	Converting
1998	AF&PA creates licensing agreement that allows non-members to participate in the SFI	Conforming		
	AF&PA introduces third-party certification option for SFI	Conforming		
1999	SFI Executive Committee creates the Interim Inconsistent Practices Reporting Protocol, which allows foresters to report perceived violations of the SFI standards confidentially (Dec)	Conforming	Southeastern standards working group proposes changing an FSC criterion to allow plantations created after 1994 to qualify for certification in the Southeast	Conforming
	AF&PA publicizes awards and implies support from Sierra Club and other environmental groups	Converting	Home Depot announces pro-FSC procurement policy	Converting
			FSC creates US Standards Committee to guide harmonization of regional standards (Sept)	Conforming

Year	Event	Category
2000	AF&PA creates Sustainable Forestry Board (Sept)	Conforming
	Lowes, Kaufman and Board Homes Corp., Centex Corp., Andersen Corp. announce pro-FSC procurement policies	Converting
	SFI and Tree Farm program agree upon mutual recognition (June)	Conforming
	GOAL program created to give members of the public confidential means of reporting perceived violations of the SFI standards (July)	Conforming
	Fifteen state legislatures endorse the AF&PA and/or SFI program	Informing
	AF&PA releases policy recognizing CSA standard as acceptable for fulfillment of SFI-membership compliance requirements in Canada[a]	Conforming
	AF&PA joins the PEFC, but does not seek endorsement for its program from the PEFC Council	Conforming
	AF&PA lobbies retailers to include SFI in procurement policies	Converting
2001	First Conservation International and then the Nature Conservancy join Sustainable Forestry Board	Conforming
	FSC auditor and other groups initiate work on SLIMF small landowner initiative	Conforming
	AF&PA creates "Office of Label Use" to administer use of SFI label by program participants	Conforming

Table 4.2 continued

| Year | SFI supporters | | FSC supporters | |
	Action	Strategy	Action	Strategy
2002	AF&PA applies for non-profit status for the SFB, with a shift of board composition from 40 percent SFI members and 60 percent other forestry interests to one-third SFI members, one-third conservation and environmental community, one-third broader forestry community (Jan)	Conforming	SCS creates Cross and Globe	Conforming
	AF&PA makes third-party verification of gatewood compliance with BMP an SFI requirement	Conforming		
	AF&PA creates partnership with NatureServe and Natural Heritage Programs to further incorporate biodiversity into 2002–2004 SFI standards	Conforming		

[a] While this policy existed in principle (AF&PAb 2000), officials with Canadian companies noted that in practice the AF&PA was less willing to formally recognize CSA certified Canadian companies.

beaten by their colleagues, so they [weren't] doing it."[21] But arguably even more important was that industry viewed the FSC regional processes as being designed to give forest companies and non-industrial private forest owners only token or minority participation. Three developments entrenched these views. First, industry officials felt limited by regional processes that could not veer from, or change, the previously developed FSC international Principles and Criteria. Second, many industrial forest companies and AF&PA officials viewed the FSC's founding meeting in Toronto in 1993 as a conscious effort to exclude most of the key players in the North American forest industry.[22] Third, the experience of JD Irving (one of the few large industrial forest companies that did choose to support the FSC) in the Canadian FSC maritime regional standard setting process reinforced industrial forest company views that the regional standards setting processes were stacked against industry interests (Cashore and Lawson 2003; Lawson and Cashore 2003). Industry decision not to participate in standards processes created a feedback loop. In the absence of industry representatives the standards likely became more stringent than they would have been if industry had participated, which led to increased industry opposition to the FSC. In the case of the Pacific Coast, for example, environmental groups and their allies, excited about the prospects of increasing regulation of private and state-owned forestlands in a way that they were unable to do in state-level public policy processes (Forest Stewardship Council 1999f), quickly came to dominate policy discussions. One participant viewed the process as a case where the "the environmentalists had the process by the tail, with the social people and the economic people bending over backward to try to placate the environmentalists."[23]

The dominance of environmental groups and their agenda within the Pacific Coast standards process led to an early understanding that the standards would not be applicable to National Forest lands. Success in the public policy arena during the 1990s in protecting National Forest lands from any kind of harvesting (described above) meant that certification was seen as inappropriate for these lands. In fact, many environmental groups came to support the "zero cut on National Forests" campaign (see, e.g., Dolcini 1997; MacCleery 1999), which they believed certification, as a standard that recognized harvesting practices, would undermine. As one observer noted: "That is really the battle that we got caught up in. A lot of the environmental chamber members don't want to see the Forest Service get certified. Most likely they are eminently certifiable."[24]

Such dynamics increased, rather than reduced, the wedge between conservation interests and industrial and community interests. Industrial groups were quick to criticize efforts to exclude National Forest lands from the Pacific

Coast standards, asserting that despite the limited amount of fiber coming from these forests, many rural communities still relied on National Forest harvests. [25] The decision to exclude National Forest lands also left many members of rural communities feeling disenfranchised with the FSC process.[26]

Certainly the industry criticism that the FSC standards were being developed unevenly bore fruit relatively early on. For example, in the US Southeast, where 40 percent of the world's plantation forests are found (Hyde and Stuart 1998), the FSC regional working group created standards governing certified plantations that were more relaxed than the requirements in FSC Principle Ten.[27] In a strategic move, and in a direct effort to conform to industrial landowners in the US South, the FSC-US board of directors approved these changes — but the FSC secretariat in Oaxaca did not. The result was a three-year stalemate within the FSC, delaying the US Southeast standards from being set and contributing to the FSC-US losing valuable time in a legitimacy race in which it had no time to lose.

At this time the AF&PA drew on its strong ties with the established forestry community (AF&PA 1999) to use informing strategies to publicize its dissatisfaction with regional FSC standard-setting processes, and to seek and receive support for the SFI from many professional foresters, non-industrial private forest owners, loggers, and even state governments (AF&PA 2001a). Perhaps anticipating the potential future need to track the sources of its fiber, professional loggers played a paramount role in distributing information regarding the SFI program to almost 79,500 landowners in 1998 (AF&PA 1999). Fifteen state legislatures "enacted resolutions of proclamations endorsing the AF&PA and/or SFI program" (AF&PA 2000a) and Forest Service executives were invited to sit on the SFI Expert Review Board. In a highly strategic conforming move, the SFI proponents made an explicit effort to recruit an Environmental Protection Agency member for the panel. AF&PA members also lobbied other government agencies for support and provided them information about the program.

SFI efforts to increase support at this time did not stop at informing the broad community or refusing to participate in the FSC standard-setting processes. The SFI and its supporters also focused directly on the one arena in which the FSC appeared to be making limited inroads — choices made by customers of forest products further down the market's supply chain. SFI strategists began by questioning the legality of commitments made to purchase FSC products, arguing that they might contravene US anti-trust and fair competition laws (Simula 1998; Grant 1989:131–134), and pressing customers, companies, and governments to adopt broad, non-specific wood procurement policies, rather than policies that would preclude the SFI. This was effective in limiting the CFPC's ability to obtain sole support for the FSC, as it meant the organization had to focus on promoting features that it deemed necessary for a credible certification program. Thus, many retailer announcements that implicitly favored the FSC did not refer to the FSC at all, but rather indicated a preference for "third party independent certified forest products" (Certified Forest Products Council 2001a). The legal issues were raised a number of times as a means of limiting the efforts of FSC strategists to gain procurement support from city councils across the country. For example, the City of Los Angeles was convinced to change its wood purchasing policy so that it would potentially recognize non-FSC certified sources. As one observer noted: "Exactly what happened was that the city of L.A. councilman advised the people drafting the standard — that was the environmental affairs department — to go back and re-draft the standard such that it would be broader. It wouldn't be just 'FSC certified' but would include programs that would meet whatever attributes the city might decide it wanted . . . industry felt that they had gotten what they wanted — which was to remove 'FSC' from [the purchasing policy]."[28]

Second, in 1998 the AF&PA began to expand its reach by allowing non-AF&PA members to become enrolled in the SFI — a not insignificant effort to promote the SFI and the amount of land under SFI authority. It did this by introducing a licensing program to "attract other champions of sustainable forestry practices (AF&PA 1999)."[29] Expanding the program to allow the participation of outside groups fostered "partnerships with diverse stakeholders" and helped "to expand the number of forest acres in [the U.S.] that [were] managed wisely and in an environmentally friendly manner" (AF&PA 1999). Key woodlands falling outside the industrial land base, such as Callaway Gardens in Georgia, and those managed by environmentally reputable organizations (e.g., the Conservation Fund) became enrolled under the SFI (AF&PA 1999). In addition to their immediate significance in increasing support, these participants also allowed the SFI to offer the public examples of SFI-managed forests that were not being intensively managed for wood production.

These efforts on the part of the SFI initially worked to heighten the resolve of FSC supporters. Environmental activists stepped up their market campaigns targeting companies further down the supply chain. In summer 1999, their efforts seemed to pay off. After two years of campaigning by the Rainforest Action Network, Coastal Rainforest Coalition, and others, the do-it-yourself (DIY) retailer Home Depot announced in August 1999 that it was going to "eliminate from [its] stores wood from endangered areas" by the end of 2002 unless certified by a credible independent third-party organization.

The Home Depot was clear that its policy preferred FSC-style certification over the SFI (Thompson 2001; Jordan 1999b). The Home Depot announcement was viewed as a major coup for the FSC, was widely covered in the US press, and resulted in a number of other companies following suit with similar announcements. For instance, Lowe's, the second largest DIY retailer in the US, released its own wood procurement policy supporting the FSC soon after Home Depot (Bond 2000), as did companies such as Kaufman & Board Homes Corporation, Centex Corporation, and the window-making giant Andersen Corporation (Carlton 2000a and b; the Forestry Source 2000). In the eyes of many FSC supporters, this breakthrough would be just what was needed to finally bring the FSC into mainstream US markets.

Yet at the same time as this potential new market pull developed, the FSC standards development process continued to be criticized by forestry professionals and companies as being too slow,[30] increasingly stringent, not appropriate for industry, and uneven across jurisdictions — resulting in, they argued, an unfair playing field that was more the result of politics than ecological differences. The strong criticism of the regional standards processes seemed to dampen the impacts of the FSC's market advances, with seemingly no industrial forest companies swayed by the Home Depot announcement to support the FSC.

The FSC recognized that its fragmented and decentralized structure was creating the perception that it was developing uneven standards, and, as a result, it created a regional standards "harmonization" committee in September 1999. This process was charged with developing a set of national indicators, which would act as a baseline standard and/or template for the development of regional certification standards (Forest Stewardship Council 2001). National indicators were to "foster a sense of trust and transparency by representing a common baseline standard" (Forest Stewardship Council 2001). Yet this process did little to increase industry support of or participation in FSC standard development.

While the harmonization process was designed to show the FSC as responsive to outside criticism, it also slowed down the development of approved regional standards — adding fuel to the argument that the FSC was overly bureaucratic and slow in making decisions. AF&PA members remained committed to the SFI and, in the two years that followed the Home Depot announcement, the organization took an astounding number of conforming actions in an attempt to make its program acceptable to retailers, while maintaining support from its industry base. Since many of these retailers' new policies required that certified products be tracked from the forest to the marketplace (chain-of-custody) and that standard setting be multi-stakeholder in nature, the AF&PA undertook conforming changes to make their program fit those criteria. AF&PA members and strategists felt confident that this approach would eventually earn the SFI support in the marketplace. According to one SFI participant, "If . . . we have third-party and we have performance measures and things are happening on the ground, we can't be ignored, we won't be ignored. So, it's interesting to me because every time . . . Home Depot says this or Wickes says this or whatever, there is always a little bit of a mild panic with some people [in the forest industry] going 'Oh my gosh, what are we going to do?' Keep working on the program, keep making it better, it will happen."[31]

One of the most important conforming strategies at this time was aimed at addressing the critique that those enrolled in the SFI program did not have to be independently audited by outside organizations, as members could either audit themselves (first party) or have the AF&PA do it (second party). To address these concerns, the AF&PA changed its policy in 1998 to allow third-party auditing; since then, most of its large industrial members have committed to or have already undergone such auditing. In a similar vein, in 1999 the "Interim Inconsistent Practices Reporting Protocol" was created to provide a "whistle blower's" hotline, allowing loggers and other interested parties to report perceived violations of the SFI standards (AF&PA 1999).

While moving to create third-party auditing procedures, the SFI continued to be criticized by environmental groups as being an industry standard that excluded other interests from participating in rule development. This concern about balanced standard setting also reached the retailers: the Home Depot explicitly mentioned the need for a balance of ecological, economic, and social interests in its 1999 procurement policy. Reflecting its fast-paced efforts at conforming to achieve legitimacy further down the supply chain and as a result of significant survey research, the AF&PA moved to address these criticisms as well. In a bold move, it created the Sustainable Forestry Board (SFB) in September 2000. The SFB was designed to increase the credibility of the SFI program by giving management responsibilities over standard revisions, verification procedures and program compliance to an independent group. Indeed, one of the first tasks taken on by the SFB was to "determine how the current SFI standard should be improved such that it can be defended as a multi-stakeholder consensus standard" (AF&PA 2000). The selection of board members had a multi-stakeholder flair; the board was set up so as to include nine (out of fifteen) non-AF&PA members that could come from interests such as environmental or conservation organizations, state and federal agencies, academia, labor organizations, foundations, customers, non-industrial private forest owners, and other forestry community organizations (Meridian Institute 2001). The remaining six participants were to be senior level AF&PA member-company executives appointed by the AF&PA Board (AF&PA 2001b). The move was clearly aimed at maintaining a prominent role for industry — much more so than permitted by FSC institutions. The SFI state level implementation committees also worked to enhance the outside influence on SFI procedures. For example, in July 2000 the Maine SFI Standards Implementation Committee (SIC) designed the SFI GOAL program, which created a process through which members of the public had the opportunity to report forest practices they believed were inconsistent with the SFI program (AF&PA 2000). Other state SICs soon followed suit, making the program national in scope.

The one area that the AF&PA had more difficulty in conforming was the requirement in the marketplace that certified wood be tracked through a "chain-of-custody system." This was because the AF&PA industrial forest companies relied on literally thousands of private forest landowners for additional supplies of fiber (see structure of forest sector section above), rendering chain-of-custody tracking a potential logistical nightmare. The SFI recognized that the first route to addressing this problem was to create an environment in which its private forest landowner sources would be able to have their own forestlands certified as well. Recognizing that non-industrial private forest landowners were ideologically opposed to FSC-style certification and likewise hesitant to support an industry standard (Newsom et al. 2003; Vlosky 2000), SFI strategists worked to create a mutual recognition agreement between the SFI and American Tree Farm system (AF&PA 2000b). For the SFI, the Tree Farm system was an opportune audience with which to ally. In fact, many AF&PA member companies had helped launch the Tree Farm system back in 1941 as a mechanism to promote reforestation on non-industrial private forest lands (Coulombe 1999).[32] By the middle of the 1990s, the system had grown to include approximately 70,000 participants. This gave it significant strategic importance. AF&PA companies recognized that the system represented one small, unified audience among the nearly 10 million non-industrial private forest owners in the US with which it could collaborate to promote its version of sustainable forestry (American Tree Farm System 2000). This was one step toward enabling the SFI to claim it had a tracking system, yet avoid an FSC-type chain-of-custody system. Furthermore, it directly addressed the cost concerns of the non-industrial owner, as the Tree Farm program required no fee for participation with volunteer professional foresters carrying out the assessments. While non-industrial private forest associations and its leaders generally supported this agreement, the process of gaining increased industry and small landowner support for Tree Farm that would facilitate and permit the SFI to develop a chain-of-custody system would require some effort. The structure of the US forest sector and the importance of non-industrial private forests made chain-of-custody issues the one key arena in which even SFI strategists were unable to make swift changes.

During this time the SFI also worked to address a major remaining criticism — that the program was not international in nature. It became a member of the Pan European Forest Certification initiative (PEFC International 2001) and even established a policy recognizing the CSA's standard as the "functional equivalent of the SFI program" (AF&PA 2000a). The AF&PA also promoted its program on Canadian forestlands in a direct effort to establish itself as the North American alternative to the FSC. [33] Furthermore, it worked to bring together industry certification programs in key forest producing and exporting countries under the auspices of the International Forest Industry Roundtable.

With a frantic pace of change now well under way, the AF&PA used informing strategies to broadly publicize the SFI's recent changes to environmental groups, retailers, and lumber purchasers and to contend that these changes meant that the program had become a credible, independent, and international alternative to the FSC. Certainly environmental group officials, even those clearly supporting the FSC, noted these changes[34]: "A year ago, we could say honestly that SFI did not have third-party verification available, now they do, we can't say that anymore, and won't say that anymore. A year ago, we could say that they don't do any field verification, I can't say that anymore and I won't say that anymore."[35]

SFI supporters also directly lobbied companies having pro-FSC procurement policies to change their policy wording to allow the acceptance of SFI-certified products. An active dialogue was kept between SFI proponents and the CFPC; one SFI proponent reported, "We want to continue that dialogue because we think that the Certified Forest Products council should recognize third-party SFI as the equivalent of FSC."[36]

These efforts led FSC supporters to call for clear wording in green procurement policies (ForestEthics 2001). A white paper was developed through the Wye River Coalition, composed of World Resource Institute, World Wildlife Fund, Natural Resource Defense Council, Rainforest Action Network, and ForestEthics, that offered clear requirements for sustainable forest management procurement policies that, if adhered to, would rule out support for the SFI in its form at that time. At the same time the white paper did not rule out support for other certification programs if they met key conditions, opening the door for the AF&PA to continue its conforming strategies in the hopes that it might gain eventual acceptance from CFPC member companies. Focusing on conditions of procurement policy rather than simple industry or landowner support for the FSC was important for moving the debate over SFM away from support for the FSC per se, and more to a series of conditions through which, conceivably, any program could adhere. This would not only open the door to the SFI in the future, but also to other groups dissatisfied with the slow pace of the FSC.

By 2001, many FSC proponents began to feel frustrated with the lack of concrete action taken by those retailers who had made public commitments toward FSC-style certification. Momentum in the marketplace was developing much slower than FSC had hoped. The CFPC planned a conference for autumn 2001 to promote FSC-certified products. The conference organizers excluded the SFI from the conference, which AF&PA members objected to. The conference was postponed because of the September 11, 2001, terrorist attack.

Recognizing that the FSC market campaigns were not moving most forest companies and that some of the early benefits claimed of FSC certification — such as price premiums — were not consistently materializing, FSC supporters worked to reduce the cost of certification for the most highly burdened group, small landowners, and began emphasizing the non-monetary benefits of participation in the program, which included improved forest management and a green image.

To address the forest sector's concerns that the program was too expensive for small landowners, FSC supporters proposed the SLIMF initiative, which allows a streamlined, less costly certification process geared toward "low risk" landowners. SmartWood also tested "umbrella certification," a form of group certification that would lower costs by certifying a "group of groups." In 2001 the Rainforest Alliance spun off the TREES[37] program as a sister program to SmartWood that would work toward creating new innovations in certification.

However, in general these innovations have been slow to change the FSC options available to landowners due to the low capacity of the FSC-US and FSC-AC to authorize the policy changes needed to facilitate innovations in a speedy way. These innovations are important not only because they highlight the fact that FSC supporters recognized that changes were needed to gain any significant degree of landowner support for the FSC, but also because of their source: the majority of innovations came from FSC-accredited certifiers rather than from the national office. Similarly, many of the key strategists in the FSC-US were WWF employees. This highlights how the fragmented decision-making structure of the FSC has plagued its efforts to seek broader support. This situation puts into sharp relief the role that core audiences play in the choices of certification programs — despite the fact that many FSC founders knew from the start that the fragmented decision-making structure would decrease the program's responsiveness, they felt that it was necessary to gain buy-in from core environmental and social groups.

Recognizing the need to show some degree of success, FSC supporters, particularly those driven by US foundations, began to focus on forestlands owned and managed by state agencies and federal lands outside of national forests. Since public forest clients are motivated by other factors than markets and the FSC market campaigns were not moving industrial companies, the FSC strategists reasoned that those motivated by non-economic interests might be more interested in participation.[38] A project undertaken by the Pinchot Institute and funded by leading US foundations sought to conduct independent FSC and SFI assessments on the same forest tracts and afterward ask the forest managers a set of questions that would evaluate and compare the standards, process, outcomes, and relevance of the SFI and FSC systems (Mater 2002). In addition to providing information for other state agencies (and private companies) considering certification, the project generated FSC-certified wood for the marketplace.[39] By 2002, the project included state or federal lands in Tennessee, Maine, North Carolina, and Vermont and Native American nations in fifteen states (Mater 2002). However, unlike the situation in the UK, it is unclear whether the decisions of state agencies to certify with the FSC created any "spillover" effects to choices made by private forest companies or non-industrial private forest landowners. These groups — from which the FSC needed legitimacy the most to create market momentum — were still firmly behind the SFI.

Indeed, more changes were being made to the SFI program to continue its quest to appeal to environmental groups and retailers as a credible certification program, on par with the FSC. In 2002 the SFI continued to address its lack of chain-of-custody system by making the verification of procurement practices a requirement of its participants. Fiber coming from company-owned lands and gatewood would be subject to third-party verification against a portion of the SFI standard. Companies were required to know all their suppliers and negotiate the right to audit these suppliers to ensure reforestation and adherence to applicable best management practices (AF&PA 2002b). To further bolster its credibility in the marketplace, the AF&PA worked extensively to provide SFI participants with an on-product label; an "Office of Label Use" was established to administer the use of the SFI label by third-party verified participants (AF&PA 2001b). Although the label was launched in 2002, it has seen limited use. Finally, the AF&PA conformed the structure of the SFB again — this time to mirror the FSC's three-chamber system even more closely. Now, one-third of SFB members were conservation and environmental community representatives, one-third was made up of SFI members, and the remaining one-third consisted of members of the broader forestry community.

These changes and others were highlighted by SFI representatives at the CFPC's conference in Atlanta, which had been rescheduled to April 2002 and reformulated to incorporate all programs, not just the FSC. In fact, the agenda for the revised meeting was jointly crafted by several leading forest products companies with several non-governmental organizations and the CFPC staff. Recognizing that support for FSC was stalling and that the SFI was stepping up its efforts to gain support, FSC strategists began to shift their own strategies from one in which the SFI was strongly criticized to one that would have the marketplace recognize the FSC as the most environmentally sensitive among certification programs, but that other programs also existed that might eventually gain some marketplace acceptance. The idea behind this approach was that it would permit the FSC to maintain viability and recognition as a credible and meaningful forest certification program in an area of the world where forest company and landowner support was limited, rather than face a less desirable outcome — the potential disappearance of the FSC as a certification program in the US context.

Measured in terms of limiting the FSC's market presence, it appears that the conforming efforts of the AF&PA have begun to meet with some success. In the 2002 annual report, the Expert Review Panel noted: "Major paper and forest products customers are responding with procurement policies that recognize companies that have achieved 3rd party certification under the SFI program. Customers that recognize the SFI program in terms of certifying forest products include 84 Lumber, Centex Homes, MASCO, MasterBrand Cabinets, McCoy's, Menards, Norm Thompson Outfitters, Payless Cashways, Pella Windows, Wickes Lumber and many others" (AF&PA 2002b). And at the same time companies such as the Home Depot that have maintained a public position giving preference to the FSC have failed to translate these policies into across-the-board purchases of FSC products.

Measured in terms of participation by environmental groups, the SFI's conforming activities also appear to have enjoyed some successes. Environmental group presence on the SFB includes representatives from the Nature Conservancy, Conservation International, and the Conservation Fund, while academic presence briefly included James Gustave Speth, dean of the Yale School of Forestry and Environmental Studies, who brought to the SFI his extremely high credibility with transnational environmental groups and decades of experience in promoting strong environmental policies in the US and globally.

Both programs' efforts to increase support have created potential problems for their "core audiences." While major industrial forest companies and forest landowners must have raised their collective eyebrows with increasing autonomy granted to the SFB, we can only imagine the private reactions when activists like Yale's Dean Speth joined the board. Yet AF&PA strategists seem to be managing rather well the SFI's need to maintain broad industrial support while moving closer to credibility further down the supply chain. Few AF&PA member companies have publicly criticized these efforts, and given the well-coordinated structure of the AF&PA, it can be assumed that all major companies were supportive of these strategic efforts.

The same cannot be said for the FSC and its supporters — where considerable criticism is mounting about the array of relatively scattered and loosely connected strategic choices made by FSC strategists and those acting on behalf of the FSC. While many FSC environmental group supporters continued or stepped up campaigns in an effort to maintain support for the program in the supply chain, the actions of a number of other groups' campaigns may inadvertently hinder FSC strategic efforts. For example, the Pinchot Institute's continued interest in facilitating certification pilot projects on state public lands (Price 2000; Mater et al. 1999) (mentioned above) caused division within the FSC's core environmental group audience.[40] While these projects were aimed at state and tribal lands, environmental groups promoting a "zero cut" policy for National Forests viewed them as a slippery slope: if state lands were being certified, would federal lands be next?[41] And efforts by supporters of the "Paper Campaign," whose purpose was to increase the sale and availability of paper products made of recycled or alternative fibers, may have had inadvertent effects on FSC efforts to gain increased acceptability from forest companies and landowners. This is because supporters of the Paper Campaign, led by ForestEthics and the Dogwood Alliance, used direct action campaigns to gain commitments from key retailers that went *beyond* FSC requirements. They demanded the immediate phaseout of all products made from a 100 percent virgin fiber (no recycled content) and all products containing fiber from US public lands and/or old growth forests. Further still, in a recent press release they criticized Domtar, an FSC-certified supplier of Staples (Forest-Ethics and the Dogwood Alliance 2002). While an understandable policy stance for those seeking to preserve the last remaining old growth forests globally, such a campaign undermines the FSC's efforts to gain support from companies whose main source of timber came from old growth regions. These efforts also gave the message that FSC certification would not provide forest companies, forest owners, manufacturers, or retailers with immunity from environmental targeting, reducing economic incentives that existed to support

Table 4.3 Forms of support given to FSC and its competitor by industrial forest companies in the US

Top 10 forest products-producing companies in the US[a]	Announcements of intention to or actual certification with[b]		Other forms of support[c]	
	FSC	SFI[d]	FSC	SFI
International Paper		✓		✓
Weyerhaeuser Company		✓		✓
Georgia-Pacific		✓		✓
Bowater		✓		✓
Smurfit-Stone Container Corporation		✓		✓
MeadWestvaco		✓		✓
Temple-Inland Inc		✓		✓
Louisiana Pacific		✓		✓
Boise		✓		✓
Potlatch[e]	✓	✓		✓
Rayonier		✓		✓
Plum Creek Timber Company		✓		✓

[a] Ranking based on information from company websites.

[b] For the SFI, all the companies included in this table had pursued and received third-party verification for some or all of their lands, as of February 12, 2002. At this time, these companies contributed 10,443,983 ha (51 percent) to the total SFI third-party area of 20,613,107 ha (AF&PA 2002c).

[c] "Other forms of support" include for the SFI participation, but not third-party certification, and for the FSC participation in a standards development process, financial support, or membership with the FSC.

[d] We include here all companies that have actually undertaken third-party verification to the SFI standard for some or all of their forest lands, as of February 12, 2002 (http://www.afandpa.org/forestry/sfi frame.html).

[e] Potlatch has had some of their poplar plantation lands (17,300 ac) FSC certified (Scientific Certification Systems 2001). This represents approximately 1 percent of the 1.5 million acres Potlatch owns (http://www.potlatchcorp.com/company/company.html).

the FSC. This diffuse and poorly coordinated environment was given a further shock when in summer 2002, the FSC-US began the search for a new executive director, leaving FSC strategic efforts even less coordinated.

It was in this context that in spring 2002, FSC auditor SCS proposed its Cross and Globe program as a new forest certification system. The program plans to certify forestry operations to the generic FSC Principles and Criteria, but it will not require operations to meet the more specific FSC regional standards. This approach is intended to address the problem of slow or nonexistent regional standard development and the uncertainty that forest companies feel this brings; however, the program worked to further undermine FSC efforts to gain support at a very sensitive time in its US history.

As a result of these developments, members of the US forest industry generally saw fewer and fewer reasons to support the FSC. As of mid-2002, the country's main forest products producers were strongly behind the SFI, with no indication that situations were about to change (table 4.3). This trend was paralleled among TIMOs who are mostly supporting the SFI and non-industrial private forest owner associations that are supporting the combination of the Tree Farm system and the SFI when they have any policy at all (tables 4.4 and 4.5). The number of FSC-certified acres in the US is indeed growing, but its success has been mainly limited to attaining market support from a small segment of public forest managers, some non-AF&PA member companies, and a handful of AF&PA members.

Conclusion

FSC strategists faced an uphill battle in US in their attempts to use converting strategies to alter the lack of initial forest company and non-industrial landowner support for their program. Converting strategies had to focus on the domestic market, and, while not inhospitable, these did not provide the same kind of pro-FSC environment that FSC converting strategies focusing on German and UK retailers did. Most features of the US forest sector further limited FSC efforts to use converting strategies, as fragmented landownership patterns saw the FSC attempting an array of conforming strategies aimed at small landowners, while large industrial companies were, for the most part, unsuccessfully wooed by attempts to conform the FSC standards regarding forest plantations. While converting campaigns continue by targeting increased numbers of forest products retailers, there are no indications that these activities are having any effect in altering landowner or industrial support for the FSC, who have turned to their own associations' programs to counter FSC efforts.

The slowness of FSC regional standards setting processes, its loose and fragmented efforts to obtain support from forest companies and those further down the supply chain, its inability to address non-industrial private forest owner and chain-of-custody issues to the satisfaction of most forest companies and forest owners, and the relative swiftness of FSC-competitor programs to adapt to the US environment worked to hinder FSC efforts in the US.

Table 4.4 Forms of support given to FSC and its competitor by timber investment management organizations in the US

Top 8 timber investment management organizations in the US[a]	Announcements of intention to or actual certification with[b]		Other forms of support[c]	
	FSC	SFI[d]	FSC	SFI
Hancock Timber Resource Group[d]	✓			✓
Forest Investment Associates		✓		✓
Wachovia				
The Campbell Group				✓
Molpus Woodland Group				✓
Forestland Group[e]	✓			
Forest Systems				✓
UBS Timber Investors				✓

[a] Companies were chosen based on reported forest area owned or leased and under management. These combine to own and manage 2,738,915 ha of forestlands.
[b] This includes certification with the FSC or commitments to become certified and commitments or actual third-party verification under the SFI program.
[c] For the SFI, "other forms of support" include participation, but not third-party certification, and considering third-party certification for non-AF&PA members. For the FSC, "other forms of support" includes participation in a standards development process, financial support, and/or membership with the FSC.
[d] Hancock Timber Resource Group has FSC certified some if its California portfolio.
[e] The Forestland Group had 13,000 ha (located in Tennessee) of its forestlands FSC certified in October 1999, which represents 14 percent of the Forestland Group's total forest area under management (176,000 ha) (http://www.forestlandgroup.com).

Although most environmental groups and pro-FSC retailers have reaffirmed their commitment to the FSC, recent actions by some environmental groups and retailers make the continued shunning of the SFI by these groups unsure.

The US case supports hypotheses 1 and 2 (*Forest companies and non-industrial forest owners in a country or region that sells a high proportion of its forest products to foreign markets are more likely to be convinced to support the FSC than those who sell primarily in a domestic-centered market [H1]; and Forest companies and non-industrial forest owners selling wood to a domestic market in a country or region that imports a large proportion of all the forest products it consumes are more likely to be convinced to support the FSC than those in a region that imports a small proportion of all the forest products it consumes [H2]*) in two ways. First, in comparison with BC, the US

Table 4.5 Support for the FSC and the SFI/Tree Farm by forest industry and landowner associations in the US

Association	No position	Supports FSC	Supports SFI/Tree Farm[a]
National Hardwood Lumber Association			✓
National Woodland Owners Association	✓		
Oregon Forest Industries Council	✓		
Washington Forest Protection Association			✓
Washington Contract Loggers Association			✓
Washington Farm Forestry Association			✓
California Forestry Association	✓		
Western Wood Products Association	✓		
Michigan Forestry Association			✓
Ohio Forestry Association			✓
Missouri Forest Products Association			✓
New Hampshire Timberland Owners Association			✓
Georgia Forestry Association	✓		
Texas Forestry Association			✓
Arkansas Forestry Association			✓
Louisiana Forestry Association			✓
Florida Forestry Association	✓		
Virginia Forestry Association			✓
Southern Forest Products Association			✓

[a] Associations that had no official position but undertook activities in support of the SFI or Tree Farm (such as participation in an SFI implementation committee) were included in this column.

case revealed that even when the FSC was able to gain support from key retailers such as Home Depot, the ability of this support to alter forest company evaluations was much weaker in the US than in BC. While part of the explanation has to do with the nature of the forest sector and the history of forestry on the public policy agenda, another important factor is that it is more difficult to obtain and maintain support from companies further down the supply chain when they operate in the same country or region as those suppliers of whom they are making unpopular demands. Many retailers do not want to take actions that might punish companies operating in the same political arena (i.e., within the same region or country), fearing that they may face significant backlash made possible by laws regulating business practices within the given country or region, and thus are more open to supporting

FSC-competitors' programs as well. The result was that many retailers came to support the SFI in addition to the FSC and that only those targeted by environmental groups with direct action used language more supportive of the FSC.

Where support did exist further down the supply chain, the other "forest sector structure" and "policy agenda" variables intersected to limit the effects of these choices. In particular, the US case, with its abundance of small landowners who provide much of the raw materials used by industrial forest companies, including most AF&PA members, also provides strong support for hypothesis 4 (*Unfragmented non-industrial forest ownerships are more likely to be convinced to support the FSC than those in a country or region characterized by a high degree of non-industrial forest ownership fragmentation*). Indeed, forest industry officials continually voiced concerns about the FSC program's chain-of-custody requirement, and SFI supporters took advantage of this criticism to claim that the requirement was not feasible in the US context, where company forestlands generally supply a small proportion of total fiber requirements, unlike in BC, where forest companies have long-term tenure over a land base that essentially meets their supply needs. This created a vacuum in which the SFI is now working to create a product tracking system in the hopes that the market might eventually accept it over a full-fledged chain-of-custody system.

The US also strongly supports hypothesis 5 (*Forest companies and non-industrial forest owners in a country or region with diffuse or non-existent associational systems are more likely to be convinced to support the FSC than those in a country or region with relatively well-coordinated, unified associational systems*). The well-coordinated and resource-rich AF&PA allowed for the development of highly strategic choices in their efforts to promote the SFI. The AF&PA was not only effective in responding quickly to the FSC threat when it developed in the mid-1990s, but it also had the experience and financing available to craft the SFI program in an optimum way, taking into consideration the results of public opinion research and the concerns of its core audience, AF&PA members. As the debate progressed and the AF&PA realized that public opinion was not, in fact, driving certification, but rather it was the preferences of retailers, the program took measures to conform the SFI to those characteristics that retailers deemed important in a forest certification program—independent of the forest industry, having third-party auditing, and international. The effect of the conforming changes on the buying preferences of retailers has been twofold: it has forced the FSC and its supporters to redouble their efforts to maintain support further down the supply chain, and it has been successful in earning the SFI support from a number of key retailers.

The US case supports and qualifies hypothesis 3 (*Large and concentrated industrial forest companies are more likely to be convinced to support the FSC than relatively small and less concentrated industrial forest companies*). Large industrial US forest companies exhibit a high degree of vertical integration, but a lesser degree than competitors in BC and Sweden, as they rely on non-industrial private forests for much of their fiber supply. Hence, hypothesis 5 is qualified because if accompanied in a region with fragmented ownership, this feature can work to solidify support for the FSC competitor, once a decision has been made not to support the FSC.

The US case supports hypothesis 6 (*Forest companies and non-industrial forest owners in a country or region with sustained and extensive environmental group and public dissatisfaction with forestry practices are more likely to be convinced to support the FSC than those in a country or region with less dissatisfaction*) and hypothesis 7 (*Forest companies and non-industrial forest owners in a country or region where access to state forestry agencies is shared with non-business interests are more likely to be convinced to support the FSC than those in a country or region where forest companies and non-industrial forest owners enjoy relatively close relations with state forestry agencies vis-à-vis non-business interests*), in two key respects. First, while there was a high level of criticism by environmental groups on private land, scrutiny was relatively low — as historical efforts by most citizens and groups were focused on National Forest lands. Second, the closer relations enjoyed between the forest industry and landowners and their state forest management agencies worked to limit industry's interests in engaging with an FSC-style system, which might see them lose considerable influence over policy choices. Hence, such a condition facilitated support for more flexible and industry friendly standards at the state level, and support for the FSC competitor, with its similarly relatively flexible approach to sustainable forest management.

And just as importantly, the US case also provides insights into issues surrounding a certification program's need to maintain core audience support. While the AF&PA has, so far, been able to keep nearly all of its members true to the SFI program despite the conforming activities undertaken to change the program from its original incarnation, the FSC already appears to face different challenges in keeping its core membership satisfied as it attempts a variety of conforming techniques to become more than a niche player for the US certification market. The difficult environment for FSC strategists in the US made it difficult for them in walking the fine line between maintaining support from their core audience and from the market in an environment in which significant program conforming efforts were necessary. And, even when changes were made to the FSC, the impacts of such strategic choices were far from preordained. In this sense, the US chapter illustrates our point made in chapter 2 — that any understanding of the way the FSC emerges and potentially gains legitimacy must explore the interaction of the identified explanatory factors with the strategic choices made by FSC supporters within this environment. Our case shows that despite best efforts, many of these choices have worked to hinder, rather than enhance, FSC efforts to achieve legitimacy in the US context.

PART **III**

Europe

5

The United Kingdom

Forest certification came to the UK earlier than any other case reviewed in this book, as Forest Stewardship Council (FSC) supporters made a strategic calculation that pursuing certification of the UK's own domestic forestlands would give enhanced credibility to their efforts to promote the FSC internationally. Initial efforts to promote the FSC not only gained the interest of UK companies down the supply chain, but the lumber retailer B&Q helped *create* the FSC, following its efforts to address whether lumber products they purchased came from sustainable sources. Yet key forest management interests, including timber growers, woodland management firms, and even government land management agencies, did not, at first, grant support to the FSC. However, after nearly a decade of strategic efforts by FSC supporters to gain support — involving a combination of market-based converting approaches and program-centered conforming tactics — all government-owned lands and some lands overseen by woodland management have supported and have become certified under the FSC. The two largest private landowner associations were marginalized as the only major groups not directly supporting the FSC, but even they took a "neutral" stance on FSC versus Pan European Forest Certification (PEFC) support. And unlike their German counterparts, the few private landowners supporting the PEFC in the UK have yet to see their program certify any UK forestlands.

This chapter reveals that, like the other cases in this book, the UK forest sector's particular position in the global economy, the structure of its forest sector, and the historical evolution of the public policy process in addressing environmental interests all worked to facilitate FSC efforts to gain support in a way that did not occur in Germany despite both countries being large importers of forest products.

The UK's position in the global economy was important in explaining the eventual support for the FSC seen there. Not only does the UK rely more heavily on imports than any other case reviewed in this book, but it is also particularly dependent on solid wood products from regions such as Brazil, Indonesia, and Malaysia, where forest degradation was and remains a serious international concern. These features meant that in the UK it was the retailer who, arguably more in tune with public demand, played a key role in certification both domestically and internationally. And unlike German publishers, who relied on pulp wood imports and who were forced to react in a way consistent with environmental groups' market campaigns, key UK retailers would, relatively early on, take *proactive* decisions, deciding on their own that certification should also be required of domestic producers. The UK domestic forest sector is also the least important to its domestic economy of any of the cases under review. This had two effects. First, it meant that UK landowners and value-added companies were not as easily able to muster backlash efforts against the FSC, be they claims of increased economic costs or loss of sovereign authority. Second, it meant that UK landowners and value-added companies would be, as we show below, more susceptible to choices by foreign competitors to support the FSC, such as made by AssiDomän, for fear of losing market share.

Four aspects of the structure of the domestic forest sector influenced FSC efforts to gain support. The first was that the government was a key landowner in its own right, and unlike Germany, the government, qua landowner, was able to make certification choices about its own forestlands somewhat independently from the decisions of private forest landowners. Second, the existence of large forest management companies who managed private forestland was also important because choices made by these woodland management firms could influence, to some degree, forest management choices on private forestlands. Third, the associational system was among the most diffuse of all the sectors reviewed in this book, making it difficult for it to develop timely strategic responses to FSC efforts. Fourth, the absence of vertical integration among forest companies meant that there were few companies that had to address both the forest management and chain-of-custody aspects of certification. This feature presented companies considering the FSC with fewer obstacles than those in other countries who both processed forest products and owned forestland.

The place in the global economy and structure of the forest sector themselves shaped the historical evolution of the public policy process in addressing environmental interests, which also facilitated direct and indirect support for FSC-style certification. Direct support came from widespread citizen concerns about UK domestic forest practices that began to emerge in the 1980s.[1] Indirect support came from the UK government itself. After the failure of the Rio Earth Summit to agree to a legally binding global forest convention and in the face of increasing domestic scrutiny, the government chose to become a global leader in the promotion of sustainable forest management (SFM). This decision worked to increase slightly the government's autonomy from traditional government-landowner clientelist policy networks, while retaining its credibility as a facilitator of stakeholder processes. And the UK government's choice to first develop sustainable forestry approaches through a multi-stakeholder process, rather than introduce legislation governing SFM, opened the door for FSC supporters to promote the FSC as the policy instrument through which SFM could be implemented. Such maneuvering made it difficult for private landowners and other forest sector interests to seek widespread support for the FSC-competitor program, the PEFC.

This chapter details this argument in five analytical steps. First, it reviews the three factors that facilitated or hindered FSC efforts to gain support in the UK. This section explores the UK's place in the global economy, highlighting both its similarities to and differences from Germany's place in the global economy, the historical evolution of the public policy process in addressing environmental interests, and key structural features of the UK's domestic forest sector. Second, it reviews the key events in the 1980s and 1990s that ultimately led to certification being an issue for UK domestic forestlands. Third, it traces initial support for the FSC and the initial "FSC-competitor" response. Fourth, it details how the three explanatory factors mediated efforts by FSC strategists to gain increased support. The conclusion illustrates how these findings support the hypotheses detailed in the introduction to this book.

Factor 1: Place in the Global Economy

HIGH DEPENDENCE ON FOREIGN IMPORTS

The UK is the most import reliant case explored in this book.[2] The UK's domestic production of industrial roundwood (logs sold in the marketplace) (FAO 2002) is six times smaller than that of Germany. These features render

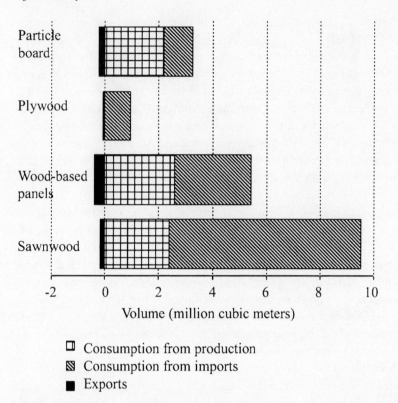

Figure 5.1. Volume (in cubic meters) of UK apparent consumption (production plus imports less exports) for various aggregate wood products in 1999. In this year, the UK purchased 46 percent of its sawnwood imports from Finland and Sweden and 53 percent of its plywood imports from Brazil, Indonesia, and Malaysia. Exports to Germany and the Netherlands represented 55 percent of its total exports. Source: FAO 2001.

the UK one of the world's largest import consuming markets in a number of key forest product categories, ranking fourth in sawnwood,[3] fifth in wood-based panel imports,[4] fourth for paper and paperboard products, and eighth in wood pulp. We distinguish these products because they help us understand, in the UK case, from where the first international pressures for certification came. Sawnwood is imported predominantly from Scandinavia; hence, early competitive pressure from Swedish forest companies (chapter 7) was important in affecting lumber sold by retailers. Many of the UK's wood-based panels (particularly plywood) come from tropical countries such as Brazil, Indonesia, and Malaysia (figure 5.1), which had become a key focus of UK consumer and environmental group campaigns in the late 1980s and early 1990s.[5] Over one-third of the UK's wood pulp imports come from Canada (326,182 metric tons) and the United States (318,880 metric tons), with an additional one-sixth

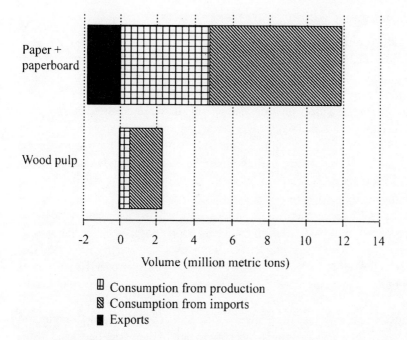

Figure 5.2. Volume (in metric tons) of UK apparent consumption (production plus imports less exports) for wood pulp and paper plus paperboard products in 1999. In this year, the UK purchased 19 percent of its wood pulp imports from Canada, 18 percent from the US, 16 percent from Brazil, 11 percent from Finland, and 11 percent from Sweden. The most significant source of paper plus paperboard products was Finland (27 percent), with another 19 percent coming from Sweden, 12 percent from Germany, and 8 percent from France. Exports to Germany, France, the Netherlands, and Belgium-Luxembourg represented 50 percent of its total paper plus paperboard exports. Exports of wood pulp were negligible. Source: FAO 2001.

from Brazil (275,211 metric tons) (figure 5.2). This meant that when BC (see chapter 3) became a target of UK activists, UK paper companies became important targets and sources of additional pressure within the UK for FSC forest products. Likewise, the UK imports half of its paper and paperboard from Finland (1,926,588 metric tons), Sweden (1,378,622 metric tons), and Germany (849,857 metric tons) (FAO 2002).

While the domestic grown forest products represent a small share of the UK market, virtually all of these products are consumed within the UK.

Factor 2: Structure of the Forest Sector

Three factors are key in understanding the influence of the structure of the UK's forest sector in influencing FSC efforts to gain support.

INDUSTRIAL FOREST-COMPANY CONCENTRATION

The forest sector in the UK is the smallest of any case in this book and, much like in Germany, there are few vertically integrated companies.[6] Local primary and secondary manufacturers of wood do not own forestlands, but instead rely on supply secured through contracts with domestic and international suppliers. Production is divided among three principal sectors, with the greatest concentration in pulp and paper and wood-based panel processors. In 2000 there were 315 sawmills, 4 pulp and paper mills, and 9 wood-based panel mills operating that used domestic industrial roundwood (Forestry Commission 2001). Sawmills and wood-based panel processors combine to consume nearly 80 percent of the domestic grown timber (Forestry Commission 2001). More than a third of the sawmills produce less than 1,000 cubic meters per annum (Balachandran and Henderson 2001), while the nine wood-based panel processors combine to consume over 20 percent of the total domestic industrial roundwood production; this represents over 50 percent of these mills' fiber requirements. For the other 50 percent of their fiber needs, these companies rely almost exclusively on sawmill co-products (residual sawdust, chips, and slabs from sawn wood production) (Forestry Commission 2001). Wood-based panels are a key component of many pre-packaged furniture, flooring, and other products sold by the do-it-yourself (DIY) retail market.[7] These structural characteristics of UK-based production are important because they separated landowner decisions from those made by processors and gave increased power to horizontally integrated parts of the supply chain, particularly the retail companies.

NON-INDUSTRIAL FOREST OWNERSHIP FRAGMENTATION

Interests other than industrial forest companies own the lion's share of forestlands in the UK. Approximately one-third of forestland (861,000 ha) is concentrated in the hands of the British Forestry Commission and the Forest Service of Northern Ireland; the other two thirds (1,929,000 ha) is fragmented among farmers, other private owners, voluntary bodies, and other government departments or agencies (figure 5.3) (Forestry Commission 2001).

At the same time, a significant number of private landowners contract woodland management firms to manage their land, in coordination with other landowners. Three firms, Scottish Woodlands, Fountains PLC, and Tilhill Economic Forestry are responsible for management of over 500,000 hectares of UK forestlands (Scottish Woodlands 2002; Tilhill Economic Forestry 2002; Fountains 2002). These landownership patterns were important because they meant that while there were few large private forest landowners to target in

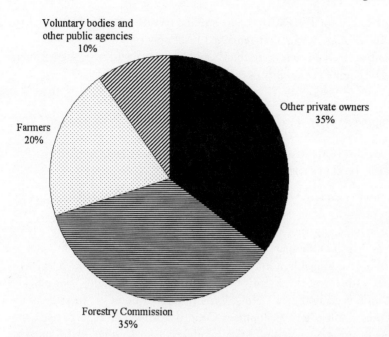

Figure 5.3. Ownership of forested lands in the United Kingdom. The figure does not include forest land ownership in Northern Ireland where the Forest Service is the dominant land-owner: it owns 73 percent (61,000 ha) of the total forest area (Forestry Commission 2001). Source: Forestry Commission, Great Britain 1998.

campaigns, there did exist a large public landowner and a concentrated group of management firms that could be targeted for FSC-style certification.

ASSOCIATIONAL SYSTEM COHESION

Various organizational attempts have been made since the early 1980s to build a more cohesive associational system to represent the entire British forest sector. These efforts first led to the creation of the Forestry Industry Committee of Great Britain (FICGB), which eventually gave way to the Forestry Industry Council, which in turn was replaced by the Forestry Industry Development Council (FIDC), whose membership comprised most of the British sector's major interests.[8] However, individual members, especially the Timber Growers Association (TGA), were left able to continue developing their own policies and strategies. The relatively loose nature of the FIDC associational system led, in part, to the solid wood sector developing its own association, the UK Forest Products Association, in 1995 (Forestry Industry Council of Great Britain 2000). These divisions within the industry made it relatively easier for the FSC to court specific members of the FIDC and to bypass those associations that remained resolutely opposed. This stands in stark contrast to the situation in the US and Germany, where strong national associations limited the ability of the FSC to pursue such strategies.

Factor 3: History of Forestry on the Public Policy Agenda

SUSTAINED AND EXTENSIVE PUBLIC DISSATISFACTION WITH FORESTRY PRACTICES

Unlike Germany, where very few significant issues arose to challenge the forestry practices of its sector, the UK forest sector had been subject to a number of concerns about forestry practices beginning most prominently in the 1980s. Much of the spark for this scrutiny had its origins in the findings of the National Auditing Organisation's report in 1986 — a report that began as a critical assessment of the sector's financial accounting methods[9] but quickly evolved into a general critique of a broad range of domestic forestry practices (Forest Industries Development Council 2001). The report paved the way for an array of equally critical analyses from leading environmental groups, such as the Royal Society for the Protection of Birds (Turner 1991), which claimed that industrial forestry practices were not financially or environmentally appropriate. Environmental groups and the general public called for preservation of the country's few remaining ancient woodlands (Tickell 2000), increased protection of endangered species and ecosystems, and greater emphasis on non-timber forest products and forest recreation and amenity values.[10] The result of these pressures was considerable. Persistent environmental pressure and public concern about forestry practices in the UK resulted in broad support for efforts to improve forest management practices, which could not be ignored by forest companies and landowners.

FOREST-COMPANY AND LANDOWNER RELATIONS WITH STATE AGENCIES RELATIVE TO OTHER SOCIETAL INTERESTS

The role of the Forestry Commission of Great Britain is also central to our analysis not just because of the forestland it owns and manages, but also for its regulatory, education, and support responsibilities, which shape forest management on private lands and sectorwide responses to emergent forest management problems.[11] Its official mandate is to protect British forests, promote reforestation and afforestation (planting areas formerly not under forest cover), and encourage and support domestic production of wood and other forest products. Statutory authority gives the commission the right to regulate

Table 5.1 Influence of explanatory variables on FSC efforts to gain support from forest companies and non-industrial landowners in the UK

Factor	Existence
Place in global economy	
Dependence on foreign markets	✓
Structure of forest sector	
Large, concentrated industrial forest companies	
Unfragmented non-industrial forestland ownership	1/2
Diffuse or non-existent associational systems	✓
History of forestry on public policy agenda	
Sustained and extensive public dissatisfaction with forestry practices	✓
Forest companies and non-industrial forest land owners share access to state forestry agencies with other societal interests	

Note: A "check mark" indicates strong presence and "1/2" indicates mixed presence.

forest practices with felling controls for private lands and forest design plans for public lands (i.e., Forest Enterprise)[12] as well as providing education and research programs to advise and assist woodland managers and owners with their management operations.[13] The commission's responsibilities were divided among three executive agencies, the Forest Enterprise that oversees forest management, Forest Research that conducts research, and the Forest Authority that deals with regulatory and grant issues (which are now the responsibility of the commission's Policy and Practices and Conservancies divisions). Both the commission's agencies and divisions developed close, somewhat paternalistic relations with the private sector during the late 1980s and early 1990s.[14] And while close ties limited early FSC attempts at converting, these relationships permitted the commission to emerge as a widely respected arena in which to develop multi-stakeholder processes over forest resource use domestically, processes that would prove integral to the FSC's success in becoming the dominant program at the supply and the demand ends of the market's supply chain.

The factors discussed above influenced the ability of the FSC to gain support from UK forest owners. While not all factors created facilitated FSC efforts (table 5.1), highly strategic choices made within this environment by the FSC and its supporters helped earn the eventual support from the government (as forest landowner) and woodland management firms, while concurrently limiting resistance from the country's private landowners.

Introduction of FSC Certification to the UK

Introduction of certification to the domestic UK forest sector was directly linked to the country's reliance on tropical wood product imports (Glastra 1999; Global Witness 1999; Matthew 2001; Ghazi 1994b). As in Germany, concerns over tropical forest depletion and degradation were addressed in the 1980s through boycott campaigns, which, by the 1990s were deemed largely unsuccessful in arresting tropical forest degradation and destruction (P. Knight 1996).[15] As Alan Knight, B&Q's leading environment official, concluded, "Our numerous, worldwide forest visits and research identified quickly that a boycott of tropical timber would not work. Such action would remove any economic incentive to protect tropical forests and do nothing to protect non-tropical forests" (A. Knight 2000:3).[16] Parallel efforts were also made through the International Tropical Timber Organization (ITTO), where the UK government and retailers pushed tropical countries to use the ITTO's International Tropical Timber Agreement to promote sustainability as much as to open access to trade.[17] Yet efforts to promote sustainable forest management through the ITTO were also hindered. Through the ITTO's "Target 2000" program, in 1990 many tropical exporting countries pledged that by the year 2000 they would only trade in timber that came from sustainable sources (Humphreys 1996b). However, many environmental groups were skeptical of this commitment and the limited resources earmarked for its implementation.[18] Tropical countries also felt that they were being unfairly scrutinized by developed countries, which led one UK retailer to assert: "One of the stumbling blocks to reaching a consensus [on sustainable forest management within the ITTO] is the producer countries' insistence — quite rightly — that wood certification be applied to *all* timbers: something which certain consumer countries refuse to accept. But we must carry on working towards achieving a level playing field" (Ankrah 1994).

The ITTO may have proven a poor institution within which to promote global sustainable forest management[19] but it did serve to emphasize and focus efforts on the use of markets, rather than traditional governmental regulations, to promote environmental stewardship. The World Wildlife Fund United Kingdom (WWF-UK) picked up on this theme and argued that companies importing wood products should not only be aware of their products' origins, but ought to also develop purchasing policies requiring their products come from forests certified to be well managed. Following the failure of Rio to sign a binding global forest convention, such an approach came to be viewed by leading environmental groups as a more effective and appropriate way to ensure that companies stopped contributing to the destruction of tropical forests in Brazil, Malaysia, and Indonesia, and ancient temperate forests in British Columbia (Schoon 1992).[20] As a result of these efforts, in 1991 the WWF-UK formed the world's first buyers group, made up of member companies committed to "purchasing substantial and increasing volumes of their wood and wood products requirements from well-managed forests" (World Wildlife Fund UK 1997).[21]

Unlike Germany however, interest in forest certification as a tool to promote sustainable forest management in tropical countries led, almost immediately, to enhanced support for FSC-style certification on the UK's own domestic forestlands. This early entry of the FSC is attributable to the UK's huge dependence on foreign imports, which led environmentally proactive retailers to blaze the certification path in the UK, rather than less proactive pulp and paper producers or associated sectors, such as publishers, further down the supply chain. And UK retailers quickly realized that if they were to demand sustainable forestry practices of their tropical suppliers, they would also need to require the same of their own domestic forest sector. As an official with Do It All, a lead DIY retailer, stated: "There is pictorial and documentary evidence to suggest that some forests in tropical countries are better managed than certain temperate forests. This kind of data prompts the question: just how righteous can northern hemisphere countries be?" (Ankrah 1994).

The decision by key UK retailers to pursue FSC certification on domestic forestlands was enhanced by a UK government that itself had begun to raise serious questions about global and domestic forest management, as well as the lack of public support for existing forestry practices. The UK government's increasing role in scrutinizing both global and domestic forestry practices can be traced to the failure of the Rio Earth Summit to produce a binding global forest convention. After Rio the UK government took a proactive role in defining and encouraging the practice of sustainable forest management. It participated heavily in European processes developed to address forest management, such as the Helsinki conference in 1993. By 1994, it had released four reports outlining its approach to sustainable forestry, sustainable development more broadly, biodiversity conservation, and climate change (Forestry Authority 1998). The government maintained an interest in a legally binding convention, but likewise recognized a need to "put its own house in order."[22] While the government of the UK had not created these initiatives with certification in mind as the solution to the global forest degradation crisis, its attention to SFM created a domestic policy environment in the UK that was more conducive to new and innovative approaches. Ironically the historically close relationships between British Forestry Commission officials and the domestic forest sector and the increasing scrutiny of the sector's own forestry practices meant that certification, as a policy instrument outside of government, was viewed more favorably by critics of domestic forest policies than any governmental response would have been (Ghazi 1994a; Turner 1991).

As a result, and in contrast to the German experience, local environmental and conservation groups quickly supported FSC-style certification as a means to address domestic practices. Early on, the UK's Soil Association developed programs to aid the FSC in its local certification efforts, establishing in 1992 the Responsible Forestry Programme that "[applied its] philosophy of sustainable resource use and [its] certification expertise to the area of forestry and timber" (Wenban-Smith 2001).

These conditions and initial support from environmental groups did not, by themselves, guarantee FSC success in its efforts to achieve widespread governing authority. Rather, it was a keen awareness of their strategic implications that allowed the FSC and its supporters, through converting and conforming strategies, to gain support for domestic certification in ways that FSC supporters in Germany were unable.

Initial Efforts to Obtain Forest Owner Support: The FSC versus Label of Origin

Until 2002, no sustainable forestry management certification program had emerged as an alternative to the FSC. Instead, those opposed to the FSC initially relied on the UK's version of a "label of origin" program, which only verified that wood was cut from UK forestlands in adherence with government regulations. The lack of an alternative gave the FSC and its supporters valuable time with which to institutionalize their program in the UK, thus making it more difficult for future competitors to undertake support-generating strategies of their own.

The FSC and its supporters were quick off the mark to establish and promote the FSC domestically[23]; for instance, the UK was home to the first FSC-endorsed national contact person (table 5.2) (Meridian Institute 2001b). As with the approach of UK retailers, the FSC supporters asserted that the UK could not require higher standards of foreign countries than it did of its domestic producers, asserting that it must promote "well-managed [forestry] both at home and *abroad*" (Tickell 1996, italics added).

The principal player in developing FSC's strategic choices was the WWF-UK, who, in conjunction with other members of the FSC steering group, first focused the bulk of their efforts in applying market-based converting strategies.[24] The WWF was able to effectively draw on its already established UK

WWF 95 Group, whose members immediately supported the FSC domestically and internationally. They recognized that once the market connection could be solidified, FSC certification would provide the "single credible, mark for all types of wood and paper from all over the world" (P. Knight 1996). The WWF worked to enhance the market pressure by expanding the 95 Group's (now the 95+ Group) membership. The WWF 95+ Group gained increased support, often through direct action campaigns led by more radical groups such as Greenpeace, Earth First!, and Friends of the Earth, who organized direct action protests at retail outlets and wholesalers' lumber distribution centers. Targets included companies such as Harrods and Timbmet, both of which imported significant amounts of tropical hardwoods (Tickell 1993; "Harrods Faces Rainforest Protest" 1993).[25] In other cases the WWF relied on its public profile and understanding of the technical issues as a means to inform companies who would already be morally predisposed to FSC-style certification.[26]

The FSC and its supporters also worked to institutionalize and facilitate anticipated market support. Before the FSC standards were officially endorsed in the UK, the FSC and its supporters worked with the Soil Association to establish the Woodmark certification scheme in 1994, which was part of the organization's Responsible Forestry Programme (Soil Association UK n.d.; Wenban-Smith 2001). The scheme had already developed a chain-of-custody process that would prove crucial in the race to get FSC-certified forest products to the market place.

Despite increasing market support for the FSC, private landowners, woodland management firms, and wood product processors remained opposed to FSC-style certification, and, like their counterparts in other countries, viewed the FSC as far too prescriptive and limiting of forest managers' discretion. Officials from these groups argued that the FSC lacked the required knowledge of the British industry to develop valid standards measuring appropriate forestry practices — feelings fueled by their realization that many FSC supporters came from outside the traditional forest management community. These groups were particularly concerned that the WWF, an organization many treated with skepticism, was spearheading FSC efforts in the UK.[27]

In addition, they voiced concern about the initial voting structure of the FSC, which institutionalized a limited role for business interests[28] and claimed that the FSC was acting as a "foreign entity," attempting to usurp rule-development authority from the UK forestry community (Timber Grower 1994b). Most landowners and forest managers generally agreed with the idea that forest management should be improved in tropical forests, but felt that having the same program applied in the UK was inappropriate due to existing government

Table 5.2 Strategic efforts to either gain or maintain support between 1991 and 2002

Year	FSC competitor program supporters[a]		FSC supporters	
	Action	Strategy	Action	Strategy
1991			WWF and 15 demand side companies create the world's first buyers' group, the WWF 95 group	Converting
			Earth First, Greenpeace, and Friends of the Earth lead ongoing direct action campaigns (protests) against retailers and wood product importers and distributors[b]	Converting
1992			Soil Association creates Responsible Forestry Programme	Converting
1994	Forestry Industry Council of Great Britain (supported by the Forestry Commission) creates the FICGB Woodmark "label of origin"	Conforming	Soil Association creates its Woodmark certification scheme	Converting
	Government releases reports on approach to international commitments (climate change, biodiversity, and sustainable forestry and development)	Conforming		
1996	Creation of Woodmark Recovery allows recycled products to carry label	Conforming	SGS Forestry and Soil Association receive FSC accreditation	Converting
			FSC begins to develop a group certification system	Conforming
1997			B&Q issues letter to suppliers indicating it intends to source only FSC-certified wood by the end of 1999	Converting

Year	Event	Classification
	WWF/Soil Association creates Local Authorities project	Conforming / Converting
	Membership in WWF 95+ Group expands to 75	Converting
	FSC and supporters participate in audit protocol talks	Conforming
1998	Membership in WWF 95+ Group expands to 87	Converting
	WWF works to break down tense relations with the forest sector through talks with the TGA	Conforming
	FSC releases guidelines for group certification	Conforming
	FSC opens discussion on revising percentage-based claims policy	Conforming
1999	FSC supporters accept the UKWAS standard (liberal provisions on chemical use permitted conditional on steering group being set up to address the issue)	Conforming
	FSC recognizes the FSC-UK standard and its equivalence with the UKWAS standard	Converting
2000	PEFC UK Ltd. established to compete with the FSC as a certifier to the UKWAS standard	Conforming
	Minister of Environment announces UK government timber procurement policy supporting certification	Converting
	FSC lowers thresholds for percentage-based claims	Conforming
2002	PEFC UK initiative endorsed by the PEFC council	Conforming
	FSC creates group chain-of-custody certification	Conforming

[a] FSC competitor programs include the FICGB Woodmark and the late-developing UK PEFC initiative.

[b] Direct action campaigns occurred throughout this period.

standards.[29] As the chair of the Timber Growers Association noted, "We have no problem with the principle of a certification scheme; what we cannot accept is a scheme administered by a self-appointed body having no legal status and which second guesses the decisions of our own Forestry Authority" (Christie-Miller 1994). Landowners and managers instead sought to defend existing practices and public forest policy processes — where they enjoyed a more prominent role in policy making and a close relationship with the British Forestry Commission.

This position meant that private forest landowners, woodland management companies, and even those responsible for choices on government-owned forestland were focused on influencing SFM within existing public policy processes and, hence, limiting their idea of certification to a program that simply identified trees harvested within Great Britain (Timber Grower 1994a). The FICGB and the Forestry Commission developed a "local wood"-labeling program: the FICGB Woodmark.[30] Launched in 1994, the program was designed to offer third-party verification[31] to end-consumers that labeled wood products coming from trees harvested according to government regulations (Certification Information Service 2001; Forestry Industry Council of Great Britain 2000; Kiekens 1997).

The decision to focus on a label of origin program and on the public policy-making process led key players within the UK domestic forest policy sector, including most private forest landowners and woodland management companies, to initially boycott the FSC standards setting process,[32] as their counterparts in most other cases in this book did. And the FSC international policy that forbade government officials from formally participating was viewed as illegitimate by a wide range of interests within the domestic forest sector.[33] However, retailers were strong supporters of the FSC-GB national initiative, with the B&Q, Homebase, and others participating in standards development and offering financial backing to the process (Scrase et al. 1998).[34] Both retailers and environmental groups were frustrated by governmental efforts to address global forest degradation and shared little faith in domestic initiatives designed to measure, rather than implement, intergovernmental sustainable forestry commitments.

The choices of private forest landowners and key value-added producers in Great Britain to promote their label of origin program and governmental policies as appropriate means for verifying sustainable forest management had two consequences. First, they led the FSC to focus increasingly on market-based converting campaigns in an effort to force private forest and governmental landowners to change their evaluations of the FSC. Second, they resulted in an increased attention by these organizations to SFM issues and the *supporting of governmental efforts* designed to define and promote them.

The British Forestry Commission contributed to the latter goal by using public policy processes and sector wide discussions to seek agreement on sustainable forestry practices for the UK. The forest sector took the lead by signing a Forestry Accord in 1996 that detailed its commitment to implementing the Statement of Forestry Principles adopted at the United Nations Conference on Environment and Development (Forest Industries Development Council 2001).

The commission also worked on the development of "a UK Forestry Standard" designed to address the UK government's commitments under the Helsinki and Lisbon sustainable forestry criteria and indicator process. Such an approach was viewed as more legitimate by the TGA than the FSC standards process. As TGA Chair Mark Crichton-Maitland (1995) reasoned, "We support the inter-governmental process started at Rio and continued through Helsinki. We believe that the principles and practices developed through the United Nations Commission on Sustainable Development (UNCSD) should be applied at a national level, taking into account the different circumstances in each country. . . . But we do not accept the principle that non-accountable bodies should be able to impose principles and criteria where national systems, in accordance with the UNCSD, exist."

The TGA thus supported parallel tracks of governmental efforts to define sustainability and its own label of origin program that would identify and label products coming from British forests. Early on, the TGA and the FICGB exerted considerable efforts to promote the program, informing TGA members about the program, and encouraging them to support it (Timber Grower 1994a; Crichton-Maitland 1996). As Mark Crichton-Maitland (1997:11) exhorted, "The FICGB Woodmark is fast becoming the accepted mark of environmental quality assurance . . . Over 4,000 forests and woodlands are signed up to the scheme already and some 8,000 packages of timber are produced every week carrying the Woodmark label. This is where our efforts must lie and I call on all TGA members who have not already done so to take out Woodmark registration."

Imposing no additional cost on participants above those associated with government regulations (Williams 1995), the program attempted to build an alternative to the FSC that would enjoy the unified support of domestic forest owners, and managers. Expanding the program to include users of recycled wood fiber through the development of the Woodmark Recovery enhanced this mission (Timber Grower 1996) and enabled mills using recycled wood to produce FICGB Woodmark labeled goods. As Ronnie Williams (Timber Grower 1996), director of the FICGB, noted: "Responsible sourcing of raw material goes beyond drawing timber from well managed forests, and includes using available recyclable materials, waste from other processes, and the re-use of other wood products whenever technology allows. This is part of our policy to promote environmentally efficient sourcing of raw material in parallel with promoting the use of timber from British managed forests. All fibre is precious if it can be used and re-used."

However, just as in BC with its government's forest policy initiatives in the mid-1990s, these joint efforts to develop policy responses were criticized by environmental groups. The label of origin was deemed inadequate as a certification program. Environmental groups asserted that the Forestry Commission and the UK government's draft UK Forestry Standard of 1996 was weak and that the FSC was the only legitimate route, "Massive clear-cuts of up to 250 acres are still allowed . . . .[It] does nothing to stop landowners converting native woods to open pasture . . . .Guidelines on the blanket afforestation of important open habitats remain hopelessly weak. And European wildlife legislation such as the Habitats and Species Directive, [are] not even mentioned . . . The key to restoring our woodlands is thus to raise the price paid for timber from well-managed forest both at home and abroad, which is only likely to be achieved by widespread certification under the Forest Stewardship Council" (Tickell 1996).

The UK government, in turn, initially addressed these criticisms by committing to enhanced procedures to measure, monitor, and ensure sustainable forest management in UK forests (Guardian 1998).

At this point a stalemate existed, not between two private-sector certification programs, but between the environmental group and retailer-backed FSC on one hand and the industry and government sponsored label of origin program on the other hand. For environmental groups and members of the WWF 95+ Group, a "local wood" labeling program was not sufficient (Ankrah 1994; P. Knight 1996). Environmental groups also found the UK Forestry Standard to be wanting, asserting that it contained few substantive requirements and that by its very nature it was unable to address international forest degradation. As a result, the FSC and its supporters stepped up their market-based efforts aimed at converting UK forest owners to the FSC.

Turning to the Market: FSC Efforts to Convert Forest Owners

The FSC and its supporters first attempted to alter support by targeting its market converting campaigns to the most receptive audiences they could find. Market research conducted by the WWF indicated that one of the most receptive and easily altered consumers of wood products was the UK's local authorities (local governments). They were finite in number, consumed a substantial quantity of wood,[35] and were exposed to public scrutiny.[36] The WWF, with the help of the Soil Association, immediately created a highly strategic local authorities project,[37] which was designed to encourage authorities to develop, and then aided them in implementing wood procurement policies that specified a commitment to purchase independently certified forest products (World Wildlife Fund United Kingdom 2001b). The WWF created a template policy "assessed to conform to international and national trade agreements,"[38] provided guidance on how to ensure a policy was properly implemented, and set up an information and assistance service for any local authority confronting wood procurement related problems (World Wildlife Fund United Kingdom 2001b; World Wildlife Fund United Kingdom n.d.; World Wildlife Fund United Kingdom n.d.). The project was successful in that it generated further market interest for FSC within the UK, which increased the pressure on domestic producers to consider FSC-style certification.

Further, and more important in terms of its impact on the size of market supporting FSC certification, the WWF 95+ Group was continuing to expand. By 1997 the group had increased from 15 to 75 companies (World Wildlife Fund UK 1997). By mid-1998 membership had increased to 87 companies whose combined sales were reported to total approximately £69 billion (Hansen and Juslin 1999:24, table 2). Specific actions of key companies were also important. In late 1997 B&Q sent a letter to all of its suppliers — including wood-based panel manufacturers in the UK — stating that it was the company's intention to only purchase wood products carrying the FSC label by the end of 1999.[39] Later, in January 1998, these intentions were made public during a meeting of the World Bank in Washington, DC where B&Q's CEO announced its changed purchasing policy and its clear deadline: the end of 1999 (Stanbury 2000; DIY 1998; National Home Center News 1998) (see chapter 3). This had serious implications for suppliers the world over: UK manufacturers, and by association UK woodland managers and public and private landowners, were issued the same ultimatum as forest companies in BC, Sweden, Finland, Malaysia, and elsewhere.[40] The impact appeared to be immediate, as many of the companies further down the supply chain joined the 95+ buyers' group, or at least began to consider options for securing FSC-certified wood and obtaining FSC chain-of-custody certification, following B&Q's commitments. These companies presumed that such support was needed or would be needed to maintain and potentially increase market share.[41]

In June 1998 market-based efforts were given a significant boost when the Swedish forest company AssiDomän, which at the time was the world's largest forest owner, announced its successful FSC certification of 3.3 million hectares of forestlands (World Wildlife Fund 1998). AssiDomän's certification meant members of the 95+ Group could soon purchase certified products from an existing international supplier, either directly or via chain-of-custody merchants, dealers, or secondary manufacturers in the UK. Indeed, soon after the company was certified the first FSC-labeled paper products appeared on the UK market (World Wildlife Fund 1998).

Added threats emerged as other supplier nations, such as Latvia and Estonia, with a 25 percent share of the UK's softwood sawnwood market (Forestry Industry Council of Great Britain 2000; World Wildlife Fund United Kingdom 2000), appeared poised to follow AssiDomän's lead. Since gaining independence from the former Soviet Union in the early 1990s, these countries had placed significant competitive pressures on private and public production of industrial roundwood in the UK, making their interest in certification even more threatening (Bills 2001).[42]

These pressures had different impacts on various components of the UK forest sector due to the individual business challenges faced by manufacturers, woodland management firms, and forest owners. Companies that owned UK processing facilities that supplied wood-based panels to B&Q and other members of the 95+ Group (such as Nexfor Ltd., formerly CSC Forest Products Ltd.) had to consider where they could obtain sufficient FSC-certified industrial roundwood and sawmill co-products to meet thresholds necessary to offer FSC products to the market. At the time, FSC imposed strict requirements on how much non-FSC wood could be included in a FSC-labeled product.[43] With panel manufacturers reliant on British industrial roundwood and sawmill co-products for 50 percent of their supply (Forestry Commission 2001c) many manufacturers were still monitoring in what direction FSC certification would go.

That UK wood panel manufacturers had a large share of the domestic market and produced domestically harvested wood products influenced forest owner support for the FSC in two ways. First, domestic manufacturers had a great deal to lose if B&Q and other retailers chose to obtain their panel products from other FSC-certified producers. Second, woodland managers and forest owners that harvested domestic industrial roundwood could not afford to lose their contracts with the panel sector. These firms feared that FSC-certified wood coming from countries such as Sweden and (potentially) Latvia and Estonia would reduce their own ability to sell their wood products in the UK. There was also the concern that if domestic owners and managers were not willing to get FSC certified, primary and secondary manufacturers would be forced to seek imported industrial roundwood from FSC-certified forests in other parts of the world. At the same time it would be in the interest of the UK domestic manufacturers to have their own domestic supply FSC certified, especially given the high thresholds set by the FSC's percentage-based claims policy.

Such recognition provided a key shift within the UK forest sector. Manufacturing firms informed forest landowners and the firms managing these lands that if they were not willing to pursue FSC certification they would lose their contracts. As Peter Wilson, the executive director of the TGA noted, "When our pan-industry umbrella group, the Forestry Industry Council of Great Britain, met to discuss the matter in May, the UK sawmillers and processors, who had been supportive of the cross-industry opposition to certification, advised that they now needed certification to safeguard UK timber's market share. These were our direct customers and their's [sic] was a view that could not be cast aside lightly" (P. Wilson 1998). Woodland management firms, the Forestry Commission's Forest Enterprise, and the TGA now recognized that their label of origin/government indicators route was failing to meet market demands, and that another route was needed.

At this time the commission itself became increasingly torn between its different interests. The commission clearly wished to enhance the economic well-being of the commercial forest sector and maintain close relations with private landowners and processors. The Forest Enterprise and authority agencies, having survived privatization, had a keen interest in maintaining their role in the management and regulation of British forests.[44] But the overall consensus-building stakeholder-participation interest of the commission was most focused on developing consensus over forest resource use, wherever that path might lead. The latter two roles began to be given greater weight as the public started to call for more attention to forest amenity and environmental values, in particular ancient woodlands and threatened and endangered species (Royal Society for the Protection of Birds 1993). As these values came to the fore, the commission gradually began to emphasize its bridge building role, focusing on addressing the divide between environmental groups and the forest sector.[45]

The FSC market pressures and increasing interest within government to address sustainable forest management created a unique conjunction of events that gave the commission an incentive to resolve certification disputes. The FSC had the label the market wanted, yet private landowners remained mistrustful of the program (Bills 2001; Forestry Industry Council of Great Britain 2000).

Changing the FSC to Increase Support

While many interest groups participated in both governmental processes and the FSC's Great Britain national initiative, environmental groups remained unconvinced that the governmental process could create rules to adequately address global forest degradation or stop its underlying causes, such as illegal logging.[46] All stakeholders became increasingly aware that certification, maybe even the FSC, would be needed to demonstrate and monitor sustainable forestry commitments. With the pending release of the UK Forestry Standard and the near completion of the FSC's Great Britain draft standard,[47] discussions began among these parties about whether a single process could not be established to recognize their compatibility (Bills 2001; Scrase et al. 1998).

Feeling confident about their compatibility, the commission contracted SGS Forestry, an FSC-accredited certifier, to carry out a comparison to determine the gaps between the requirements of the UK Forestry Standard and the FSC principles and criteria (P&C) and the FSC-GB draft standard. The report made a number of recommendations and noted some substantive differences between the FSC (both the FSC P&C and the FSC-GB standard) and the UK Forestry Standard, such as attention given to social impacts and the FSC's policy against the use of chemicals.[48] Nonetheless, its most important general conclusion was that "it would be possible to produce an audit protocol which included the requirements of both the United Kingdom Forestry Standard and additional requirements of the FSC-GB Standard" (SGS Forestry 1997:10, 6).

With this information in mind, and recognizing it could play a key role, the commission decided to act as an arbiter, bringing together all the groups involved in debates over certification and the overlapping issue of setting sustainable forest management standards for British forests (Goodall 2000).[49] It was well positioned to play this role as private landowners, managers, and manufacturers as well as other groups involved in certification viewed the commission as a relatively neutral body.[50] Further, by providing resources to fund meetings and accommodation expenses for participants, the commission was able to engender a fruitful environment in which to reach consensus.

Irrespective of the commission's motives for facilitating these talks, the process that resulted provided the FSC access to interest groups that had refused to participate in the FSC-UK working group (Scrase et al. 1998). In particular, the TGA, the FICGB, and the Ulster Timber Growers Organisation, and individual companies (such as Kronospan) (Forestry Commission 1999) — all of whom had refused to or chosen not to participate in the FSC-UK standards process — took part in the commission-led talks starting in September 1997. By February 1998, agreement had been reached by fifty delegates that work should begin on developing a United Kingdom Audit Protocol (the name given to the process that developed into the UK Woodland Assurance Scheme) contingent on it being endorsed in some manner by the FSC (Forestry Commission, Great Britain, 1998). FSC and its supporters (the 95+ Group and environmental groups including the WWF and Friends of the Earth) recognized the strategic opportunities the process provided. Retailers had a keen interest in securing domestic wood with an FSC label, and environmental groups were concerned that continued controversy over UK forestry would limit the FSC's ability to have impacts on forestry practices in more controversial forested regions.[51] Most FSC supporters viewed these developments favorably and began to strategize about how they could help facilitate and institutionalize the FSC. These actions were not coordinated or authorized by the commission, but the UK Audit Protocol amplified their potential for success since the process deflected attention away from forest sector contentions that the FSC had no right to develop an independent forestry standard for the UK or British forests. Among the most notable of these actions was that FSC supporters now held out an olive branch to its TGA opponents, admitting that their market-focused converting campaigns did not focus enough on how the FSC might adapt to increase support. As Steve Howard, WWF-UK's senior forest officer noted, "I agree that the FSC did not get it quite right in the beginning. We approached it in a very much campaigning fashion. We are currently five years on from that. We are now beginning to talk and to get the right things in place and to address everybody's needs. It's a shame we did not do that sooner" (Timber Grower 1998).[52]

The UK Woodland Assurance Scheme (UKWAS) resulted in compromises typical of multi-stakeholder processes. Environmental groups felt that the interests of landowners and managers had been given too much weight (especially with respect to the use of chemicals), but they also felt these concessions were reasonable if they could help FSC efforts internationally (illustrating that the international effects of FSC certification were still a significant factor affecting FSC supporters' strategy and choices within the UK).[53]

These developments paved the way for the June 1999 release of the (UKWAS), which was heralded as a monumental achievement in resolving certification debates (Tickel and World Wildlife Fund for Nature 2000; Forestry Commission 1999). The commission had played a central role, not only in generating consensus over a broad forestry standard for the UK — endorsed by the full gamut of governmental, non-governmental and private interests — but in making available a mechanism through which its Forest Enterprise might validate, independently, the positive role it played as a steward of British forests.[54]

But the most significant aspect of these developments was not the agreement itself, *but the strategic environment it created for the FSC as the only certification program operating in the UK that could offer immediate auditing and certification of forestlands*. With the Soil Association and SGS accreditation as

FSC certifiers already in place, and the Soil Association's already established role as a certifier of UK organic agriculture production,[55] these certifiers were able to legitimize the FSC-UK-based auditing as a vehicle for the UKWAS process. This institutionalization had significant effects on the UK government's own position, as a landowner, on forest certification. While official commission policy was support for all independent certification measured against the UKWAS standard, with the FSC being the only operational program in the UK at the time, by default, the Forest Enterprise achieving certification meant the FSC gained a strong foothold, marginalizing later attempts of the PEFC to compete in the UK certification arena.

By participating in the UKWAS standard, the FSC was able to sidestep the mistrust of private forest landowners and managers on the FSC national initiative and then emerge as the only program able to certify forestlands. The implications of this approach were important. For example, on July 28, 2000, the government's Ministry of Environment released a timber purchasing policy that gave preference to third-party certification, specifically mentioning the FSC as an appropriate and dependable scheme (Shotton 2000).[57] And while this policy responded directly to pressure from environmental groups calling for an end to importation of illegal timber and products from unsustainably managed forests, it had the indirect effect of further legitimizing the FSC within the UK. Not only was certification on the mind of the commission, a government department, but also sustainable forest management was gaining attention at the ministerial level.

Dividing and Conquering

Having a standard in place acceptable to the broad range of interests in the UK forest sector, FSC supporters quickly worked to translate this effort into direct and indirect support for the FSC. As the only program that had an auditing system in place and could grant chain of custody, FSC supporters quickly worked to use these features to institutionalize the FSC as the dominant certification program in the UK. Soon, woodland management firms, the Forest Service of Northern Ireland, and the British Forestry Commission's Forest Enterprise sought FSC certification (Forestry Commission, Great Britain, 1999a and b) (table 5.3). They did this not because they felt that the FSC was a more appropriate standard of sustainable forestry, but because the market demand was for an FSC label.

Manufacturers however continued to voice complaints about the FSC. They asserted that the FSC percentage-based claims thresholds were still too high by processors, owing to the complexity of their fiber inputs.[57] In response, the program moved to conform to these concerns; it reduced the percentage of FSC fiber required in a product,[58] in recognition that the existing policy was causing "problems and obstacles for relations with forest owners, and practical financial difficulties for some forest managers, industries and dealers" (Forest Stewardship Council 1999e).[59] The FSC was able to disentangle these percentage-based claims issues from forest management ones owing to the lack of vertical integration in the UK forest sector: most companies concerned about chain of custody did not own forestlands and thus their principle concern was the proportion of certified wood inputs they would need to obtain from the timber market. This, combined with concentrated public ownership and management of forestlands, meant the FSC found itself again able to avoid, in the short term, the concerns of individual private landowners and their limited participation. It did so by setting thresholds that required manufacturers to source little additional FSC materials beyond wood supply available from government certified lands. Thus, the FSC was able to increase significantly the quantity of certified products[60] in the UK's domestic retail sector. Recognizing the value of this conforming avenue, chain of custody remains a policy issue the FSC has continued to tweak. Early in 2002, a policy allowing group chain of custody (multiple processors certifying through a single audit) received early interest and participation in the UK. Two sawmills and five merchants joined the country's first group, whose manager, David Ogg, forecasted membership to expand to between 100 and 150 participants by the middle of 2003 (World Wildlife Fund 2002). His forecasts proved close to the mark; by September 2003 over 150 companies from all along the supply chain were certified members (Independent Forestry 2003).

At the same time the FSC continued to make conforming overtures to private owners by introducing programs to reduce their costs associated with diseconomies of scale (UKWAS Steering Group 2000). Woodland management firms received FSC certification as resource managers (see Scottish Woodlands n.d.), offering lower cost certification options for private landowners. In some instances the cost of the certification was entirely waived where it was unlikely certification would have proceeded otherwise.[61] Yet, such conforming strategies failed to move the opinions of most private landowners that the FSC system was costly and contrary to what they deemed legitimate approaches to regulating forest management (see tables 5.3 and 5.4).

But arguably the most strategic and shrewd maneuvering was the choice by the FSC-UK working group to submit the FSC-UK draft standard[62] to the international FSC for endorsement, rather than the UKWAS standard. They chose to *cross-reference their own standard* with the UKWAS standards to show how and why they were equivalent (Jenkins 1999). In October 1999 the

Table 5.3 Forms of support given to FSC and its competitor by woodland management firms and the public forest management agencies in Great Britain and Northern Ireland

Land owner/manager	Announcements of intention to or actual certification with[a]		Other forms of support[b]	
	FSC	PEFC	FSC	PEFC
Tillhill Economic Forestry	✓			
Scottish Woodlands Ltd.	✓		✓	✓
Fountains plc.	✓			
Forestry Commission Forest Enterprise	✓			
DANI Forest Service	✓			

[a] Fountains plc manages 150,000 ha of forestlands in the UK, some of which (3,255 ha) have been certified with the FSC (http://www.fountainsplc.com and http://www.certifiedwood .org). Till Economic Forestry manages approximately 200,000 ha in the UK, operates as a resource manager (7,817 ha), and oversees a group certification scheme (4,145 ha) (http:// www.fscoax.org and http://www.qualifor.com). Scottish Woodlands Ltd. manages approximately 170,000 ha in the UK and operates a group scheme representing a total of 27,851 ha as of October 9, 2001 (http://www.scottishwoodlands.co.uk/fsc.html). Forestry Commission Forest Enterprise has its Scottish, English, and Welsh forestlands certified, totaling 891,496 ha (http://www.fscoax.org). DANI Forest Service has 75,000 ha certified (http://www.fscoax.org).
[b] "Other forms of support" include participation in decision-making processes (i.e., FSC-UK working group or PEFC UK), and financial contributions. The Forestry Commission (and later DANI) participated in the FSC-UK process (as an observer). In early 2003 the Wales Forestry Commission agreed to provide a second staff member to FSC-UK until 2006.

FSC board endorsed the FSC-UK standard with the recognition that the UKWAS standard was completely equivalent. Certification could occur to either standard, but the FSC label would only be available to operators who were certified by an FSC-accredited certifier (Jenkins 1999). Such an approach institutionalized the FSC label in the marketplace, but at the same time permitted the FSC and its supporters to honestly claim that the standard had been developed and supported by private forest landowner representatives and their associations.[63]

In just a few short years the FSC had mustered a significant reversal of fortune — virtually all government forestlands were to become FSC certified and woodland managers were now offering a process for their clients to become FSC certified. Private forest landowners and the TGA were now isolated

Table 5.4 Forest and farm owner association support for the FSC and its late-developing competitor program, the PEFC

Organizations	Support for Certification	
	PEFC	FSC[a]
Forestry and Timber Association[b]	✓	✓
Scottish Landowners Federation[c]		
National Small Woods Association[c]		

Note: Support is measured as participation in the rule development or governance of the certification program at the national or international level. For the FSC this includes membership in the organization or participation in the FSC-UK national initiative. For the PEFC this includes membership in PEFC UK Ltd.

All these organizations participated in and endorsed the UKWAS standard.

[a] Prior to the creation of the FTA, no landowner organizations directly supported the FSC; however, the FIDC did join the FSC as a member of the economic chamber (http://www.fscoax.org). The TGA (now FTA), therefore, had a direct connection to the FSC given its membership in the FIDC.

[b] The FTA was created in the beginning of 2002 as the merger of the TGA, the Ulster Timber Growers Organisation, and the Association of Professional Foresters. Its 2,500 members represent both growers and forestry professionals. Prior to its inception, the TGA and Ulster Timber Growers Organisation merged in the summer of 2001 and at that time were supporting the PEFC, exclusively. In May 2002, the new association formally stated that it supports certification to the UKWAS standard and will encourage its members to become certified. It has not stated a preference for one program over another (i.e., the FSC versus the PEFC); rather, it supports the existence of multiple programs so as to foster competition in the provision of certification services (http://www.forestryandtimber.org/).

[c] Neither organization has an official position on certification but many of their members are also members of the TGA (now FTA) that was supporting the PEFC.

in their opposition and outmaneuvered by FSC interests. And in what they treated as rubbing salt in their wounds, the FSC added a caveat to its more relaxed percentage-based claims policy — continual improvement would be required with increasingly stringent thresholds developing over time (hence increasingly limiting market share for landowners who were not FSC certified), and they added a requirement that non-FSC-certified wood in labeled products be from FSC-defined non-controversial sources (Forest Stewardship Council 2000b).

This set the stage for the TGA action. In what it deemed as actions reflecting the interest of many UK landowners, the TGA supported the introduction of the PEFC system for UK forests in 2000. Officials with the TGA and some other landowners were especially concerned that without an alternative certifier offering an on-product label, the FSC could decide to revoke its recognition of the UKWAS standard, breaking the present connection between the UKWAS and the FSC label. TGA chairman and head of PEFC UK argued: "For those of you who have already gone through a certification process in response to demands from your customers, I would advise watching developments closely so that at the time of the five year renewal — or indeed before — you are able to consider new options. For those who have yet to seek certified status, I would suggest waiting a little longer, as you may find that a single operation will give you simultaneous access to a choice of labels, and at a lowered cost — partly as a result of competition but also because of the development of more user friendly, but nonetheless highly credible, schemes" (Yull 2001).

The PEFC UK worked with the UK Forest Products Association and the Wood Panel Board Federation to develop and promote a chain-of-custody system as they recognized that the FSC monopoly in this area was a key factor in FSC success (PEFC UK 2001) and worked hard to argue that they, too, could now audit and implement the UKWAS standard.

The environmental community immediately raised concerns that this effort to promote the PEFC threatened to undermine the credibility of the FSC. Fern, Friends of the Earth, Greenpeace, and the WWF decried the lack of consistency among PEFC standards in different European countries (Liimatainen and Harkki 2001). These environmental groups also began to fear that some might argue that the PEFC and FSC were the same, since both FSC and PEFC certifications drew on the UKWAS standard as an appropriate benchmark for sustainable forest management. Environmental groups and their supporters feared that such a situation might lead the broad public to believe that global sustainable forest management could be achieved through certification by either FSC or its competitors (Ozinga 2001).

The campaign by FSC and its supporters to criticize the PEFC UK has been largely successful in maintaining FSC support from woodland management firms as well as overall support from the 95+ Group[64] and the UK government qua landowner. Owing to the FSC success in being the first certification program to institutionalize itself into UK's forest certification market, efforts to promote the PEFC as a legitimate alternative beyond private forest landowner audience have proven unsuccessful.

By the middle of 2002, the political landscape had successfully undergone further change with the emergence of a new association, the Forestry and Timber Association that merged the Timber Growers Association, the Association of Professional Foresters, and the Ulster Timber Growers Organisation. Unlike the strong opposition held by TGA officials toward the FSC, early accounts report this new association supports a middle ground position, where both FSC and PEFC are valued. Independent of this, evidence today indicates little on-the-ground support for the PEFC: even though the PEFC Council has officially endorsed the UK national PEFC initiative, no UK forests have sought certification under its scheme. And at the close of 2001, at the WWF 95+ Group's tenth anniversary event, the participation of Prime Minister Tony Blair as a speaker and his support for the innovative approaches being pursued by the 95 Group to prevent forest degradation and loss are a telling illustration of the dominant position held by the FSC (World Wildlife Fund 2002). Supporters of the FSC in the UK had succeeded, as of 2002, in doing what their counterparts in Germany were unable—they had sidestepped vestigial private forest landowner opposition and may have generated a unique road toward an opportunity for FSC institutionalization.

Conclusion

The story of forest certification legitimacy dynamics in the UK reveals the importance of understanding how a sector's place in the global economy, the structure of the domestic forest sector, and the history of forestry on the public policy agenda intersect to provide opportunities for FSC strategists to mix market-based efforts aimed at converting forest owners to support the FSC, and FSC efforts to conform to concerns of forest owners and companies along the market's supply chain. The FSC was also able to draw on increasing concerns about the UK domestic forest management practices and on a government increasingly worried about the state of the global and domestic forests. Conforming was needed to augment the appeal of the FSC, yet due to the strategic venue created by the government's Forestry Commission, and the near absence of vertical integration, the specifics of these strategies were such that the FSC core audience remained convinced of the program's moral appeal.

The story reveals support for hypothesis 2 (*Forest companies and landowners selling wood to a domestic market in a country or region that imports a large proportion of all the forest products it consumes will be more susceptible to converting strategies by the FSC and its supporters than in a region that imports a small proportion of all the forest products it consumes*). The FSC, through a moral boomerang effect, was able to argue in favor of an "even playing field" to convince members of the 95+ Group to apply their purchasing polices both to international and domestic suppliers. These retailers were likewise less susceptible to political backlash from a relatively small domestic forest sector. Market converting, however, was not by itself effective because most private forestland managers, as in other countries, refused to budge in their opposition to the FSC.

What tipped the scales toward the FSC was in fact the increasing concern about existing public policy approaches to forest management regulation, lending support to hypothesis 6 (*Forest companies and landowners will be more susceptible to FSC converting strategies in a country or region with sustained and extensive environmental group and public dissatisfaction with forestry practices than in countries or regions with less dissatisfaction*). And even the UK government's increasing scrutiny of forest management ironically reduced the credibility of its regulatory agency, paving the way for its role as a consensus builder to help generate indirect support for the FSC — support that was further reinforced when the government, as a landowner, chose to pursue FSC certification on its own forestlands. The latter, in particular offers support to hypothesis 4 (*FSC converting strategies will be more effective in those countries or regions characterized by relatively unfragmented non-industrial forest ownerships, than in those countries or regions characterized by a high degree of non-industrial forest ownership fragmentation*). Divisions in the forest sector's associational system also help explain the environment in which the FSC, unopposed for years, was able to carefully craft a strategic approach. This illustrates the power of hypothesis 5 (*FSC converting strategies will be more effective in those countries or regions characterized by fragmented or non-existent associational systems, than in those countries or regions characterized by well-coordinated industrial and landowner associational systems*).

However, unlike the situation seen in British Columbia, the degree of criticism (not as pronounced) and the nature of these concerns (some concern over domestic species loss and the conservation of ancient woodlands, but high concern over illegal and unsustainable practices in forests of trading partners) meant the pull toward converting was not as strong. Conforming was needed and was facilitated by the absence of vertically integrated firms, making possible separate strategies to address the mostly independent interests of the processing and growing segments of the domestic sector, a situation providing evidence supporting the causal mechanism proposed by hypothesis 3 (*FSC converting strategies will be more effective in those countries or regions dominated by large concentrated industrial forest companies, than in those countries or regions where industrial forest companies are relatively small and not concentrated*). Finally, while the forest sector did not view public forest policy as faultless, the relative close ties between the commission and forest sector interests predisposed landowners and companies to instill trust in the UKWAS process, owing in part to its connections to the government. Support for existing public policy and the eventual conforming the FSC undertook on specific policy issues in order to ensure the UKWAS standard gained full acceptance illustrates hypothesis 7 (*Forest companies and landowners will be more susceptible to FSC converting strategies in a country or region where they share access to state forestry regulatory agencies with non-business interests, than in countries or regions where forest companies and landowners enjoy relatively close relations with state forestry agencies vis-à-vis non-business interests).*

The UK case highlights two key dynamics important for understanding the development of non-state market-driven governance systems. The first is that government departments and their agencies and officials can play an important role in promoting non-state market-driven governance. This case supports our point in the introduction that governments can act in many ways that are consistent with non-state market-driven governance. The mere fact that government actors are participating does not mean that the "non-state" aspect of non-state market-driven is gone. Government can act as a facilitator of processes, as a landowner and as a supporter of non-state market-driven governance, which adds legitimacy to non-state market-driven rather than diminishing it. If the UK government had gone the route of implementing SFM through governmental policy, then it most certainly would have removed the "non-state" aspect of non-state market-driven and would have rendered an analysis of stakeholder evaluations less important, since the law would have required certifications to take place. However, for the reasons noted above, the government did not take this route, and the FSC was able to capitalize on the policy environment to build greater support for its conception of forest certification. In this sense the UK case provides a glimpse into how "cognitive legitimacy" might be achieved by non-state market-driven systems — through the traditional "routinization" path envisioned by Suchman when supporters find some means of making their system a part of day-to-day business life.

Second, the UK case reveals that international efforts and campaigns can strongly influence domestic policy, providing a glimpse into the process through which the international-domestic interaction of "two level games" (Putnam 1988) may also influence non-state market-driven dynamics. What is clear is that the emergence of non-state market-driven governance, and the type of approach to regulation it embraces, is far from inevitable. Rather, strategic choices are facilitated and debilitated by the economic, structural, and domestic public policy legacies within which they are deployed.

6

Germany

*[The certification discussion was slow to come to Germany] because it
bumped up against a confidence that said 'by God, we're Germans — we
developed sustainability!' Sustainability is such a taken-for-granted concept here.*

> — *Landowner and member, Bavarian Landowner Association,*
> *Bavaria, Germany, April 20, 2001*

Although those concerned about tropical and temperate rainforest destruction had long targeted German importers of wood products from these regions, the debate over forest certification on Germany's own forestland arrived
relatively late. However, once certification did emerge as an important theme
for the domestic forest policy community, the competition for legitimacy developed into one of the fiercest struggles examined in this book. As the Forest
Stewardship Council (FSC) began to gain incremental support from some
forest owners in the mid-1990s, forest landowner associations responded ag-
gressively with the Pan European Forest Certification system (PEFC), which, as
of winter 2003, was emerging as the preferred, albeit not unanimous, certification program of those forest owners who chose to participate in certification.

Similar to the scenarios presented in the other chapters, the German forest
sector's place in the global economy, the history of forestry on the domestic public

policy agenda, and the structure of the forest sector significantly influenced efforts by the FSC and its PEFC competitor to gain or maintain support. Germany's position as a large importer of forest products was important because it first focused German attention on forest destruction in the tropics and harvesting practices in some temperate regions, such as British Columbia. But efforts in Germany to scrutinize its *imports* had a boomerang effect (Keck & Sikkink 1998), as domestic environmental groups and foreign suppliers argued that it was unfair to subject imports to different standards than domestic products.

The external pressure on domestic forests to become certified was more of a moral "fairness" issue in Germany; however, since the processing of most domestically produced fiber occurred within Germany, there were few foreign purchasers of German wood products that environmental groups could enlist in their market campaign. This situation forced environmental groups to focus much more on the domestic scene. And unlike the cases of BC and Sweden, the history of forestry on the public policy agenda within Germany and the structure of the domestic forest sector generally worked to hinder FSC efforts to gain legitimacy, while facilitating the development of the PEFC as the key FSC-competitor program. The history of forestry on the public policy agenda worked against FSC converting strategies because there was little widespread domestic criticism of German forest practices. In fact, the forest sector was generally seen by the forest policy community and the German public as practicing sustainable forestry according to long established principles of "German forestry" (Johnson 1993). Efforts in the 1980s and 1990s by domestic environmental groups, such as Greenpeace Germany, were important for increasing public support for the creation of more natural protected areas, but these groups' wider critique of German forestry practices failed to spark significant concern among the German public. And perhaps more importantly, environmental group efforts to force forestry onto the public policy agenda led many landowners and forest managers to view environmental groups as marginal players — a feature that hindered later FSC efforts to gain broad landowner support.

The low level of vertical integration in the forest sector as well as a fragmented landownership base further hindered FSC converting strategies because there were few large landowners to target and relatively diffuse companies along the supply chain, which limited domestic-oriented market campaigns. These dynamics meant that FSC supporters' efforts to sow seeds of support would be planted on relatively infertile domestic soil. And even where some of these seeds managed to grow, the nature of Germany's forest sector made it difficult to translate this support into a significant market share, limiting the impact of purchasing commitments made by a handful of domestic retailers. These same domestic features made it easier for leading landowner associations to gain widespread support among landowners and state forest management agencies for the PEFC, further placing the FSC on the defensive as it attempted to shore up its limited, but not insignificant, support.

The balance of this chapter proceeds in four analytical steps. First, it provides an overview of the three factors that mediated and directed the efforts on the part of the FSC to gain forest owner support. Second, it describes the roots of certification in Germany and the evolution of forest owner support for the FSC and PEFC certification programs. Third, it carefully reviews how the strategic choices pursued by the proponents of each system were influenced by the three factors. The chapter concludes by specifically reviewing the German case's support for the explanatory hypotheses raised in the introduction.

Factor 1: Place in Global Economy

DEPENDENCE ON FOREIGN MARKETS

Approximately 40 million cubic meters of raw logs are harvested in Germany each year. Of this amount, nearly all are sold and processed locally (Hofmann et al. 2000), although many finished products are exported, primarily to other European Union countries. Roughly two-thirds of wood grown in Germany is destined for sawmills, while one-third is turned into pulp and cellulose derivatives (Statistisches Bundesamt 2001). The fact that most wood harvested in Germany is also processed there is an important feature because it meant that there were not many foreign purchasers of German primary wood products for environmental groups to target — a tactic that was highly effective in regions such as BC. As a result, environmental groups would have to target domestic purchasers of German wood products. This is a more difficult task because domestic purchasers, fearful of political backlash, are less likely to demand that their own domestic landowners change their practices than they are to demand that foreign landowners alter their behavior.

Germany imports large amounts of wood products from both tropical and temperate rainforests (European Forest Institute 2001) and is the world's second largest importer of pulp (Verband Deutscher Papierfabriken 2000). Germany's position as a major importer of tropical and temperate rainforest products meant that the demands for sustainable forest products from certain segments of the German industry, such as from publishing and paper companies, were a larger factor in forcing support for the FSC in other regions (as the BC chapter illustrates) than in Germany domestically (figures 6.1 and 6.2). However, the country's relatively high level of imports from countries with

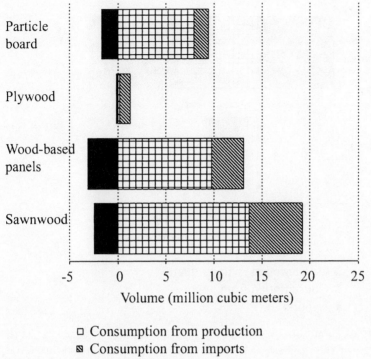

Volume (million cubic meters)

□ Consumption from production
☒ Consumption from imports
■ Exports

Figure 6.1. Volume (in cubic meters) of German apparent consumption (production plus imports less exports) for various aggregate wood products in 1999. In this year, Germany purchased 34 percent of its sawnwood imports from Sweden, Finland, and the Russian Federation and 44 percent of its plywood imports from Finland, Brazil, and Indonesia. Exports to the Netherlands, the UK, Italy, France, Belgium-Luxembourg, Austria, Denmark, and Spain represented 87 percent of its sawnwood exports, 55 percent of its plywood exports, and 75 percent of its particleboard exports. Source: FAO 2001.

a large amount of FSC-certified forests — and similar forest products — has meant that forest owners in Germany have increasingly come to fear that fledgling demand for FSC products by German retailers will result in their loss of market share to foreign imports.

Factor 2: Structure of the Forest Sector

Three factors are key in understanding the structure of the German forest sector: a low level of industrial forest company concentration, fragmented non-industrial landownership patterns, and strong associational system cohesion.

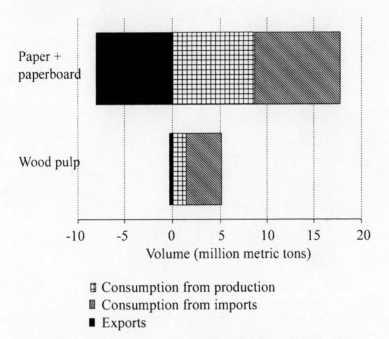

Figure 6.2. Volume (in metric tons) of German apparent consumption (production plus imports less exports) for wood pulp and paper plus paperboard products in 1999. In this year, Germany purchased 24 percent of its wood pulp from Canada, 20 percent from Sweden, and 18 percent from Finland. Sweden and Finland also accounted for 39 percent of German paper plus paperboard imports. Exports to the Netherlands, the UK, Italy, France, Belgium-Luxembourg, Austria, Denmark, and Spain represented 48 percent of wood pulp exports and 55 percent of paper plus paperboard exports. Source: FAO 2001.

INDUSTRIAL FOREST COMPANY CONCENTRATION

Unlike Canada, the US, and Sweden, there is low vertical integration within the German forest sector (Hofmann et al. 2000). Many small and medium-sized manufacturing enterprises exist, with large enterprises limited mainly to capital-intensive branches of the sector, such as paper or wood-based paneling production (Schraml and Winkel 1999). The sector is characterized by the absence of any large companies with horizontal integration at the landownership level. When combined with the lack of vertical integration within the German sector, ostensibly there was a relative scarcity of large, easy-to-target forest companies, making it difficult for FSC supporters to exert pressure on individual companies in the supply chain. This book and existing research have shown such pressure to be one important factor in changing firm-level behavior (Sasser 2001; Gereffi, Garcia-Johnson, and Sasser 2001).

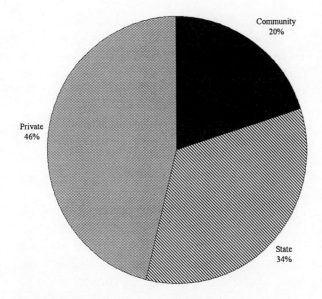

Figure 6.3. Ownership of forested lands in Germany.

The combination of the large number of small landowners and low vertical integration also makes it difficult to track wood along the supply chain. Since tracking of wood is a key requirement of the FSC, this feature made it difficult to translate the FSC's limited support among forest landowners to purchasers further along the supply chain.

NON-INDUSTRIAL FOREST OWNERSHIP FRAGMENTATION

Three major types of forestland ownership exist in Germany: land publicly owned and managed at the state level[1]; land publicly owned and managed at the community level; and non-industrial private land. In 2000, non-industrial private land provided 31 percent of wood harvested in Germany (Statistisches Bundesamt 2001). The percentages of forestland in each of these categories are shown in figure 6.3.

In general, forestland ownership in Germany is highly fragmented[2] (Schraml and Winkel 1999), although land concentration is increasing slightly and state-owned forests tend to exist in larger contiguous pieces (Schraml and Thode 2000). Community and privately owned forests are composed of relatively small tracts. Ownership of German private forestlands is divided among 450,000 landowners, whose average ownership size is 8 hectares (European Forest Institute 2001).[3] Nearly 10 percent of the private forest area in Germany is in blocks of less than 1 hectare in size, on which no forestry practices occur

(Hofmann et al. 2000). Landowners have tried to minimize the disadvantages of their status as small owners through the creation of approximately 1500 voluntary "forest enterprise cooperatives" (*Forstbetriebsgemeinschaften*), which coordinate forestry and marketing activities (Arbeitsgemeinschaft Deutscher Waldbesitzerverbände 1997).

ASSOCIATIONAL SYSTEM COHESION

Germany has a diverse array of forestry associations that are relevant to our story; these include private landowner associations, processor associations, and the overarching quasi-state German Forestry Association (*Deutscher Forstwirtschaftsrat*).

The German Forestry Association is a group of professional foresters, forest scientists, and forest landowners of all types whose members deliberate over broad national forest policy issues. So influential is this body that the association's policies and recommendations are often enacted into German law (Schraml and Winkel 1999). This semi-autonomous association "provides a forum for discussions for representatives of all ownership types, including professional associations and scientists. The results of the council's deliberations are usually passed on the government for implementation" (Schraml and Winkel 1999). The most powerful member of the association, based on the number of votes, is the Working Group of German Forest Owner Associations (*Arbeitsgemeinschaft Deutscher Waldbesitzerverbände*) (Schraml and Winkel 1999).

The Working Group of German Forest Owner Associations promotes the rights of private forest owners, and sometimes municipal forest owners as well, by providing the umbrella under which fourteen state-level associations are united. This association is regarded as being as powerful as state forestry agencies in Germany (Schraml and Winkel 1999). It not only has a large level of influence on German forest policy through its prominent role in the German Forestry Association, but also is a major source of forestry information for private forest owners: more than half of private forest owners surveyed reported that they gather nearly all of their forestry information from landowner associations and do not search out other sources (Eklkofer and Suda 2000). This well-organized associational structure — and the formal venue for communication between private forest owner associations and state agencies through the German Forestry Association — is key to our analysis because it created an institutional setting that allowed the myriad forest landowners to develop proactive alternative strategies to the FSC in conjunction with those agencies with the resources to do so.

Various segments of the German forest product processing sector are grouped under associations. The Association of German Paper Producers and Association of German Magazine Publishers (*Verband Deutscher Papierfabriken* and *Verband Deutscher Zeitschriftenverleger,* respectively) have played an important role in bringing certification to Germany by responding early on to environmental group pressure to ensure that their fiber imports were coming from sustainable sources.

Factor 3: History of Forestry on the Public Policy Agenda

PUBLIC DISSATISFACTION WITH FORESTRY PRACTICES

With the exception of controversial "treetop thinning" (*Waldsterben* or "forest die-back"),[4] there are arguably no items on the German forestry public policy agenda that are perceived by the general public or the forest sector as environmental crises. Those forest management issues that have reached the policy agenda are narrow in scope and have centered on technical issues, such as pest management or the effects of acid rain. In general, German foresters, landowners, and the general public have a strong sense of pride in German forestry, which they view as very "ecologically friendly" (Rametsteiner 1999; Schraml and Winkel 1999). The long and celebrated tradition of German forestry developed following the reconstruction of vast areas of forest that were destroyed in the eighteenth and early nineteenth centuries. It is known for its rational and scientific approach to forest management (Huette 2000), where professionals are entrusted with making sound decisions. Until recently, this tradition was based on single-species, even-aged forest management (Spathelf 1997) and embraced the concept of a "sustained yield" of timber. This approach became widely adopted in the mid-1900s in most other countries and regions reviewed in this book (Johnson 1993) and was the prevailing forestry paradigm until the 1960s, when the timber-focused "sustained yield" concept came under increasing criticism by conservation biologists, scholars, and nongovernmental organizations as failing to accommodate changes in societal values and increasing scientific evidence about fragile forest ecosystems. Such criticism led to calls for a more integrated approach and conscious decisions in the US, BC, and Sweden to move away from the German school of forestry approach to forest management.

Similar critiques were made within Germany as well (Spathelf 1997), but unlike other cases reviewed in this book, the German forestry tradition was largely seen as being compatible with, even necessary for, the inclusion of new science and values. Thus, when environmental groups and other actors, such as members of the Working Group on Natural Silviculture (*Arbeitsgemeinschaft*

naturgemäße Waldwirtschaft), began to criticize the German forestry paradigm for being too narrow, state forest managers and private landowners
could point to formal and informal changes in German forestry as evidence it
was able to accommodate and address such concerns. Changes to German
forestry laws prohibiting clearcuts, requiring reforestation, and placing restrictions on the use of certain types of machinery (Schraml and Winkel 1999)
were offered as evidence that the German approach was rational and adapt-
able and encouraged German forest managers to see themselves as environmentalists (Schraml and Winkel 1999). These changes were made in tandem
with the older focus on timber sustainability — the latest data showing that
German forest owners continue to harvest far less than they are growing
(Hofmann et al. 2000).

As a result of this history, most members of the German forest policy community tend to view groups that are critical of German forestry as fringe
players. Such a view became more entrenched when Greenpeace Germany
championed its forestland conservation agenda by criticizing existing forest
management practices in the early 1990s. Greenpeace promoted a "process
protection" concept in which it called for the protection of natural processes
to take precedence over wood production (Hoeltermann 2000) and suggested
that forest owners set aside 10 percent of their forestlands from any harvesting
at all (Spathelf 1997). Such campaigns led to the perception on the part
of many in the German forest sector that environmental groups were anti-
forestry. The German forest sector's animosity toward these domestic environmental groups increased further following the 1992 Rio Earth Summit, whose
Agenda 21 initiative called for a stronger role for conservation groups in
traditionally forestry-dominated discussions (Huette 2000; Weber 1999). The
result of these events was that German environmental groups were seen by the
forest sector as promoting a strong forestland preservation agenda. Such views
made it easier for PEFC supporters to paint the FSC as being on the "conservation" side of the forestry/nature conservation debate.

FOREST COMPANY AND LANDOWNER RELATIONS WITH STATE
AGENCIES RELATIVE TO OTHER SOCIETAL INTERESTS

While they have not been faced with addressing forestry conflicts of the
magnitude faced by state agencies in BC or the US, state forest agencies have
historically been very influential — and successful — in promoting the German
forestry tradition among private forest landowners and the general public
(Schraml and Winkel 1999). State and municipal forestry agencies are important for the advice and services they provide private landowners: three-
quarters of private forest owners in Germany use these agencies as a primary

Table 6.1 Influence of explanatory variables on FSC efforts to gain support from forest companies and non-industrial landowners in Germany

Factor	Existence
Place in global economy	
Dependence on foreign markets	✓
Structure of forest sector	
Large, concentrated industrial forest companies	
Unfragmented non-industrial forestland ownership	
Diffuse or non-existent associational systems	
History of forestry on public policy agenda	
Sustained and extensive public dissatisfaction with forestry practices	
Forest companies and non-industrial forest land owners share access to state forestry agencies with other societal interests	

A "check mark" indicates strong presence.

source of information on forestry issues (Eklkofer and Suda 2000). This direct connection between state and private forest owners, and the high status that foresters in Germany have among private forest owners and the general public, has meant that policy choices taken by state forest management agencies — including decisions about certification programs — are usually respected by municipal and private landowners and appear to influence their own evaluations. This has meant that the decision by many state agencies to support the PEFC program, undertaken for reasons described below, have heavily influenced the story of certification in Germany.

The factors discussed above have influenced the ability of the FSC to gain support from German forest companies and landowners (table 6.1).

Initial Certification Developments

The domestic debates in the late 1990s over the certification of German forests were directly framed by Germany's earlier interest in the certification of tropical and some temperate wood imports to Germany. And the debates over certification of wood imports were themselves influenced by the more blunt boycott campaigns initiated in Germany of tropical wood in the late 1980s. We start our story with these earlier struggles because they explain how certification arrived on the domestic policy agenda and framed the international influence as a moral one, with no discernible economic pressure.

The destruction of tropical forests became a major focus of German environmental groups in the late 1980s. Highly visible campaigns brought the conflict to the general public, and environmental groups used pressure tactics and protests to cause many large retailers in Germany, such as OBI, to stop selling tropical woods and wood products (OBI 2001). Many state and municipal procurement policies also were changed at this time to ban the purchase of tropical woods (World Wildlife Fund Germany 2000a).

Environmental group campaigns then expanded to the preservation of ancient temperate forests as well, moving scrutiny to German purchasers of forest products coming from regions such as the coastal temperate rainforests of BC (Stanbury 2000). By mid-1993, Greenpeace Germany had convinced four of Germany's largest publishers to stop buying pulp or paper from BC suppliers who logged in ancient temperate forests (Stanbury 2000). At this time the majority of German environmental groups were still campaigning for a policy of boycotts and import bans and were skeptical of market-based initiatives such as the *Initiative Tropenwald,* the attempt by the German tropical timber importing industry to create a label for tropical timber originating from sustainable sources (European Forest Institute 2001).

Soon, however, German environmental groups began questioning the effectiveness of boycott campaigns when fears were raised that tropical forest owners might actually cut more wood to compensate for the loss of the more lucrative foreign markets, thus worsening, rather than mitigating, forest destruction (European Forest Institute 2001). Environmental groups, both in Germany and internationally, began to reorient their strategies, attempting to encourage sustainable forest practices, rather than simply boycotting destructive ones.

This growing international desire for an alternative to boycotts, in addition to the failure of the United Nations Conference on Environment and Development in Rio to achieve a binding global forest convention in 1992, as we detail in the introduction to this book, provided the context in which the FSC was created. However, the idea that FSC certification could be required of timber imports to Germany did not immediately become a widespread topic of interest within the German forestry community, nor was it favored by key German environmental groups such as Greenpeace, who feared that certification might undermine their own campaigns. Instead, many of the German forest products importers that had been targeted in earlier campaigns carried out surveys of their suppliers and, in some cases, developed their own forest management standards for imported products (European Forest Institute 2001). The German paper and publishing industry was one of the first sectors to make concrete demands of its suppliers; in 1995, the Association of German Paper

Producers and Association of German Magazine Publishers jointly issued a position paper that outlined their stance on a credible certification program in greater detail. They were interested in purchasing products certified under a widely recognized, global program (Verband Deutscher Papierfabriken 1998) that could be implemented rapidly (Verband Deutscher Zeitschriftenverleger 2001).[5]

This early action was important because it resulted in a key part of the German forest sector identifying what it believed to be essential requirements of a credible forest certification program for German imports: that it be international in scope, created by third parties, and recognized by dominant environmental groups. These requirements were important because although most wood importing companies in Germany did not require official adherence to the FSC, there was no other international program that matched these criteria. The position paper thus gave some important industry credibility to the FSC approach and likewise also set a standard to which FSC competitor programs would be compared. Still, at that time certification debates in Germany were clearly focused on international issues and on imports, with the German government taking an active interest at the European Union and United Nations level in promoting efforts to ensure that tropical imports came from sustainable sources. As one member of the German Forestry Association explained, "This . . . was a time when the [forest certification] topic was not even brought up in Germany. I experienced an intense, lively discussion at the international level, and complete silence on the national level."[6]

The strong attention to imports and non-domestic concerns about sustainability set the stage for the domestic certification debate. Tropical governments increasingly viewed the tropical forest-focused *Initiative Tropenwald* as an invasion of their sovereignty and as an effort by northern countries to limit scrutiny of their own forestry practices (Klins 2000). The more German retailers scrutinized tropical imports, the more they opened themselves up to charges that they were not practicing domestically what they preached internationally. German environmental groups also began to feel a growing moral obligation to examine domestic forestry practices more closely.[7]

The immediate result of this sense of obligation on the part of German environmental groups was not yet to promote the FSC within Germany. Rather, German environmental groups helped create two distinct domestic forest certification programs in 1996 that were heavily influenced by their conservation agenda. The first of these was the creation of the stringent *Naturland* program,[8] which modified and extended the *Naturland* standards for organic foods to forestry. It was supported by Greenpeace Germany, the German branch of Friends of the Earth (*Bund für Umwelt und Naturschutz*

Deutschland), and Robin Wood, three of Germany's most prominent environmental groups (Reinold 1998). The second domestic forest certification scheme created at this time was the *eco-timber* program of the German Society for Nature Conservation (*Naturschutzbund Deutschland*), which also had a strong conservation focus (Klins 2000). Although these two labels and their approaches reflected the conservation values of a small group of forest owners, the vast majority of German forest managers and owners rejected the programs as impractical, an affront to private property rights, and tools that promoted environmental groups' "radical forest preservation" agenda. German forest owners also felt little widespread public pressure for change and believed that the public continued to support existing forest practices as being sustainable.

What German forest owners did come to realize was that growing international criticism of northern countries concerning labeling and certification demands of their imports meant that they, too, would have to act to verify that domestic production was also sustainable. As a result, in 1996 the German Forestry Association and the Working Group of German Forest Owner Associations created their own "Label of Origin" (*Herkunftszeichen*) labeling program, which focused on simply identifying wood harvested in German forests and stood in contrast to the standards-based environmental-group initiated programs described above. German private forest owners and state forest managers believed that this limited approach would be enough to prove that their products came from well-managed forests. As one private landowner and member of the German Forestry Association said: "With the Label of Origin we're giving the consumer a signal that he's buying a product that hasn't been transported great distances, and that — as he is able to confirm during his walks in the forest — comes from sustainable forestry."[9] The label soon appeared on forest products throughout Germany.

The clear split between conservation-oriented, prescriptive environmental group programs and the landowner and state agency program, which failed to require any change in the status quo of German forestry, resulted in a certification stalemate at the domestic level. The Label of Origin plan failed to gain environmental group acceptance, and even the media began to expose the program's lack of standards beyond already existing state and federal laws. At the same time, the eco-timber and Naturland programs were either ignored or steadfastly opposed by most forest owners and state forest management agencies. Germany's environmental groups gradually came to accept the credibility of the FSC for domestic forest production after acknowledging the sustained resistance of forest owners to their own certification programs and witnessing the support the FSC received from industrial forest companies in Sweden and

BC and the increased interest it was given from the public in the Nordic countries. A strategic decision was made in 1997 by most German environmental groups involved in the creation of the domestic certification programs to throw their support to the FSC domestically, opening the door to a fierce competition between environmental group and forestry factions within Germany. The actions undertaken by these factions between 1995 and 2002 are outlined in table 6.2.

Initial Domestic Support for the FSC

Environmental groups first began their efforts to bring FSC certification to Germany through ad hoc converting strategies, targeting large German companies that purchase relatively large amounts of German wood fiber, such as the publishing company Axel Springer Verlag and Germany's largest mail order company, Otto-Versand. These companies agreed to demand third-party verification that the forest products they were purchasing — including those from domestic sources — came from sustainable sources. However, the response down the supply chain at the level of forest landowners and state forest managers was not favorable; their position, for the most part, was to reject *any* certification program that was more prescriptive than the Label of Origin plan. The most outspoken critics of the FSC were private forest landowners, who, unlike BC companies who eventually but reluctantly agreed to the requirements of purchasers like Otto-Versand and Axel Springer Verlag, decided to take direct action against the domestic companies that began demanding certification. In what one private forest owner described as an act of "desperation," on December 4, 1997, German landowner association members took to the streets with their counterparts from Sweden, Finland, Austria, and France and demonstrated against these companies for not supporting what they saw as already sound German forest practices.

Just as some tropical countries had complained earlier that Germany was imposing its standards on them, many forest managers in Germany similarly viewed the FSC as a foreign-based program that was performing *Fremdbestimmung* — or "outside control."[10] These views were buttressed by the proud tradition of German forestry noted above and were especially acute because so much support for the FSC was coming from the UK and the Netherlands, places where German forest owners generally believe that, where forests still exist, forestry is done poorly.[11]

The fears of Fremdbestimmung, the associated focus on private property rights by forest landowners, and the fragmented nature of private forestland worked to limit the impact of choices made by German purchasing companies

Table 6.2 Strategic efforts by the FSC and FSC competitors between 1995 and 2002

Year	FSC competitor program supporters		FSC supporters	
	Action	Strategy	Action	Strategy
1995			VDP and VDZ issue joint position paper outlining requirements for credible certification program that promotes FSC	Converting
1996	German Forestry Association creates "Label of Origin" for domestic wood products	Conforming	Environmental groups create *Naturland* and eco-timber forest certification programs	Converting
1997			German FSC Working Group founded	Converting
1997			WWF and others create *Gruppe 98* FSC buyers group	Converting
1998	Private forest owners demonstrate against companies supporting certification	Converting	FSC regional standards group eliminates set-aside requirement for all private and some public forest owners	Conforming
			FSC regional standards group accepts proposal that strengthens private forest owner power within FSC voting structure	Conforming
1999	PEFC Germany created	Conforming		
2000	Forest ministers of Bavaria, Baden-Württemberg, and Thuringen unsuccessfully pressure OBI CEO to retract commitment to FSC	Converting	FSC conducts model project to demonstrate certification requirements and costs to landowners	Informing
2001	PEFC introduces chain-of-custody procedure	Conforming	FSC conducts model project to demonstrate innovative group certification options	Informing
	PEFC pursues partnerships with SFI and CSA	Conforming		
2002	Private landowners protest outside OBI stores	Converting	FSC conducts model project to demonstrate innovative group certification options	Informing

further down the supply chain on small forest landowners. While it is difficult to distinguish whether the lack of market impact along the supply chain was stronger than moral opposition by small landowners, it is clear that the moral opposition was based not so much on the fact that the FSC had stricter rules, but that the rules were initiated from international environmental groups (Janssen 1998). These concerns were supported by German state forest management agencies, some of which were not necessarily opposed to the FSC because of the program's substantive rules per se, but rather because the proposed rules were not emanating from traditional governmental agencies. One influential private landowner even speculated that if state forestry agencies were to promulgate forestry laws of equivalent rigor to the FSC, landowners would likely accept them with little resistance.[12]

Opposition to the FSC was further exacerbated by Germany's "nature conservation versus forestry" conflict described above. The groups that the forest sector perceived to be radical proponents of forest preservation were the same organizations that promoted the FSC. This resulted in fears that the FSC would become another vehicle for those groups wishing to preserve forestland — a fear made stronger by the FSC's international criterion (Criterion 6.4) that a certain percentage of certified forests be preserved as no-forestry areas.

Signs of FSC Support

Despite the relatively unified resistance on the part of German governmental and private forest managers to the third-party certification concept in general and the FSC specifically, the FSC did manage to incrementally build its support from key German wood products purchasers and retailers noted above. More importantly, FSC also began to achieve support from a handful of public land agencies some of whose political masters included German green party officials (World Wildlife Fund Germany 2000b) and an even smaller, but highly symbolic, number of private landowners.

This growing domestic and international interest in the FSC led the Association of German Paper Producers and Association of German Magazine Publishers to encourage the German forest sector to support and participate in the creation of the FSC Germany initiative (Verband Deutscher Papierfabriken 1998). The Association of Municipalities of Rheinland-Pfalz (*Gemeinde- und Städtebund Rheinland-Pfalz),* a member of the German Forestry Association, likewise attempted to bring the discussion of FSC certification into forestry circles. However, the vast majority of forest managers did not heed these calls, with some viewing the small number of landowners who supported the FSC as "traitors."[13]

Given this initial, limited influence within the German forest policy community, environmental groups that supported the FSC turned to more coordinated and systematic market-based converting strategies in their quest to increase support for the FSC. Led by the World Wildlife Fund, FSC supporters helped create the German buyers group (Gruppe 98) in 1997 as a means of focusing and identifying market support for the FSC. The group soon gained membership from existing FSC supporters, such as retailers OBI (e.g., OBI 2001), Praktiker, and Otto-Versand, as well as from a limited number of other purchasers of wood products along the supply chain who had not previously committed to the FSC. Still, the effects of these converting strategies for increasing market pressure along the supply chain were mixed. Some of the large retailers joined the German FSC buyers group to justify resuming tropical wood imports[14] and were less inclined to transfer this pressure to their domestic suppliers. While buyers group members were required to formally support FSC certification for their domestic suppliers, many of these companies made it clear to German landowners that they viewed German forestry practices as sustainable, thus limiting the impact of their FSC commitments on domestic forest landowners. As expected, German private forest owners and forest managers were highly critical of the creation of the Gruppe 98, arguing that it was an illegitimate tactic. The result was that this market pressure threat served to reinforce, rather than break down, forest manager opposition. As one private forest owner described: "Essentially they said 'either you submit to FSC or we — the environmental groups and buyers groups and companies that are associated with the FSC — will refuse to buy German forest products in the future.' That was quite drastic, right from day one."[15]

In October 1997, in the midst of these events, FSC supporters in Germany moved forward in their operations, formally establishing a German working group that was charged with developing specific standards based on FSC's international principles and criteria. A few public land agencies chose to participate, such as the Association of Municipalities of Rheinland-Pfalz and state forestry agencies of Schleswig-Holstein, Saarland, and Hamburg. Through its creation of the German Forest Initiative (*Forstinitiative Deutschland*) in 1997, the Association of Municipalities of Rheinland-Pfalz hoped to rally forest owners to support the FSC and make changes "from the inside"; the standards that the association created as a basis for negotiation upon joining the FSC would become crucial to the final FSC standards development process.

The legitimacy that FSC achieved from this limited number of agencies is best categorized as moral, since they chose to join the FSC because its focus on social consensus and dialogue on forestry issues was consistent with their own philosophical values. The source of these values can be partly traced to the

existence of governments in many of these regions that included a strong
Green Party element (World Wildlife Fund Germany 2000b). Support from
this small number of publicly owned forestlands resulted from political processes and had little to do with market pressure from retailers further down
the supply chain. A key exception to private forest landowner opposition to
FSC support at this time was found with the Working Group on Natural
Silviculture, a group of landowners that place a high importance on "ecologically friendly" forestry practices.

The first draft of the FSC regional standards for Germany was released in fall
1998. The German governmental and private forest owner majority that boycotted the standards-setting process predictably criticized the final FSC standards, arguing that they went beyond the requirements of the ten international
FSC principles (Ripkin 1999), were stricter than the FSC standards in nearby
neighbor Sweden, and were inappropriate for the German context. These
views occurred despite the clear conforming strategies that the FSC working
group undertook to address two contentious issues for private forest landowners: rejection of some conservation groups' proposed 15 percent set-aside
area requirement and rejection of a decision-making structure that would have
resulted in the FSC's environmental and social chambers outvoting the economic interests when changes to the FSC's substantive standards were made.
Instead of accepting the proposed 15 percent set-aside rule, the German FSC
went in the opposite direction, with proposed standards that *waived* the set-
aside rule all together for *all* private landowners and for public forest managers
with less than 1000 hectares (Forest Stewardship Council Germany 1998).
Similarly, the FSC working group accepted a proposed rule that the FSC standards could only be revised and changed if at least two-thirds majority agreed,
directly addressing landowner fears that less than that amount would mean
that the economic chamber, in which landowners' views were represented,
could easily be outvoted by social and environmental interests.

These conforming efforts did have the effect of maintaining support from
the minority of private forest landowners that had decided to become involved
in the FSC working group process. An official from the Association of Municipalities of Rheinland-Pfalz, which was active in the FSC standards setting
process, explained the process: "We discussed and negotiated the standards
for weeks — no, months — with Greenpeace, with WWF, with all of the organizations. Then, when we realized 'ok, this is acceptable from a forestry perspective' . . . we decided to stick with the FSC working group. We were the first
landowner association to do so."[16]

However, while these conforming strategies maintained FSC's limited support from maverick landowners, they had the opposite effect among the

majority of public and private forest landowners — who now became increasingly concerned that, in the absence of an alternative program, the FSC could theoretically earn a monopoly as the only comprehensive certification system in Germany[17] (PEFC Germany 1999b). Landowners and managers worried that if this happened, FSC standards would be constantly revised, made stricter, and most importantly, shift rulemaking away from state management agencies and powerful private landowner associations. Interactions with landowners and managers who were disgruntled with the FSC in other European countries reinforced this view. It was at this time that the German Forestry Association and the Working Group of German Forest Owner Associations reluctantly decided to stop defending the Label of Origin program and created the PEFC, which they believed would be seen as a more credible program to purchasers further down the supply chain, but would still maintain landowner control over certification rules. Supporters of the PEFC also believed that its creation would stop any further "bleeding" of limited public and private landowner support over to the side of FSC. As an official from the Association of the Municipalities of Rheinland-Pfalz explained, "Many people in Germany say that PEFC probably never would have been created if we hadn't cooperated with the FSC . . . since [our participation] created a pressure to develop another system as an alternative . . . The voices in German forestry who said 'we have to create [a program] of our own' became louder, and those voices who said 'we don't need that at all' became quiet at that moment. . . . Up until that point it had been the strategy to say that there were absolutely no forest owners who participate in the FSC, and therefore the FSC wouldn't gain any momentum."[18]

The decision to create the German PEFC was a conscious choice by German landowner and forestry associations to act in tandem with their counterparts in other European countries, who, equally threatened by the FSC, had established the PEFC as an overarching European-wide umbrella organization that allowed for the mutual recognition of landowner-initiated national forest certification programs.

In BC, the recognition that the FSC was gaining some support and was becoming a credible program resulted in forest companies deciding to influence the FSC from within. In Germany, this similar realization led to the creation and promotion of an entirely new "FSC-competitor" program, designed to succeed where previous industry labeling schemes had failed and created in conjunction with other forest owners in Finland, Norway, and Austria. The decision to create a new program, rather than work to force changes within the FSC, was clearly influenced by the longstanding and proud tradition forest managers have had in Germany, the lack of direct domestic market pressure, and forest landowner distrust of environmental groups and their motives. The latter was most certainly exacerbated by the historical debate between forestry and conservation group interests described above. Despite the FSC standards' waiving of the set-aside rule for private landowners and despite many modern conservation groups in Germany shifting from forest preservation strategies to a more holistic approach and compromise-oriented solutions (Hofmann et al. 2000), most forest managers still saw the FSC as a part of the old acrimonious preservation debates (Klins 2000).

The PEFC was formally established in June 1999, and its standards were based on criteria identified at the Helsinki[19] and Lisbon Forest Ministers Conferences in 1993 and 1998, respectively (PEFC International 2001b). From the start, the program was explicitly designed to address forest managers' criticisms of the FSC: PEFC contained a decision-making process dominated by non-industrial private forest landowners, and the program created an assessment system that has a heavy focus on management planning and the certification of regions rather than individual forest management units. The setting of national-level standards was left to national working groups.

The German Forestry Association provided the PEFC's first office and executive director, and, reflecting the close involvement of state agencies, the PEFC was granted funds in its initial years for public relations purposes from the *Holzabsatzfond,* a state agency that is responsible for promoting the use of forest products.[20]

The German national PEFC standards were created in August 1999 and recognized by PEFC International in July 2000. These standards explicitly recognized current German forestry as being "sustainable" (although the program purported to go beyond state regulations), which led to immediate and strong support from a relatively large number of German forest landowners. While the degree to which private forest owner associations and state forestry agencies influenced each other's attitudes toward the competing certification programs is arguable, the close relationship between the two groups most surely influenced the aligning of their positions. What is clear is that the support of most state agencies had a leadership and *supply* effect. As the producer of nearly half of Germany's raw logs, the state agencies' forest certification choices directly influence the amount and type of certified product that is available (Janssen 1998).

Although much of the language used by many state agencies and landowner associations when promoting the PEFC seems to indicate that the PEFC resonated with their management values and that they were granting the PEFC moral legitimacy, caution is warranted in making this claim. Most supported the PEFC for pragmatic reasons as they saw it as the "lesser of two evils." One strong supporter even called it, ironically, "the program that nobody wants."[21] If it were not for the perceived need to combat the FSC, most forestland owners would prefer no certification program at all.

Efforts to Increase Support: 1999–2000

Once the debate over forest certification in Germany turned from foreign to domestic forests and the two FSC and PEFC factions were established, the continuing competition between the two programs for forest certification rule-making authority resulted in a highly public, often acrimonious, competition for legitimacy. While the actions taken by German supporters of the FSC involved conforming strategies at the outset of the debate, the hardening of competing factions meant that its conforming strategies were viewed as arrows in the battlefield for support, rather than as peace offerings. With FSC conforming strategies increasing, rather than reducing, opposition to FSC among most forest landowners and state agencies, the FSC was forced to rely increasingly on its rather weak market pressure campaign.

Supporters of the PEFC have worked in two directions to shore up support and place FSC supporters on the defensive: increasing support for the PEFC among state, community, and private forest landowners, and increasing efforts to gain support from parts of the supply chain that are traditionally FSC territory. PEFC supporters were able to use the close state forest manager and private landowner relationship to their advantage. PEFC supporters worked to gain public statements in support of the PEFC from many forest and environment ministers (e.g., Hessisches Ministerium für Umwelt, Landwirtschaft und Forsten 2000); and prominent private landowners (PEFC Germany 2000b). As a result of these "informing" efforts, landowners have come to understand that state forest management agencies, for the most part, support the PEFC, which has influenced and/or reinforced private forest landowner choices to support that program. According to one forestry official, "If I was to put it all on a map, you'd see some forestry districts where all communities are certified with the FSC. Those regions would be where the forester supports FSC and says that it's good. And everywhere else on the map, there would be a few random dots for the communities that are certified with FSC — those are ones that opted for FSC certification even against the choice of the forestry office."[22]

While state agencies strongly influence the certification decisions of community and private landowners, a reciprocal relationship also appears to exist, with the opinions of private landowners reinforcing, or influencing, the policies of state forest management agencies. A number of private landowner and state agency officials have argued that state agencies certify their forests with the program that is preferred by the private landowners in order to show "solidarity." If state forests were to certify under a different certification program than that chosen by private landowners, state agencies could have seriously damaged their relationship with private landowners.

With the support of most state agencies and private forest landowners, PEFC supporters have used conforming strategies in the more difficult task of convincing forest product buyers that their program is also a credible approach to guaranteeing forest sustainability. They have worked to place considerable domestic political pressure on retailers with pro-FSC stances to accept the PEFC as well, which appears to have served to weaken domestic demand for FSC in some instances. In this respect, the case of the OBI home improvement chain is informative. In early 2000, forest ministers of Bavaria, Baden-Württemberg, and Thuringen wrote to the company's chief executive officer and pressured him to retract his commitment to purchase FSC certified products, using a veiled threat of a forest owner boycott against OBI if a retraction was not made (World Wildlife Fund Germany 2000c). While these actions did not result in changes to OBI's commitments to purchase FSC products, this type of pressure has had direct impacts elsewhere. For example, the decision by the managers of the Expo 2000 in Hanover to use FSC-certified products in the Expo was reversed after they came under pressure from powerful private landowners. Such domestic political pressure reveals the importance of understanding from where market pressure further down the supply chain is coming. If the support is outside the domestic political jurisdiction (the way European retailers and publishing companies were outside the BC domestic scene), support for the FSC is easier to maintain because companies are shielded from domestic "backlash" pressure. However, companies that are operating in the same country are not as shielded to charges that they are hurting their own country's industry, rendering, everything else being equal, a less direct and less durable form of pressure.

PEFC efforts to gain or maintain support have involved a mixture of converting and conforming tactics. They have attempted to convert retailers and purchasers by placing political pressure on them to accept the PEFC. On the other hand, they have conformed to retailers and other purchasers of their products by agreeing to many of the requirements of certification demanded by Gruppe 98 and other German retailers who had persisted with their support of the FSC (PEFC Germany 2000c). For instance, the PEFC has accepted, rather than fought, the terms of a "credible" certification program that was established early, as noted above, by German paper companies and magazine producers: international, third-party, and recognized by relevant environmental

NGOs. Unlike its counterparts in the US, the PEFC has introduced a chain-of-custody procedure (e.g., Rettenmeier Holzindustrie Wilburgstetten 2001), and as of January 2003, PEFC chain-of-custody certificates had been issued to 172 companies (PEFC Germany 2003). The PEFC has also boosted the rigor of its auditing procedure (PEFC Germany 2001a) and has pursued partnerships with other national certification programs, such as the Canadian Standards Association (CSA) and the US Sustainable Forestry Initiative (SFI) in order to create an international presence (PEFC International 2001b). The PEFC has turned its attention toward gaining mutual recognition with the international FSC (PEFC Germany 2001c) — a position that German FSC supporters have rejected (Forest Stewardship Council Germany 2001b) and tend to view ultimately as a takeover attempt, since such recognition would essentially remove incentives to firms to support the more onerous FSC.

However, not all PEFC supporters have bought in to the conforming strategies being undertaken at the FSC secretariat level. In May 2002 German private forest owner associations again took to the streets and protested at OBI stores Germany-wide, claiming that OBI's policy of preferring FSC-certified wood products put domestic forest owners at a disadvantage against FSC-certified competitors abroad. OBI representatives took care to clarify the retailer's stance, stating that OBI's goal is not to put German forest owners at a disadvantage or to import FSC-certified wood from abroad, but rather to "help the German forest sector be able to market its forest products internationally" (Forest Stewardship Council Germany 2002a).

While German environmental groups and companies such as OBI appear to be unimpressed by these strategies, the PEFC's efforts are having some limited success among some German forest product buyers. The Haindl paper company, a supporter of the FSC for foreign imports, now accepts both FSC and PEFC wood, and participated in the development of PEFC Germany standards. In 2001 the Association of German Magazine Publishers declared that the PEFC filled its requirements for a credible certification system (PEFC Germany 2002a). Axel Springer Verlag has decided not to put the FSC label on its paper products even though it would qualify under the percentage-based claim requirements of the FSC, citing logistical difficulties in using the label in the complex paper industry and urging the FSC to consider mutual recognition with other programs (Taiga Rescue Network 2002). These are important developments because, as we detailed in chapters 1 and 2, success for the "FSC-competitor program" only requires that it be recognized as a credible program to purchasers of certified wood, rather than being considered the only credible program.[23]

These changes to the PEFC have led some analysts to conclude that the differences between the PEFC and FSC are no longer significant (Thoroe 2000; Thierme 2001), while others have contended that significant differences remain (Ozinga 2001). PEFC proponents have used the former studies to buttress their own informing strategies, sending out the message that the two systems are becoming more similar (PEFC Germany 2000a, 2001b), while the latter studies have been promoted by FSC supporters with informing strategies, to emphasizing to conservation group supporters and Gruppe 98 members that substantive and irreconcilable differences still exist between the two programs (Forest Stewardship Council Germany 2001b). Major points emphasized by FSC supporters include the manner in which non-forestry industry groups can participate in decision making, whether certification occurs at the regional level or individual forest management unit level, and the more prescriptive content of FSC forest management rules versus the flexible and discretionary PEFC approach.

PEFC success at the landowner level and efforts aimed at the retailer level to have them accept the PEFC as an appropriate certification program forced the FSC and its strategists on the defensive. The FSC redoubled its efforts to inform like-minded forest owners that it existed, and it turned back toward the market-place in the hopes of converting at least some forest owners to support the FSC. Strong efforts also began at this time by FSC supporters to pressure state agencies to promulgate pro-FSC procurement policies, still focusing in many cases on the government's commitment to protect remaining old growth forests abroad by taking a pro-FSC stance (Forest Stewardship Council Germany 2002). However, given the domestic structural features noted above, this FSC move has had little effect in increasing support from state forest agencies and private forest owners, and it is uncertain how well it will do in maintaining existing levels of support.

One of the major problems the FSC continues to face in relying on market-based efforts is that even by maintaining retailer support for the FSC, companies along the supply chain between retailer and the landowner are not as convinced. For example, many sawmill owners believe that if they did support the FSC they would be seen as "traitors" by the landowners and state agencies on whom they rely for raw materials. As one private forest owner noted: "The debate is very emotionally charged . . . if a large sawmill was to get FSC chain-of-custody certified it would probably be viewed by the state forest agencies in Bavaria or Baden-Württemberg as a type of betrayal. And of course no large sawmill wants to get in the bad books with as important a wood supplier as the big state forest agencies."[24] Many ambivalent sawmill owners in Germany are staying out of the certification debate until they know who the certification winner will be. In addition to being concerned about hurting their relationships with forest owners and companies down the supply chain, they do not want to implement, for economic reasons, chain-of-custody requirements for a program that might fail in the marketplace.[25]

FSC efforts to convince sawmill owners to support the FSC were further hindered by the low level of vertical integration and fragmented land base. This feature made it difficult for the FSC, as there were few large sawmill companies they could directly target for support through direct action campaigns. The fragmented sawmill ownership hinders FSC efforts in both directions along the supply chain. Market pressure from the retailer is increasingly weak as it travels down to the forest owner. Similarly, it is difficult to track the "chain of custody" of the wood up to the retailer from the minority of landowners who are supporting the FSC (Becker 1999). The impacts of these supply chain dynamics is a catch-22 for the FSC: landowners say there is not enough demand for FSC wood to make certification under that program worthwhile, while retailers say there is not an adequate supply of FSC-certified wood available to stock their shelves. As one FSC supporter and private forest owner described: "The situation is like this: on the one hand is the Gruppe 98 [members], who say that they want FSC wood. On the other hand is a small group of landowners who dared to get FSC certified . . . But what's missing is the connection — the chain. It's useless for me to be certified, and it's useless that OBI would like to have FSC wood, if there's no certified sawmill in between."[26]

The lack of domestic supply and the clear gap in the German supply chain has actually resulted in demand from some Gruppe 98 members being filled not by domestic producers, but by FSC-certified imports from Poland, Brazil, and Sweden (World Wildlife Fund Germany 2000a). Stuck between market converting strategies that are currently only resulting in limited forest manager support for the FSC and conforming efforts that seem to have had the unintended impact of *increasing* support for the PEFC, FSC supporters in Germany have been largely forced into a holding pattern — hoping that their market pressure, if maintained long enough, will eventually convince landowners who are largely supporting only the PEFC to support the FSC as well. As one FSC supporter explained, "[Our detractors] will have a hard time holding us back, because we'll always find a private or state owner, or especially a community forest owner, who will get FSC certified . . . there will always be some brave individuals who will do it . . . That's happening right now. We will have increasing numbers of individuals from the sawmill industry who . . . will suddenly notice that they're losing business. That companies like OBI are buying their FSC wood in Poland."[27]

Table 6.3 *Certification choices of state forestry agencies in Germany*

State	Certification Program	
	PEFC	FSC
Baden-Württemberg	✓	
Bayern	✓	
Berlin		✓
Brandenburg	✓	
Bremen		
Hamburg		✓
Hessen	✓	
Mecklenburg-Vorpommern[a]	✓	
Niedersachsen	✓	
Nordrhein-Westfalen	✓	✓
Rheinland-Pfalz	✓	
Saarland[a]	✓	✓
Sachsen	✓	
Sachsen-Anhalt	✓	
Schleswig-Holstein[a]	✓	✓
Thueringen	✓	
Total state, community, and private forest area certified	5,899,713 ha	422,985 ha

Note: Support is indicated by check marks; total certified area includes all ownership types.
[a] These states were writing forestry plans in preparation for PEFC certification.
Sources: Forest Stewardship Council Germany 2002b; PEFC Germany 2002b.

The impact of such a holding pattern in increasing support for the FSC in Germany is far from assured, with the PEFC appearing to increase, rather than reduce, its support among landowners and gaining some support among retailers.

The trends in support for the two programs that are reviewed in this chapter are summarized in table 6.3, which presents the area of lands certified under each program. While not a comprehensive measure of support given that FSC certification generally takes longer to achieve, the data attest to the higher number of state forestry agencies that have chosen to certify their lands under the PEFC. As of September 2002, 422,985 hectares were certified under the FSC in Germany (Forest Stewardship Council Germany 2002b), compared with 5.9 million hectares certified under the PEFC (PEFC Germany 2002b).[28]

Conclusion

The domestic problem definition and structure of the forest sector have not only worked to limit the impact of FSC's market-based efforts to convert forest owners to support them, but the same features also created favorable conditions for landowner support for the PEFC. As a result, the PEFC was able to focus its efforts on gaining support from the very purchasers of German forest products from which the FSC had already gained commitments. By focusing at the level of the retailer, the PEFC was able to put the FSC on the defensive. FSC and its supporters are now spending much of their efforts shoring up support among retailers and purchasers of domestic forest products, including vigorously fighting PEFC efforts to call for mutual recognition with the FSC. The FSC is now in a holding pattern, being forced to rely on maintaining its support among retailers and state procurement agencies in the hopes that this will eventually lead to increased support from forest landowners. While predictions are difficult to make, the evidence presented in this chapter indicates that increased support for the FSC is far from assured.

The case reveals support for hypothesis 2 (*Forest companies and nonindustrial forest owners selling wood to a domestic market in a country or region that imports a large proportion of all the forest products it consumes are more likely to be convinced to support the FSC than those in a region that imports a small proportion of all the forest products it consumes*); evidence exists that Germany's role as a major importer of forest products permitted the FSC to gain attention domestically after importing companies began demanding FSC-certification of foreign suppliers. This led domestic environmental groups and other critics to contend that, for moral "fairness" reasons, German forest product buyers should make the same demands of FSC-certification of their domestic suppliers as well. The country's high amount of imports from countries with a large area of FSC-certified forest, such as Sweden and Poland, has also facilitated converting strategies; forest companies and owners in Germany fear that their market share will diminish if domestic companies begin preferring FSC-certified products from these countries over PEFC-certified or uncertified domestic sources.

Likewise, the German case shows strong support for hypothesis 4 (*Unfragmented non-industrial forest ownerships are more likely to be convinced to support the FSC than fragmented non-industrial forest ownerships*). We recommend more research be undertaken to see whether the fragmented land-ownership patterns that minimized acceptance for the FSC were as important a factor as the strong ideological positions against FSC-style certification. There was strong support for hypothesis 5 (*Forest companies and non-industrial forest owners in a country or region with diffuse or non-existent associational systems are more likely to be convinced to support the FSC than those in a country or region with relatively well-coordinated, unified associational systems*), where strong associations with close connections to state agencies meant that landowners could articulate—collectively—alternatives to the FSC. In addition, hypothesis 3 (*Large and concentrated industrial forest companies are more likely to be convinced to support the FSC than relatively small and less concentrated industrial forest companies*) is supported by the difficulty that FSC supporters faced in transmitting market pressure at the retail level to the forest owner, mainly due to resistance of small- to mid-sized saw-mills operating in the middle of the supply chain.

Implicit support was illustrated for hypothesis 6 (*Forest companies and non-industrial forest owners in a country or region with sustained and extensive environmental group and public dissatisfaction with forestry practices are more likely to be convinced to support the FSC than those in a country or region with less dissatisfaction*), since the lack of any perceived serious crisis on the domestic public policy agenda meant that there was no interest in using the FSC as a way of addressing public scrutiny. In fact, the lack of a perceived crisis in the public policy sphere worked to marginalize environmental group efforts to change the nature of forest policy development, undermining FSC (and PEFC) efforts to gain support.

This chapter also highlights how the FSC and FSC-competitor certification programs change in their efforts to gain support beyond their initial "core" audiences. As such, it reveals the indirect impacts of the FSC in influencing the emergence of and changes to alternative programs. As one FSC supporter explained, "Looking back, I think that the most important impact of the FSC [in Germany] is not the measurable results, such as 'we've certified so-and-so many hectares per year and marketed so-and-so many cubic meters of wood' . . . In reality, our biggest impact is what the PEFC has done as a reaction to the FSC . . . Despite all of the PEFC's shortcomings, it's caused a new dialogue within the old established forestry community; they're asking questions like 'what is sustainable forestry?' . . . And even if the changes in forest practices [caused by PEFC] don't go far enough for us at the FSC . . . there still are changes, and, cumulatively, they are incredibly important. But most important is the mental change, the fact that [forest owners] have learned that they must justify [their forest practices] and really think about . . . whether or not to log a wetland, for example. This movement wouldn't have happened without the FSC in Germany."[29]

Whether the emergence of FSC-competitor programs are simply meant to limit the scope and implementation of the FSC conception of certification, or whether they actually signal a meaningful attempt by the forest industry and forest landowners to address sustainable forest management, is a topic we return to in the conclusion of this book. What is clear from the German case is that the ability to gain support from forest owners is far from deterministic. Much of the outcome of this struggle is dependent on the ability of the FSC and FSC-competitor programs to recognize, and capitalize on, the environment in which they make their strategic choices.

7

Sweden

Interest in the use of forest certification as the route for developing domestic sustainable forestry standards emerged early in Sweden, when support for the Forest Stewardship Council (FSC) began gaining momentum in some of the country's key foreign export markets. The early entry of certification was important because it meant that, in the absence of any alternative program, the initial dynamics facing forest companies and forest landowners were either to support FSC certification, or not support certification at all. Like other cases reviewed in this book, forest certification was promoted by key environmental groups, who saw certification as a faster and more responsive arena than traditional governmental processes in which to develop clear and comprehensive rules governing sustainable forest management. And like other cases reviewed in this book, the forest industry and forest landowners were initially reluctant to support FSC-style forest certification. Yet FSC supporters, through active market-based campaigns, were able to attract, by early 1996, both industry and private forest landowners (whose membership had mixed feelings) to the development of Swedish FSC standards. By September 1997, major Swedish industrial forest companies came to support FSC-style certification on their forestlands, while non-industrial private forest owners, reflecting trends in the US, UK, and Germany, ultimately decided not to support the FSC, deeming the program inappropriate for their forestlands. It was these very same landowners who would end up promoting and supporting forest certification under the Pan European Forest Certification (PEFC) program (see table 7.2).

Similar to the scenarios presented in the other chapters, the Swedish forest sector's place in the global economy, the history of forestry on the domestic public policy agenda, and the structure of the forest sector significantly influenced the ability of FSC strategists to gain forest owner support. Sweden's position as a large exporter of forest products was important because it required firms and forest landowners to evaluate seriously demands from supply chain pressures in UK and German markets, where forest products purchasers, as we discussed in chapters 5 and 6, were making commitments to support the FSC. The structure of the Swedish forest sector was crucial in three key respects. First, highly integrated associational systems — themselves products of Sweden's longstanding "corporatist" approach to public policy making (Boström 2003) — meant that large forest companies and private forest landowners each had their own well-structured and organized associations that had a long history of speaking for their members with a single voice. This feature is important because it works with other factors to accelerate support or opposition toward the FSC. That is, the existence of a well-integrated Swedish Forest Industries' association facilitated industry-wide support for the FSC as members fell in line with their association, but similar associational features of the dominant private forest landowner association worked to hasten landowners' eventual opposition to the FSC. Hence, the unified approach to policy development characteristic of industry and landowner associations created potentially fertile ground for FSC strategists — but only if the seeds were planted carefully, as the soil was equally fertile for FSC-competitor programs. Second, the existence of a handful of forest companies with a relatively high degree of horizontal and vertical integration meant that these companies would become identifiable targets for environmental group boycott campaigns (similar to experiences of forest companies in BC). Reduced transaction costs associated with horizontal and vertical integration meant that FSC-style certification was also easier for them to implement. Third, and related, the role of private forest landowners meant that the industrial sector was not as horizontally integrated as in BC. Indeed, industrial forest companies' dependence on non-industrial private forest owners put these companies in a position not dissimilar to their US counterparts. This feature would mean that FSC strategists could not rely solely on converting strategies in their efforts to maintain support from non-industrial private forest owners. Yet, we show below that FSC strategists, for the most part, overlooked this feature, believing that once non-industrial private forest landowners had agreed to participate in FSC standards development processes, they would not back away. So confident were FSC strategists and forest companies (who themselves were eager to respond to foreign market pressures) that they moved ahead when non-industrial private forest owners withdrew from the FSC process. Owing to the size of the forest companies, Sweden became the largest single source of FSC-certified forestlands, comprising as of September 2002 approximately one-third of the FSC global total, and the country was thus hailed as an FSC success.

But a fundamental problem still confronted those promoting FSC-style certification in Sweden: without support from private forest landowners, the forest industry would have a difficult time tracking certified wood to the marketplace, as it would be mixed with non-certified sources. When the FSC did recognize the need to address this issue, it first did so by lowering its "percentage-based claims" (the amount of FSC fiber required to permit the use of a label), but Swedish industry, recognizing the temporary nature of such an approach, sought more permanent solutions. It worked with Swedish environmental groups and forest owners in 2001 to first identify remaining differences between the systems, and to see if a Swedish standard might be created that would allow forest owners either to participate in the FSC or to recognize their own approach as equivalent. If successful, the payoff would be important, as it would allow all of Sweden's forest products to be easily tracked along the supply chain all the way to the end consumer.

Finally, the history of forestry on the public policy agenda was important because it created a policy arena conducive to FSC efforts to gain or maintain non-industrial forest owner evaluations toward supporting their program. Increasing societal concern over Swedish forest practices in the 1980s and 1990s led the government to establish that environmental goals have equal weight to production goals. Following existing approaches, it left implementation of these objectives to the regional boards. And it was dissatisfaction by some environmental groups with these regional boards (and the county forestry boards that preceded them) that led them to see the FSC as a more appropriate means with which to put environmental protection and economic development on an equal footing. The tripartite corporatist policy-making style of the FSC already mirrored traditional approaches to public policy development in Sweden, making it arguably the smoothest fit with traditional and widely accepted approaches to public policy development. In addition, public policy initiatives aimed at privatization and deregulation in the early 1990s were key enabling factors — government privatization of AssiDomän, the dismantling of timber pricing controls, and Sweden's entry into the EU all worked to create a climate hospitable to market-based solutions to environmental problems.

The balance of this chapter details these arguments in four analytical steps. First, it provides an overview of the three factors that influenced the ability of the FSC and its supporters to gain forest owner support. Second, it describes the roots of certification in Sweden and the evolution of industrial forest company and forest owner support for the FSC. Third, it details the strategic choices made by FSC supporters and their ultimate decision to focus largely on market-based strategies aimed at converting forest companies and non-industrial landowners to support the FSC. It reviews the eventual emergence of the PEFC as a preferred alternative for non-industrial private forest owners, who were frustrated by what they felt were inadequate efforts by the FSC to conform its program to meet their concerns. The conclusion assesses the validity of the hypotheses detailed in chapter 2, reviewing how the seven specific factors worked to mediate the success and failure of strategic choices taken by the FSC and its supporters.

Factor 1: Place in Global Economy

DEPENDENCE ON FOREIGN MARKETS

Sweden is highly dependent on export markets. Figures 7.1 and 7.2 reveal that export sales are concentrated in four countries: Germany, the UK, the Netherlands, and Denmark. These features are important for two reasons. First, as we reveal below and in the UK, German, and BC chapters, some of the most intense pressures for improving forestry practices came from these countries, rendering Sweden particularly vulnerable to market-based converting. Second, Sweden's market share in these countries was, and remains, far from guaranteed — indeed the current and future roles of Eastern Europe and Russia in providing fiber to these same markets are being taken seriously by Swedish forest companies, which have been keenly interested in devising strategies addressing this emerging competition.

Factor 2: Structure of the Forest Sector

Three factors are key for understanding the structure of the Swedish forest sector: the degree of concentration of its forest industry (assessed in terms of number of industrial forest owners, the horizontal integration of

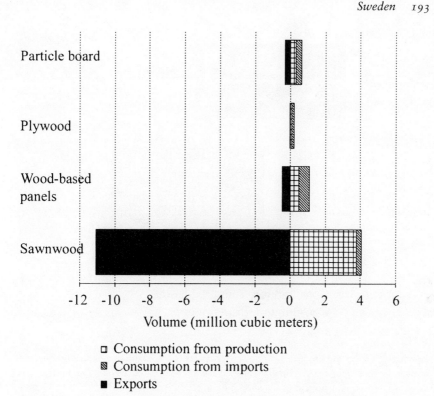

Figure 7.1. Volume (in cubic meters) of Swedish apparent consumption (production plus imports less exports) of various aggregate wood products in 1999. In this year, Sweden purchased 35 percent of its sawnwood imports from Finland, Poland, the Russian Federation, and the US and 62 percent of its plywood imports from Finland, the Russian Federation, and Poland. Exports to the Netherlands, the UK, Germany, and Denmark represented 53 percent of its sawnwood exports, 56 percent of its plywood exports, and 44 percent of its particleboard exports. Source: FAO 2001.

industrial forest landownership, and the vertical integration of processing facilities), the high degree of fragmentation of non-industrial lands that provide industrial forest companies with additional fiber, and the relatively cohesive and representative forest sector associations.

INDUSTRIAL FOREST COMPANY CONCENTRATION

Five large industrial Swedish forest companies together own and manage one-third of Sweden's industrial forestlands (figure 7.3), and control approximately 95 percent of pulp and paper processing capacity (van Kooten, Wilson, and Vertinsky 1999:160).[1] Their lands produce about one-third of the

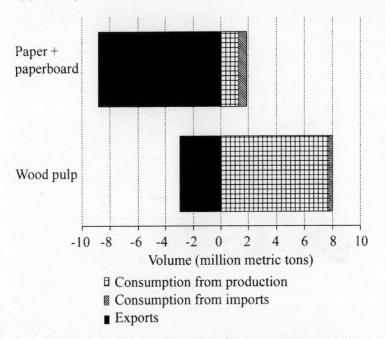

Figure 7.2. Volume (in metric tons) of Swedish apparent consumption (production plus imports less exports) for wood pulp and paper plus paperboard products in 1999. In this year,
Sweden purchased 34 percent of its wood pulp imports from Finland and 10 percent of its
paper plus paperboard imports from Germany. Exports to the UK, the Netherlands, Italy,
France, and Germany represented 68 percent of its wood pulp exports and 60 percent of its
paper plus paperboard exports. Source: FAO 2001.

total harvest volume (160). These features make industrial forestland ownership fairly concentrated, exposing it to a greater degree to FSC supporters'
international market campaigns.

The largest company of these five, Sveaskog, is also its most volatile.[2] First
known as "Domänverket" (the Domain Authority), the publicly owned company was instructed in 1979 to behave as if it were a private forest company. In
1992 it was converted to corporation status under Swedish law and renamed
"Domän Aktiebolag (Domän AB). In the fall of 1993 Domän AB bought another government owned forest company AB Statens Skogsindustrier (ASSI)
and changed its name again to AssiDomän. In 1994 it acquired a failing
northern pulp company and then became privatized (though the government
continued to maintain 51 percent of the shares). In 1999 AssiDomän split off
600,000 hectares of its forestland to create a new publicly traded company,
Sveaskog. Yet by 2001 AssiDomän bought back all of Sveaskog outstanding
stock and the entire company was renamed Sveaskog in 2002. This same year

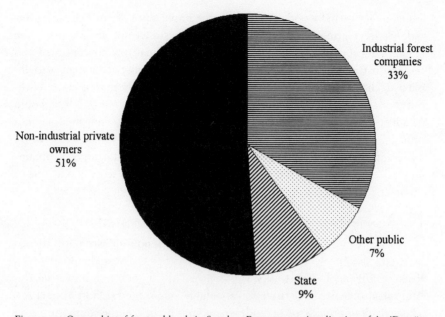

Figure 7.3. Ownership of forested lands in Sweden. Recent re-nationalization of AssiDomän to create Sveaskog has shifted approximately 13 percent of the total forest landbase from private to public ownership. However, we provide these historical patterns because they existed when active efforts to gain support were under way and thus are more important for our historical analysis than the patterns that now exist. Source: Swedish Forest Industries Federation 2001.

the company was re-nationalized, though it was again instructed to behave as a private company.

While the industry is highly concentrated, the fact that two-thirds of forest-lands are not in industrial hands makes Swedish companies less horizontally integrated at the level of forest ownership than companies in BC. The ownership issue is further complicated by the Swedish industry "wood swapping" system, whereby timber harvested on one company's lands is sent to the nearest processing facility where that particular quality and species of wood is required, regardless of ownership. Hence, while a company may, on paper, own enough forestlands to meet more than half its fiber needs from its own harvest, the location of its mills relative to company lands determines the extent to which its wood is processed in its own facilities. This is particularly important for mills in regions with a predominance of non-industrial forestlands, as wood swapping means these mills face very limited access to

company harvested wood. Thus, while the importance of non-industrial forestlands in creating supply for forest companies does not directly affect the
overall FSC converting logic associated with the relatively concentrated and
large industrial company features, it does affect the ability of companies to implement chain-of-custody tracking of certified wood to the marketplace. Without similar commitments to the FSC from non-industrial forestlands, it would
be difficult for companies to track wood through their processing facilities, not
only due to the extent of non-industrial ownership, but also due to the wood
swapping system devised by the industry itself.[3] And the difficulties in obtaining FSC commitments from non-industrial forest owners were in part due to
our next structural feature, high fragmentation of non-industrial forestlands.

NON-INDUSTRIAL FOREST OWNERSHIP FRAGMENTATION

Private forest ownership makes up the bulk of non-industrial forestlands
in Sweden and provides over 60 percent of the fiber harvested annually (van
Kooten, Wilson, and Vertinsky 1999:160). The 345,000 non-industrial private forest owners make these forests highly fragmented (Elliott 2000:179).
The density of private forest landownership is highest in the southern region
where productivity is higher, and smallest in the northern region where industrial ownership dominates. There is a handful of large non-industrial
forest owners—notably the Church of Sweden, which gives each of its eight
dioceses authority over forest management choices and strategies. The share
of government-owned forestland has fluctuated with the privatization and re-
nationalization of Sveaskog. The re-nationalization of Sveaskog in 2002 increased the government owned share from 9 to 22 percent.

These features are important for our story below because the national association was strong enough to ensure that forest landowners acted with a single
voice, but it also impacted the choice to enter the FSC standards-setting process, which was pushed by private landowners in the south, and the choice to
leave, which was strongly favored by private landowners in the north.

Factor 3: History of Forestry on the Public Policy Agenda

PUBLIC DISSATISFACTION WITH FORESTRY PRACTICES

In the past decades, forestry practices in Sweden have gained the attention of the general public—both within Sweden and its key export markets. A
history of disputes had developed, traced back to the 1970s and revolving
around the appropriateness of regulations governing forestry practices. Importantly, the concern in Sweden was not on maintaining vast tracts of old

growth forests, as it was in the US and BC, largely because Sweden had few remaining old growth forests.[4] This feature moved the problem definition more toward minimizing the impacts of forest management on the natural environment, making forest preservation a component of the policy reform agenda, rather than it being the dominant theme.

Public attention in the 1970s was at first placed on concerns that timber harvests were occurring at an unsustainable pace. But evidence that forests were not being harvested faster than their growth rate shifted attention toward the ecological integrity and protection of flora and fauna in Sweden's forests. The governmental response in 1974 was to alter slightly the 1948 Forestry Act by recommending to forest owners that they incorporate environmental concerns into their forest planning, and, more substantively, by making it mandatory for owners to notify the Board of Forestry when cutting any forests over half a hectare in size (Eckerberg 1987, cited in Elliott 2000:170). Changes were again made to the act in 1979, but these were motivated largely by (incorrect) assumptions that over-harvesting continued (van Kooten, Wilson, and Vertinsky 1999; Elliott 2000) and that timber production remained the primary value. Indeed, changes to the 1979 act limited the use of fines for non-compliance with any environmental prescriptions that County Forestry Boards might announce, focused on the boards' role as educators, and supported their close relations with their industrial and forest owner clientele (Ekelund and Dahlin 1997; Elliott 2000:170).

The limited legislative or administrative responses governing the forest sector and increasing concern within Swedish society over the environment generally (Frizzell and Pammett 1997) led to increased conflict over forest resource use management in Sweden in the 1980s (van Kooten, Wilson, and Vertinsky 1999; Elliott 2000). Swedish environmental groups, led by the Swedish Society for Nature Conservation, raised a number of specific concerns — including the use of clearcutting and exotic species in forest management and chlorine bleach in pulp production (Elliott 2000:171) — and called for increased forest preservation (to permanently remove forestland from the commercial land base) (Fridman 2000). With perceived limited responses from the Swedish forestry agencies or government in the 1980s,[5] Swedish environmental groups began their first efforts to work with their counterparts in the UK and Germany in an effort to use Swedish reliance on foreign markets to force change domestically (Elliott 2000).

Three coinciding events in the late 1980s and early 1990s facilitated FSC's strategic efforts to gain forest owner support. The first two events increased attention to environmental aspects of Swedish forestry, while the third worked to limit the role of government in developing prescriptive resolutions to these issues. The first was the United Nations' Commission on Environment and Development, known as the Rio Earth Summit, which gave Swedish environmental groups an international forum in which to raise and promote their concerns about Swedish industrial forestry (Gamlin 1998). The second was a range of scientific studies that documented the negative effects of forestry practices on a range of forest species (Elliott 2000) — all of which worked to increase societal concern about Swedish forestry. The third, however, was an increasing salience in Sweden for non-regulatory approaches to public policy amid increasing trade liberalization. This was sparked by the election of the reform and market-oriented Liberal Democrats in 1991, who for three formative years replaced Sweden's historically dominant governing party, the Social Democrats, and by Sweden's entry into the European Union in January 1995-as conditions of membership included limitation on budget financing and a range of economic integration measures.

The new government sought to address the increasing interest in the environment while minimizing a heavily prescriptive state-dominated approach. It did so in two ways. First, following a comprehensive multi-stakeholder process[6] — not dissimilar to FSC-style processes — it crafted new forestry legislation in 1994 in which environmental goals were required to be given the same weight as economic ones when developing forest policy and making forest management decisions (Klingberg 2002; Elliott 2000, chap. 6; van Kooten, Wilson, and Vertinsky 1999).

At the same time it encouraged market-oriented and flexible approaches for achieving these goals. As the Royal Swedish Academy of Agriculture and Forestry (2001) explained: "The new forestry policy, which took effect in 1994, was influenced, above all, by the wish for greater liberalization in the business sector and the need for greater attention to be paid to conservation issues in forestry. The policy structure was changed from one characterized by the imposition of regulations to one based on management by objectives. Two overriding goals of equal status were formulated: one for production and one for the environment. . . . One aspect of the thinking behind the new policy is that more room should be afforded in the market for different forest products and services."

This legislation varied markedly from the approach in BC (where the government required that its Forest Practices Code and biodiversity goals not reduce the annual cut by more than 6 percent), and its requirement that environmental goals be given equal weight to economic goals was strongly supported by environmental groups. However, with the regulations, rules, and standards left to lower levels of implementation (the national and regional forestry boards), struggles over Swedish forest policy did not subside — they were simply redirected to other arenas (Boström 2003). And it was at these levels that environmental groups continued to feel that long entrenched industry and agency relations worked to limit implementation of the dual goals. The policy process had indeed opened up to include new interests, but largely at the national goal-setting arena rather than at the lower levels charged with implementing the goals.

The second way that the government sought to address the environment while minimizing the state-dominated approach was to embark on a "protected areas" effort, whereby it sought to increase the amount of forestland in forest reserves (Royal Swedish Academy of Agriculture and Forestry 2001). This would be an important move as forest companies could use these initiatives in their efforts to satisfy the high conservation requirements of the FSC.

The result of these initiatives was that by 1994 the Swedish public policy approach worked to create an environment that strongly supported FSC efforts to promote its style of forest certification. The policy vacuum provided by the flexibility in the 1994 act proved important, as it would open the door for the FSC as a way to implement these goals outside of existing state-centered processes. The recognition of broader forest conservation goals by the government, and even its efforts to address them, provided a helpful environment for FSC strategists wishing to convert forest companies and forest owners toward supporting the FSC. The FSC and its supporters could argue that being certified under the FSC would provide recognition and evidence that companies were indeed meeting these dual goals.

FOREST COMPANY AND NON-INDUSTRIAL LANDOWNER RELATIONS WITH STATE AGENCIES RELATIVE TO OTHER SOCIETAL INTERESTS

Historically the Swedish approach to forestry policy development has been highly institutionalized. It begins with Parliament providing broad framework legislation and goals, which is then reviewed by Cabinet and sent to the National Forestry Board for the development of specific regulations. Regionally based boards of forestry are then in charge of implementing the regulations on the ground, directly interacting with industrial forest companies and private forest owners (Elliott 2000:176). (Following the 1994 Forestry Act, the twenty-four county forestry boards were amalgamated into eleven regional boards.[7]) This system was created to develop a rational and technical approach to forest management and led to forest companies and private forest owners enjoying a close relationship with the county boards and the board of forestry. This close relationship came under scrutiny, first during the 1970s as the "first wave of environmentalism" resulted in Swedish citizens paying more attention to domestic forestry practices. And by the mid-1990s the relatively closed

Table 7.1 Influence of explanatory variables on FSC efforts to gain support from forest companies and landowners in Sweden

Factor	Existence
Place in global economy	
Dependence on foreign markets	✓
Structure of forest sector	
Large, concentrated industrial forest companies	1/2
Unfragmented non-industrial forestland ownership	
Diffuse or non-existent associational systems	
History of forestry on public policy agenda	
Sustained and extensive public dissatisfaction with forestry practices	✓
Forest companies and non-industrial forestland owners share access to state forestry agencies with other societal interests	1/2

Note: A "check mark" indicates strong presence and "1/2" indicates mixed presence.

business and government relations in the forest sector had largely dissipated (particularly at the national level), so that business did not feel it was risking anything by participating in the tripartite environmental-social-business struc-ture of the FSC.

Sweden's place in the global economy, the structure of its forest sector, and the history of forestry on the public policy agenda worked to create an arena conducive to FSC efforts to convert forest companies and non-industrial forest owners to support the FSC (table 7.1). However, the existence of important and fragmented non-industrial private forests meant that FSC strategists did face important obstacles. Our story below shows that failure of FSC strategists to place enough attention on conforming their own program to meet the needs and demands of non-industrial private forest owners opened the door for a competitor program to emerge.

Initial Efforts to Obtain Forest Company and Non-industrial Forest Owner Support

Supporters of the FSC, initially led by the World Wide Fund for Nature (WWF), were extremely eager to see the FSC introduced in Sweden. As early as 1992, the WWF began to reach out to companies in Sweden with hopes of finding receptive audiences ready to endorse the FSC as a model for stan-dardizing the practice of sustainable forestry globally. For instance, the WWF informed AssiDomän of the program in 1992, and in the following year an official from the company participated in the FSC's inaugural meeting in Toronto in 1993.

Back in Sweden, the chapter-based Swedish Society for Nature Conservation (SSNC) — the largest environmental group in Sweden — joined WWF's efforts, and together they helped launch the creation of an informal "reference" group, established in 1994 to begin work on forest management standards for Sweden based on FSC principles and criteria (at the time it was not an official FSC national initiative as not enough of its participants were FSC members) (Elliott 2000). These early environmental supporters also immediately turned to their networks in Germany and the UK — Sweden's top two forest products export markets — in the hopes of convincing customers of Swedish forest products to support the FSC process. Like other cases reviewed in this book, most forest companies and non-industrial forestland owners were initially hesitant to participate in this process.[8] Instead, and like other forest sectors described in this book, industry attempted to create a short-lived industry alternative, known as the Nordic Forest Certification Program system.[9]

But just as the Nordic system was fledging, market campaigns targeted at AssiDomän and Korsnäs were heating up. The Dutch company Tetra Pak, which purchased liquid paperboard for making drink cartons, itself under pressure from environmental groups and their market campaigns, communicated to both AssiDomän and Korsnäs that it would be asking its suppliers to provide independent verification that their forestlands were managed sustainably (Elliott 2000:189). Both general market concern about Swedish industrial forestry practices and specific attention on Swedish industrial companies led AssiDomän and Korsnäs to see if certification could be used both as a way of accomplishing the environmental and economic goals required by Swedish legislation, and to transform negative economic attention into positive gains.[10]

Meanwhile the FSC established its national contact person in Sweden, and the reference group established in 1994 to work on an initial set of standards was disbanded in favor of a preliminary FSC working group set up in November 1995. This new group included representatives from WWF, SSNC, Friends of the Earth, Greenpeace, the Church of Sweden, the Forestry Society (Skogssallskapet), and representatives of the Sami people (Elliott 2000:189). Now Swedish industrial forest companies and forest landowners faced a clear choice: participate with the FSC and earn market recognition, or refuse to participate and face increasing boycotts and international scrutiny.

The choice was not preordained. Efforts within the industry association were led by AssiDomän, which, as the largest of the five dominant forest companies and the least dependent on private forest owners for industrial supply, had the most to gain and the least to lose by supporting the FSC. Some industrial forest companies that relied more heavily on non-industrial private forest owners hesitated, and choices over what to do were not made overnight. Likewise the Swedish federal forest owners' association was ambivalent about certification, since its members asserted that they had long been advocates of sustainable forest management. There were also internal differences. Non-industrial private forest landowners in the north were much more hesitant, largely owing to their specific concerns about FSC Principle Three, which required certified owners to consider and duly address the interests of those holding customary and indigenous rights on certified lands (Forest Steward-ship Council 1999c). This concern was owing to a longstanding dispute be-tween forest owners and the Sami people, whose customary rights to have their reindeer herds forage on private forestlands was recently recognized in legislation, although the details of what this was to mean remained in signifi-cant dispute (Elliott and Schlaepfer 2001; Elliott 2000). Many forest owners in the north felt that their lands were being negatively impacted by the reindeer herding practice, which now involved moving reindeer by trucks, rather than on foot, from area to area (Klingberg 2002). On the other hand, landowners in the south, especially the large landowner cooperative Sodra, which operated its own mills and did not face the same diseconomies of scale as other more fragmented ownerships, were more inclined to participate. This was par-ticularly true since the Sami issue was the exclusive concern of landowners in the north, as reindeer herding does not take place on forestlands in the south.

There was also strong concern about the FSC governance structures. Non-industrial private forest owners felt marginalized in an economic chamber that, in their view, was too dominated by retailers and other companies that did not own or manage forestland.

The FSC and its supporters continued efforts at building market demand. Beginning in 1995, the WWF buyers group in the UK, as part of a restructur-ing effort, began to exert more far-reaching and specific demands for the FSC (see chapter 5). The group was renamed the 95+ Group, and some of its members began to issue more concrete purchasing preferences for the FSC. Two prominent members of this group, the UK do-it-yourself retailer B&Q and the supermarket chain Sainsbury, announced that they would give a pref-erence for FSC certified wood; later B&Q even gave the end of 1999 as the deadline for all its suppliers to be FSC certified (DIY 1998). This helped tip the scale for AssiDomän (1996), which, along with Korsnäs, announced in 1995 it would undergo pilot testing of FSC certification (Olsson 1995). By the end of 1995 AssiDomän told its industry counterparts that it intended to break with tradition and participate in the FSC working group even if the others did not. This announcement tipped the scales for the other members of the industry association, who, evaluating the pros and cons, acquiesced to AssiDomän and agreed to have the association participate.

After they formally announced their decision in February 1996 (Barker 1996) Sweden's industrial forest companies followed two tracks, focusing efforts on negotiating creating a workable FSC standard and implementing FSC certification on their forest lands. Indeed, Stora Skog AB (now Stora-Enso) was the first company to earn an FSC certificate in October 1996 for its Ludvika forest lands even before the standards setting process had been completed. And when the standards were endorsed by the FSC's international office in May 1998, Stora Skog became the first company in the world to operate under an approved FSC national standard.

The industry decision to join the FSC working group had an immediate impact on the still deliberating forest owners who, concerned about the potential of increasing market action by Greenpeace, finally agreed to participate as well, though they were clearly more reluctant than other members of the working group. Those representatives in the north were still opposed to joining while those in the south, whose lands were not affected by the reindeer herding issue, and who felt the market pressure more directly owing to their ownership of large pulp mills, saw greater gains from participation.[11] The strong tradition of speaking with one voice meant that the landowners from the north reluctantly accepted the Swedish Forest Owner federation's decision. Since support was so tenuous, a high degree of strategic effort would be required throughout the negotiation process if landowners were to sign on to any formal agreement. With both industry and landowner associations on board, the industry-led Nordic initiative faltered, lacking support from domestic organizations and international markets.

Attempting to Maintain Support

With the forest industry and private forest owners now participating in the development of Swedish FSC standards, strategic efforts by the FSC and its supporters shifted from market campaigns to developing the most appropriate standards with which to achieve environmental and social goals. The governmental policy of requiring companies to give up some of their commercial forestlands to protected areas facilitated these negotiations; industrial forest companies reasoned that since they would have to do it anyway, it made sense to use this requirement to their advantage in shaping certification standards and gaining market recognition. Industry participants also reasoned that since national policy required that they give equal weight to production and environmental goals, and given that conflict existed with lower level interpretations of how to do this, certification represented an opportunity to demonstrate that companies were addressing these goals.

There is no question that negotiations were intense — and that just because the time was ripe for FSC-style certification, the situation or path chosen was in no way inevitable. Issues such as clearcut use and size, where and how much productive forest area should be set aside, and whether exotics and chemicals were to be permissible were all treated seriously by those involved in negotiations. However, the most contentious issue concerned the historical conflict between the Sami people and the private forest owners. Private forest owners were willing to recognize the Sami people's legal rights to graze on their lands, but were not willing to support broader "customary" rights proposed by the Sami and written into draft standards. Indeed, non-industrial private forest owners reasoned that if they had agreed to the FSC standard that recognized these broader customary rights it would hinder their court challenges that sought judicial clarification (and a limiting of) these asserted customary rights.[12] The Sami were quick to argue that the FSC standard must contain support for what they argued was part of their cultural heritage. Private forest owners were equally adamant that the Sami's use of twentieth-century technology in herding its reindeer could not fall under customary rights. The FSC process was not equipped to address this dispute, and in the end, most participants supported the Sami position. While the Sami issue alone might have been enough to have the forest owners leave the process, private forest owners also identified other grievances including specific management rules and the limited role for private forest owners within FSC governance structures. The private forest owners revisited their support of the FSC process and, led by those in the north, announced their withdrawal in March 1997.[13] As Klingberg (2002) explained: "We found, however, a lack of understanding of our ownership pattern, for our small-scale forestry practice and culture, for our complicated network of supply lines to saw mills and pulp mills. Also, on some principal issues, primarily concerning forestry in the mountain region where relations to the reindeer herders are complicated, we were completely overruled. We had to leave the FSC process."

While industry representatives were not surprised, the original supporters of FSC in Sweden, especially environmental groups, were taken aback. They were confident that the market pressures were enough to keep companies and landowners in the process and reasoned that the marketplace would eventually bring them back. Industry was less certain about this and believed that the converting strategies focusing on market pressure alone might not be enough and that conforming approaches that addressed landowner frustrations and concerns might be necessary.

What is clear is that the consequences of non-industrial private forest owner withdrawal on FSC efforts to gain or maintain support from forest owners in Sweden and elsewhere in Europe were enormous. Its significance was that it paved the way for the Swedish owners to initiate alternatives to the FSC — first through their support of the systems-based Eco-Management and Auditing System (EMAS) (van Kooten, Wilson, and Vertinsky 1999:173) and then, when markets failed to recognize this, the creation of the PEFC. The PEFC was immediately attractive to private forest owners and, as we see below, is becoming increasingly attractive to industrial companies that have become frustrated with the FSC and its product tracking requirements, which limit their ability to sell certified products in the marketplace. As one environmental group official involved in the process explained, the failure to maintain non-industrial forest owner support at this time was "a huge strategic mistake,"[14] as it led to the creation of a viable certification alternative to the FSC. And once created, FSC strategists were then required to wage campaigns not only to gain support for the FSC, but to explain why their program was more appropriate than the alternatives.

Negotiations on the FSC standards proceeded without non-industrial private forest owners, and after intense debates in which industry itself threatened to leave (Elliott 2000), an agreement was reached in September 1997 (Swedish Forest Industries Association 1997). The agreement had strong support from the Swedish Society for the Conservation of Nature and the WWF. The opposition Greenpeace Sweden (which was opposed to specific rules in the standard, rather than the idea behind the FSC itself) did not seem to have much effect and, when combined with the private forest ownership opposition, implied a central "middle of the road" approach to achieving sustainable forestry. Industrial forest companies went from being the targets of boycotts and market campaigns in the UK, Germany, and the Netherlands to being the toast of the environmental community. Not only did industrial forest companies commit to becoming FSC certified, but by 1998 they had all met their commitments (Anonymous 2000a), giving Sweden the largest share of the global FSC-certified land area (Anonymous 2001a).

The choices by Swedish industrial forest companies to support the FSC were clearly influenced by market pressures and signals. AssiDomän (1996) immediately explained its support for the FSC as responding to perceived market advantages and to the demands of purchasers of its products along the supply chain that it be recognized as conducting environmentally friendly sustainable forest management (Moller 1996). A report written by the Royal Swedish

Academy of Agriculture and Forestry (2001) (of which the Swedish Forest Industries Association is a leading member) also explained that the FSC choice was largely an economic one: "There is an increasing need for players on the timber market to be able to provide their customers both with assurances of the quality of forest products and with hard information on the direct or indirect environmental impact of the products. One way in which this is done is through forest certification, which offers a standard by means of which the producers can communicate with the consumers. Being major exporters of wood and wood-based products, Swedish enterprises have been at the forefront of certification schemes for the forestry sector as part of the drive to promote timber as a sustainable raw material."

Post Agreement

Industrial forest companies responded directly to the market pressure in the most proactive way of the five cases reviewed in our book. They responded just as the B&Qs and Sainsburys of the world had required and just as the environmental community had demanded. Their commitment to corporate social responsibility was perceived as extremely high, but a key problem loomed in undertaking full implementation of the FSC. With over half of their fiber coming from non-industrial private forests, and given the "wood swapping" system and the importance of small independent sawmills as a source of chips for large pulp and paper mills, Swedish industrial forest companies faced considerable challenges transforming certified timber from their own lands into FSC-labeled products ready for markets (Editorial 2001). These companies faced a conundrum. The FSC percentage-based claim requirements for solid wood and fiber and composite products were in conflict with the realities of the Swedish production system. Changing this system threatened to significantly increase the costs of operations, and without a clear benefit of getting FSC wood to buyers, companies were reluctant to push too hard in this vein. For instance, chain-of-custody tracking would mean giving up on the wood swapping system, which was developed to reduce transportation costs. This was especially so for industrial forest companies such as Korsnäs, which, owing to the swapping system, relied on an above average share of input fiber from private forest ownerships.[15]

This meant that while Swedish forest companies were no longer boycotted or suffering market campaigns, they were not able to provide significant amounts of FSC-certified wood to their external markets. It was an issue to which they knew they must expend considerable attention. They did so in two ways: first, they lobbied the FSC secretariat and board to change its "percentage-based claims" policy so that they could claim that a portion of their output came from FSC-certified forests (percentage-in-percentage-out approach). While the percentage-in-percentage-out approach was not approved, the FSC conformed to the pressure by reducing downward its thresholds. Initially, in its 1997 policy, FSC labeling was only permitted on solid wood products containing no less than 100 percent certified wood, while chip, fiber, and component products required 70 percent FSC certified content (by dry weight). In 2000, a much more complex policy was released which, to industry's delight, reduced thresholds on both solid wood and pulp, but also forbade products containing *any* wood from controversial sources.[16]

At this point FSC strategists (both within the FSC and its environmental group supporters) appeared to reason that the market pressure would still be enough to address the chain-of-custody issue and were thus reluctant to use conforming strategies to gain non-industrial forest owner support. Non-industrial private forest owners were nonplussed with this approach and responded with their own proactive campaigns designed to defend their approach to forest management. On December 4, 1997, a coalition of Swedish, German, Finnish, Austrian, and French non-industrial private forest owners demonstrated in front of the German publishing house, Springer's Hamburg headquarters, to publicly criticize its support of the FSC. These forest owners met with and presented the president of Springer their "manifesto" and explained that family forests had for generations managed their forests sustainably.

They also embraced the European Union's procedural Eco-Management and Auditing System (EMAS) approach (van Kooten, Wilson, and Vertinsky 1999:172), which was increasingly popular among European companies (Kollman and Prakash 2001).[17] However, when these initiatives did not seem to have much impact on certification pressures, private forest owners in Sweden undertook the single most important choice affecting certification politics in Sweden: they helped create and were central players in developing the PEFC, which was formally released in November 1999 (Reuters News Service 1999).

As described in chapter 1, the PEFC was created by private forest owner associations who, while critical of the FSC approach, came to recognize that forest certification was increasingly attractive to the European market and that Label of Origin plans were not enough. The PEFC was set up as a mutual recognition framework for endorsing national certification systems developed in its members' countries. Each national initiative has discretion in setting up its own program; the Pan European Criteria and Indicators and Operational Level Guidelines for Sustainable Forest Management developed in Helsinki and Lisbon act as principles for the system, but are not defined as requirements

Table 7.2 *Strategic efforts to either gain or maintain support between 1994 and 2002*

	FSC competitor program supporters		FSC supporters	
Year	Action	Strategy	Action	Strategy
1994			WWF and Swedish Society for Nature Conservation set up "reference" group to develop FSC standards for Sweden	Converting
1995		Conforming	Foreign customers in Sweden's top export markets begin releasing buying policies stating preferences for FSC wood	Converting
1997	Nordic Forest Certification Program developed	Conforming	B&Q issues letter to suppliers indicating it intends to source only FSC-certified wood by the end of 1999	Converting
1998				
1999	Swedish landowners help establish the PEFC	Conforming	FSC opens discussion on revising percentage-based claims policy	Conforming

Year	Event	Status
2000	Forest landowners participate in Stockdove process	Conforming
	FSC releases revised percentage based claims policy that reduces threshold requirements for labeled FSC solid wood, chip, fiber, and component products	Conforming
	WWF leads media information campaign to discredit PEFC	Informing
	Stockdove process initiated to compare and bridge gaps between the Swedish PEFC and FSC standards	Conforming
2001	FSC Sweden undertakes revision of its standard to comply with FSC requirement for five-year revision process; companies work from within to bring standard in line with PEFC	Conforming
2002	PEFC proposes addressing Stockdove recommendations in revisions of its standards to begin in 2004	Conforming

that must be met by national initiatives (PEFC International 2001a). This discretion and an approach that gives greater weight to the interests of landowners than does the FSC are key components of the program (Liimatainen and Harkki 2001; Ozinga 2001). As Klingberg (2002), who was head of the Swedish Forest Owners' Federation at the time, explained: "In the PEFC national differences are respected. That in itself is an element of diversity. The Swedish PEFC-scheme is independent, based on local involvement. This is a grass roots, bottom-up approach respected by all involved and not a top-down approach, which might be seen as distant, detached and directorial."

The Swedish landowners and their PEFC national initiative immediately undertook significant conforming strategies (table 7.2) targeted at diffusing the international market pressures for the FSC by creating the most wide-ranging and strict standards of any PEFC national initiative. Aside from systems and governance issues, the key substantive difference between the PEFC Sweden and the FSC Sweden was the issue of reindeer herding rights. While the stringency of the Swedish PEFC rules posed difficulties for other national initiatives in the PEFC system, it was recognized by the PEFC board that these standards were needed to compete with the FSC in Sweden.[18] The PEFC policy on chain-of-custody tracking also conformed to the Swedish industrial forest companies' critique of the FSC system — it did not require that every piece of wood or every bit of fiber produced from certified forests be tracked through the mill, but rather followed a percentage-in-percentage-out approach in which a mill could be certified to sell a percentage of its product with the PEFC label equal to the percentage of certified product that went into the processing facility. This was a significant difference and was created to make it easier for landowners to support the PEFC (by making it easier for them to get recognition in the marketplace), but was also appealing to industrial forest companies for the reasons noted above.

It is somewhat ironic that the Swedish PEFC looks more like the FSC than do most other PEFC initiatives (Editorial 2001),[19] as the creation of the PEFC system as a response by dissatisfied Swedish forest owners would eventually spell significant difficulty for FSC strategic efforts to gain support, not only in Sweden, but in Germany (see chapter 6) and elsewhere.

FIGHTING THE PEFC

The creation of the PEFC and its particular form in Sweden forced many FSC supporters onto the defensive. The FSC and its supporters saw that they needed to turn from earning support from industrial forest companies to maintaining support from these companies as well as their customers, who were now confronted with two systems that, in Sweden, looked substantively *similar.* A series of reports were commissioned by WWF comparing the FSC to the PEFC, in which it was asserted that the PEFC represented the status quo (Environmental News Service 2000), did not create inclusive governance systems (Lindahl 2001), and permitted ecologically poor practices to continue (Anonymous 2001d, 2000b; Meek 2001; Liimatainen and Harkki 2001). An advertising campaign was launched by the FSC that was primarily aimed at companies along the supply chain who were being wooed by the PEFC, while popular press advertising strategies were initiated by the FSC in the UK and Germany.[20]

The PEFC and its supporters responded, arguing that the comparison studies were seriously flawed. They initiated their own comparisons (Confederation of European Paper Industries 2000), asserting that on a range of values the PEFC performed better than the FSC (PEFC 2000). Meanwhile, private forest owners launched their own protests in front of a large German do-it-yourself retailer with pro-FSC policies, asserting that they, too, were a grassroots movement created within Europe. They also asserted that, unlike the FSC, their rules were generated from "within Europe" rather than originating from Mexico (where the FSC was headquartered until January 2003).[21] The PEFC also embarked in broad stakeholder consultations (PEFC 2000), arguing that it was more representative of broader interests than the FSC.

At this point the Swedish forest companies maintained support for the FSC and also watched PEFC and market developments. B&Q, recognizing these dynamics and maintaining their desire to claim they sold certified wood, startled the certification world with a position paper noting that the company was considering accepting products from the Finnish Forest Certification System (A. Knight 2000) — a PEFC national initiative — with the qualification that these products remain unlabeled. And while this system has been strongly opposed by FSC strategists (Liimatainen and Harkki 2001), since it certified virtually all commercially productive forestlands in Finland, it provided B&Q with an important source of certified products.

With the market apparently showing greater flexibility for non-FSC systems, FSC supporters became slightly less intransigent. They recognized that they needed to shore up remaining support and realized that increased conforming strategies might be key. Indeed, all of the Swedish FSC participants were by this time supportive of the percentage-in-percentage-out approach advocated by Korsnäs and the industry in general, but such conforming efforts were stalled at the international level, where officials worried about the precedent this would create for markets in other regions.[22] Arguably the most important example of FSC strategic conforming was its support of the Swedish "Stockdove" process — an effort that attempted to model, at least in part, the UK Woodland Assurance Scheme (UKWAS). FSC supporters hoped that such a process might be able to gain them indirect support from non-industrial forest owners, as they had accomplished in the UK (see chapter 5). The key difference between Stockdove and UKWAS, however, was that the Stockdove efforts were initiated after the PEFC had established itself — leading to significantly more obstacles in gaining indirect support than had existed in the UK.[23]

The Stockdove Process

The Stockdove process was created in 2000 with strong support from the Swedish Forest Industries Federation, whose officials recognized the need to bring private forest owners on board in order to gain increased market access and who were asserting that the PEFC Sweden and FSC Sweden were already very similar (Swedish Forest Industries Federation 2000). The process was also supported by the SSNC and the WWF, whose strategists appeared to be highly aware of the strategic environment in which they operated, and the Swedish Forest Owners' Federation, which was open to seeing if the creation of a Swedish forestry standard might lead to mutual recognition of PEFC Sweden with the FSC (Aulen and Bleckert 2001). If successful, this would allow Swedish industrial forestry companies to provide FSC-certified wood in the marketplace, gaining them ultimate consumer recognition and providing retailers with the certified products that they had long been demanding.

The Stockdove process commissioned a report to identify those standards that the PEFC would have to make more rigorous or otherwise address to be compatible with the FSC and to identify those standards that the FSC would have to make more rigorous or otherwise address to be compatible with the PEFC. It found that the PEFC would have to increase the rigor of seventeen standards, while the FSC would have to increase the rigor of four. The Swedish Federation of Forest Owners immediately addressed three of the seventeen points in the PEFC program and put in motion a process for revising its standards in 2004 to address most of the other points. And while the Sami issue is still present, Swedish landowner officials have indicated that Swedish government efforts to resolve this dispute, including providing compensation to landowners, may increase the possibility that the major substantive obstacle to recognizing the two systems could be overcome.

Likewise, the FSC is in the middle of revising its standards, and industry officials are using the Stockdove process to assert that there could be some standards that the FSC might revise downward, in the hopes of achieving broad consensus on Swedish standards and ultimate recognition for Swedish certified products in the marketplace. As Swedish industrial forest companies explain in their recent annual report (Swedish Forest Industries Federation 2000): "Until now, the development of certification has been characterised by competition between different systems. This process is now entering a new and less confrontational phase, where a number of professional certification systems are learning to coexist in a state of mutual respect, known as 'mutual recognition' ". The report went on to assert that substantive differences were minimal: "The differences between the Swedish standards are in fact limited, and the intention [of the Stockdove process] is now to simplify the certification process by striving to achieve agreement on a joint Swedish standard. It will then be up to individual forest owners to select the certification that offers highest market credibility" (Swedish Forest Industries Federation 2000).

While mutual recognition is usually a forbidden term within the FSC, officials with the FSC secretariat seem to recognize the importance of Sweden in providing FSC-certified wood. Given this and the similarities between the two systems, the idea that FSC and PEFC Sweden might mutually recognize at the *national* level does not seem to present significant difficulties.[24]

The strategic maneuvering between the FSC Sweden and PEFC Sweden, as well as within each system, is a highly delicate process. Certainly PEFC officials are leery of supporting another UKWAS process that might inhibit PEFC growth; however, there is support for some kind of resolution that would give market credibility to both systems. Likewise, FSC officials want to make sure that any resolution sees the FSC label and program dominate, since they believe that, at least outside of Sweden, the PEFC rules represent the status quo rather than improve sustainable forest management.

Where Swedish Forest Certification Is Headed

By fall 2002 both sides were engaged in highly strategic thinking and maneuvering. The FSC secretariat's efforts to decentralize decision making were expected to help facilitate a "made in Sweden" solution, potentially permitting the Swedish national initiatives to undergo strategic mutual recognition with the PEFC there, while preserving the uniqueness of the FSC globally. Meanwhile, companies like Korsnäs, frustrated with the FSC's refusal to adopt a percentage-in-percentage-out policy over chain of custody, are rethinking their certification options. And the Swedish retailer Ikea has adopted a four-step process to its forest certification policy, in which the FSC is recognized as the best, but not the only credible, certification system in the marketplace.

What is clear is that by not conforming to private forest owners during negotiations back in 1996, the FSC must do much more conforming in 2002

Table 7.3 Support granted to FSC and PEFC by Swedish industrial forest companies and the industry association

	Support for certification	
Industrial forest companies	FSC	PEFC
AssiDomän AB	✓	
SCA Skog AB	✓	
Stora Skog AB	✓	
MoDo Skog AB	✓	
Kinnarps AB	✓	
Korsnäs AB	✓	
Swedish Forest Industries Association	✓	

Source: Elliot 2000, company websites, personal interviews and communications

than would have been the case six years prior, when no FSC competitor existed. Despite these movements, by the end of 2002 choices by forest companies and non-industrial forest owners regarding forest certification had not (yet) changed (tables 7.3 and 7.4).[25]

Conclusion

The Swedish case illustrates how the place of a country's forest sector in the global economy, the structure of its domestic forest sector, and the history of public policy approaches to forest resource use conflicts shaped and constrained the strategic choices available to the FSC and its supporters in their efforts to convince forest owners to support their style of forest certification. It illustrated support for hypothesis 1 (*Forest companies and non-industrial forest owners in a country or region that sells a high proportion of its forest products to foreign markets are more likely to be convinced to support the FSC than those who sell primarily in a domestic-centered market*) in that the FSC and its supporters were able to undertake a variety of market-based converting approaches aimed at wooing industrial forest companies into the FSC fold. Support was evident for hypothesis 3 (*Large and concentrated industrial forest companies are more likely to be convinced to support the FSC than relatively small and less concentrated industrial forest companies*), through the ease with which converting strategies were used to gain pragmatic support from a handful of large concentrated Swedish forest companies The case illustrated nicely the importance of hypothesis 4 *(Unfragmented non-industrial forest ownerships are more likely to be convinced to support the FSC than*

Table 7.4 Forms of support given to FSC and its competitor by Swedish non-industrial forest landowner associations

	Support for certification	
Landowner associations	FSC	PEFC
National Federation of Swedish Forest Owners		✓
Södra Skogsägarna		✓
Mellanskog		✓
Norrskog		✓
Norra Skogsägarna		✓
Norrbottens Skogsägare		✓

Source: Elliot 2000, company websites, personal interviews and communications

fragmented non-industrial forest ownerships). Swedish landowners eventually balked at FSC converting strategies, given their higher per hectare costs of undergoing a certification assessment. That they numbered in the hundreds of thousands made them less susceptible to direct targeting by the FSC.

The narrative provides mixed support for hypothesis 5 (*Forest companies and non-industrial forest owners in a country or region with diffuse or non-existent associational systems are more likely to be convinced to support the FSC than those in a country or region with relatively well-coordinated, unified associational systems*), since a well-integrated industry association decided to accept the FSC. At first glance the strength of the industry association and the success of the FSC in gaining industrial forest company support appears at odds with our predictions. While there was an internal debate about whether to support the FSC, ultimately the association acted in unison and all companies ended up supporting the FSC. In this case, the presence of a well-integrated industrial forest company association intersected with the "place in global economy" and "large company" factors to reverse this feature's hypothesized direct effects. (We return to the issue of how direct effects may become reversed in chapter 8). The direct effects of this feature were, however, illustrated in the experience of the private forest owners, as their well-integrated associational system permitted them to be active members in creating the PEFC as a legitimate alternative.

The story also provided support for hypothesis 6 (*Forest companies and non-industrial forest owners in a country or region with sustained and extensive environmental group and public dissatisfaction with forestry practices are more likely to be convinced to support the FSC than those in a country or region with less dissatisfaction*), as industrial forest companies (and even initially landowners) saw the FSC as potential shield from future criticism. Finally, hypothesis 7 (*Forest companies and non-industrial forest owners in a country or region where access to state forestry agencies is shared with non-business interests are more likely to be convinced to support the FSC than those in a country or region where forest companies and non-industrial forest owners enjoy relatively close relations with state forestry agencies vis-à-vis non-business interests*) received partial support. Multi-stakeholder processes at the national level paved the way for a new forest policy that required that industry and forest owners address equally environmental and economic goals, while fairly closed public policy networks at the Forest Practices Board and forestry board level turned attention toward implementation of these goals by way of the FSC.

At the same time the Swedish case illustrates that even in a region quite conducive to efforts to convince forest companies and non-industrial forest owners to support the FSC, strategists who ignore factors that are not conducive ignore them at their own peril. Failure to conform more to non-industrial private forest owner concerns in 1996 largely explains the rise in the PEFC in Sweden and beyond. These developments, in turn, explain the more defensive mode that the FSC has been on both in Sweden and abroad since that time. The significance of this empirical account is that it illustrates how we can identify factors that shape and mediate FSC efforts to gain forest owner support, but that prediction is difficult. The chapter also illustrates the need to understand what happens when forest companies and non-industrial forest-land owners choose to influence the FSC from within. By becoming involved in the standards process and bringing traditionally close governmental allies to the table, forest company officials were able to influence from the inside FSC rules that would have arguably been stricter if not for their active participation. In chapter 8, we explore further the Swedish and BC cases where forest companies participated from within versus forest companies influencing the FSC from the outside through participation in the industry alternative, the Sustainable Forestry Initiative.

PART IV

Private Authority and Sustainability

8

Competing for Legitimacy

Forest certification programs have presented the world of policy analysis with one of the most provocative and startling institutional designs since governments the world over first began addressing the impacts of human activity on the natural environment. Traditional political struggles over the use of the world's forest resources have not been subsumed by these institutional designs but they have changed the arena and the rules of the game through which these struggles occur. This book has carefully documented five different cases where the Forest Stewardship Council (FSC), an international forest certification program with widespread environmental group support, has competed with industry- and/or landowner-initiated certification programs for rule-making authority. The outcomes of this struggle are critical to understanding the ability of forest certification to address some of the most important problems facing global sustainable forest management. The struggle in the forest sector also provides important lessons to non-state market-driven certification programs that are emerging in other sectors, including mining, tourism, coffee, food production, and fisheries.

A key finding of this project is that the competition for "legitimacy" does not just affect which certification program comes to be accepted by forest companies and landowners, but also the way each program *changes* from its original conception in order to earn broader support along the market's supply chain. The degree of change required to gain this support depends in large part on a region's place in the global economy (dependence on foreign markets), its domestic forest sector structure (industrial forest company concentration, non-industrial ownership fragmentation, and associational system cohesion), and the history of forestry on the region's public policy agenda (sustained and extensive dissatisfaction with public forest policy and relationship between state forest agencies and their company/landowner clientele).

These findings have two important implications. First, the influence of the FSC on sustainable forest management is not simply through its own rule development, but also on the impact it has had on the way competing programs develop their decision-making processes and their procedural and substantive rules. While our specific research question focused on divergent levels of support for the FSC, answering this question necessarily shed light on the interaction between the competing systems. While it is doubtful that FSC competitors will ever end up looking the same as the FSC (after all, their raison d'être was to offer what they perceived to be a more appropriate approach for forest companies or landowners), the important issues to understand are how their rules do change (appendix 1) and what the impacts are of their own program's rules in promoting sustainable forest management and environmental stewardship. This book sheds light on the former, and future research needs to explicitly and systematically address the latter.

Second, any comparison of existing standards at one point in time misses the fact that rule development is not static. With certification emerging and taking on new forms and approaches, descriptive comparisons of certification program standards become quickly out of date — especially during the current phase, where the programs are clearly in the process of institutionalizing. While comparisons of these dynamic standards are useful for providing context, our cases reveal that the more important question at this time is to understand the *processes* through which change occurs and where support for certification might be headed.

This chapter addresses these issues in three steps. The first section assesses the impact of our identified structural features in shaping the support granted by forest companies and forest owners in our five regions following FSC efforts to alter their initial (negative) evaluations. We review here our argument and hypotheses about the identified factors' direct effects and explain how key choices made in the beginning of these efforts to gain support reinforced the strength of some of these factors, while redirecting the influence of others. Recognition of this reinforces the need to understand the timing and sequence of choices made at the initial stages of efforts by FSC strategists to gain forest company and forest owner support. We argue that recent break-throughs in path dependency research in political science and the concept of "increasing returns" offer important directions for future research and refine-ment of our argument. And recognition of the fluidity and contingent nature of the emergence of these non-state forest certification systems leads us to identify additional factors that are both dependent and explanatory — as ini-tial strategic choices influence the ability of future efforts to gain forest owner and forest company support. Drawing on our empirical cases, we identify in this section three additional factors that appear to influence FSC efforts to gain forest company and forest owner support and that appear worthy of future exploration.

A second section reflects on what our cases mean for the emergence of non-state market-driven governance as an alternative form of regulatory authority to traditional Westphalian sovereign authority and the relevance of these find-ings for understanding the emergence of the non-state market-driven phenom-enon in other sectors. A third section reflects on the implications of these trends on the ability of forest certification to address and promote global sustainable forestry.

The strength of our approach was in identifying those features that influ-enced the ability of the FSC and its proponents to gain "legitimacy" from forest companies and forest owners. What soon became clear during our re-search was that very different struggles that emerged over forest certification rule-making authority in our five study countries or regions revolved more around the *strategies* used by the FSC and its competitors to achieve legitimacy from forest companies and landowners than the *type* of legitimacy given. When forest companies and forest owners chose to support the FSC, they tended to do so for economic interest-based reasons or what we referred to in chapter 2 as falling under the category of "pragmatic legitimacy." The other "moral" and "cognitive" legitimacy granting distinctions simply did not gener-ally capture firm-level evaluations in support of the FSC. What did describe very well the competition for certification legitimacy were "converting" efforts (where the FSC was able to exert outside pressures to convince forest com-panies and forest owners to support their program) and "conforming" strat-egies (where the FSC changes aspects of its own program in hopes of gaining support) that took place in all our cases. Using these distinctions to character-ize active efforts to gain legitimacy revealed very different abilities in convert-ing forest companies and forest owners to support the FSC (figure 8.1).

The country or region's place in the global economy, the structure of the forest sector, and the history of forestry on the public policy agenda all played

HIGH   LOW
(Converting)   (Conforming)

◄──►

British   Sweden   United   Germany   United
Columbia   Kingdom   States
Canada

Figure 8.1. The combined effects of factors facilitating the FSC across cases.

key roles in our cases and all worked to influence forest company and forest owner evaluations of the FSC once active campaigns to convince forest companies and forest owners were under way (table 8.1). At the same time, we noticed in our cases that the perceived direct effects identified in our hypotheses were, in some instances, reversed once active legitimacy achievement campaigns were underway. We detail these points by reviewing our three broad factors and the seven hypotheses to shed light on the independent effects they had in each of the cases (table 8.3) and to clarify how and why their hypothesized direct effects were sometimes reversed. We then introduce and discuss a matrix (table 8.4) synthesizing these findings and turn to path dependency as an analytical launching point for future research.

Factor 1: Place in the Global Economy

Hypothesis 1: Forest companies and non-industrial forest owners in a country or region that sells a high proportion of its forest products to foreign markets are more likely to be convinced to support the FSC than those who sell primarily in a domestic-centered market.

Hypothesis 2: Forest companies and non-industrial forest owners selling wood to a domestic market in a country or region that imports a large proportion of all the forest products it consumes are more likely to be convinced to support the FSC than those in a region that imports a small proportion of all the forest products it consumes.

The evidence from our cases illustrates the importance of these hypotheses in understanding the strength of market pressure facing forest companies. The regions or countries most reliant on exports to foreign markets (BC and Sweden) were clearly the most influenced by market pressure. When industrial forest companies made choices to support the FSC in some way, they did so because they took the pressure seriously and believed that by supporting the FSC some kind of positive economic benefit would accrue (be it market access

Table 8.1 FSC ability to alter evaluation by hypothesis and case*

Case	Place in the Global Economy	Structure of the Domestic Forest Sector			History of Forestry on Public Policy Agenda		✓/6
	H1 or H2	H3	H4	H5	H6	H7	
BC	✓	✓	✓	✓	✓	✓	6
Sweden	✓	1/2	X	X	✓	1/2	3
UK	✓	X	1/2*	X	✓	X	2.5
Germany	✓	X	X	X	X	X	1
US	X	1/2	X	X	X	X	.5

Notes: H1: high dependence on foreign markets for exports; H2: high dependence on imports; H3: concentration of forest industry; H4: low level of non-industrial forest fragmentation; H5: fragmented forestry associations; H6: long history of unresolved forestry conflict; H7: industry shares access with non-business interests.

* The factor's effects described by each hypothesis do not have equal weight as we elaborate and explain in our case studies. We use the numbers simply to synthesize and present a general guide to understand the cumulative effects of each of the individual effects described by our hypotheses.

** H4 Non-industrial forest land in the UK is distinguished from concentrated government ownerships and fragmented private forest owners.

or a more abstract corporate social license to operate). We also noted that those countries (UK and Germany) that rely more heavily on foreign imports also facilitated FSC efforts, but usually for moral suasion and credibility reasons (it was hard to argue what was good for the foreign goose was not good for the domestic gander). While this factor did facilitate FSC efforts, it did not appear to be as strong as the economic pressures that faced forest companies and owners in BC and Sweden.

And in the US, which had the least relative dependence on imports *and* exports, we witnessed the lowest degree of forest company and forest owner support for the FSC, despite widespread market-based converting efforts on the part of the FSC and its supporters. What became evident was that environmental groups had a much easier time gaining support from retailers for their cause when the forestry problem they were addressing resided outside of the political system in which the retailer was based. This was certainly true in the

BC and Swedish cases, and it also proved relevant in Germany and the UK where commitments from large wood product purchasers not to buy imports from unsustainable sources also affected (though to a lesser degree) certification choices made by domestic forest companies and landowners. The US case revealed that it was easier to secure the commitment of a Home Depot to prefer FSC wood by focusing on "endangered" forests in BC and the tropics rather than on problems with domestic forestry practices.

Our exploration of these hypotheses is revealing about the influence of economic globalization in assisting efforts by environmental groups to force upward environmental standards. The cases strongly support Bernstein and Cashore's argument that the "downward" "race to the bottom" effects of economic globalization can be reversed by efforts to link access to these markets with environmental performance requirements (2000). As a result, our analysis challenges those environmental critics who contend that economic globalization always has negative consequences. While much more research needs to be done to understand how the upward and downward effects intersect, there is no question that environmental groups have used the power of the global marketplace to force companies and forest owners to make choices they otherwise would not have made.

Factor 2: *Structure of the Forest Sector*

Hypothesis 3: Large and concentrated industrial forest companies are more likely to be convinced to support the FSC than relatively small and less concentrated industrial forest companies.

Hypothesis 4: Unfragmented non-industrial forest ownerships are more likely to be convinced to support the FSC than fragmented non-industrial forest ownerships.

Hypothesis 5: Forest companies and non-industrial forest owners in a country or region with diffuse or non-existent associational systems are more likely to be convinced to support the FSC than those in a country or region with relatively well-coordinated, unified associational systems.

All five cases revealed the importance of the structure of each country or region's forest sector in understanding FSC efforts to increase support. Developments in countries or regions with and without large, concentrated industrial forest companies, fragmented non-industrial forest ownerships, and well-integrated associational systems all provided support for the hypothesized direct effects of these factors in mediating strategic efforts by the FSC and its supporters. Intersecting effects also occurred that had the impact of reversing the direct effects described by these hypotheses. The direct effects occurred most notably in BC, the UK, and Germany, and intersecting effects occurred in Sweden (where a strong association did not correlate with the predicted direct effects) and the US (where the existence of large industrial concentrated companies did not correlate with the predicted direct effects).

In BC and Sweden, large concentrated forest companies were relatively easy and identifiable targets for campaigners focusing largely on UK and German retail markets. This increased the tendency of both BC and Swedish companies to positively evaluate the FSC; in the case of BC, it pushed them away from solely supporting the competitor program (the Canadian Standards Association [CSA]), and in the case of Sweden, it caused them to withdraw support from the stalled Nordic Forest Certification competitor program. The absence of large, concentrated industrial forest companies in the UK and Germany made these regions much less fertile for direct targeting market campaigns by environmental activists. Instead, activists relied on targeting bigger companies down the supply chain, but this was a second best option and, owing to the fact that most of these companies were in the same country (place in global economy), was made more difficult by charges that the FSC was a "foreign" organization and hence inappropriate for domestic forestry. It was the US case where the hypothesized direct effects about large, concentrated forest companies *were not supported* by the story. With few exceptions, the existence of large, concentrated industrial forest companies in the US did not facilitate FSC efforts. Instead, market-based converting efforts by the FSC in the US had the effect of industrial forest companies supporting more vigorously the FSC competitor program, the Sustainable Forestry Institute (SFI). Of course, since we identify seven factors with direct effects that push in different directions, it is logical that not all will be able to strongly influence the dependent variable (forest company and forest owner choices to support the FSC). What is important is to understand better which factors "trump" other factors and under what conditions the hypothesized direct effects may intersect with other factors to create unpredicted outcomes.

In the US case it appears that the role of non-industrial forestlands as a source of most of the country's fiber actually trumped the direct effects of industrial forest company concentration. The fragmentation of non-industrial forestland made it difficult, if not downright impossible, to convince most landowners to support the FSC. This feature alone would make chain of custody extremely difficult for industrial companies who might otherwise have been open to supporting the FSC. As a result, the "large, concentrated" factor ended up pulling in the opposite direction that we argue would have been the case had it operated by itself. Instead, and because of intersecting effects, industrial forest company concentration actually worked to hasten opposition to the FSC and increase support for the Sustainable Forest Initiative.

Recognition of this highlights the need to understand the role of non-industrial private owners in facilitating or debilitating FSC efforts. In the case of Sweden the importance of non-industrial private owners and their eventual opposition to the FSC worked to limit slightly industry's commitment to the FSC. Given the non-industrial private forest landowner issue, Swedish forest companies first looked to the FSC to *conform,* by pushing the program to alter its percentage-based claims approach. Later they went so far as to urge the FSC to reach out to the landowner-initiated program, the Pan European Forest Certification (PEFC), in an effort to develop a "made in Sweden" system that Swedish forest owners could deem appropriate.

The UK case is a good example of how the effects of fragmented non-industrial private ownership were very real, but concentrated government owned lands and highly strategic maneuvering by FSC officials worked to downplay their significance. We demonstrated in the UK chapter how FSC strategists carefully read conditions in this country, allowing them to gain indirect support from most landowners in the UK.

Finally, the hypothesized direct effects of a cohesive associational system did influence as predicted forest company and forest owner choices in BC and the UK (low associational system cohesion aided FSC converting strategies) and Germany and the US (high associational system cohesion limited FSC converting strategies), where well-developed associational systems helped companies and landowners to develop strategic alternatives to the FSC. We observed mixed results for the hypothesized direct effects in Sweden. The associational system cohesion for non-industrial private landowners was high, allowing them to vigorously create and defend the FSC competitor, the PEFC. However, the choice of Swedish industrial forest companies was not as expected. Indeed, the Swedish forest industry's associational system cohesion ultimately worked to enhance support for the FSC, rather than limit it. Again, the explanation for this has to do with understanding the role of intersecting effects during the early stages of FSC efforts to gain support. The effects of Swedish companies being highly exposed to foreign markets and also being large and concentrated ended up, ultimately, trumping the direct effects of associational structure. Once the association reversed its opposition toward the FSC, however controversial it may have been within the association, the associational structure ended up facilitating support for the FSC since once the association made a choice, all of its major forest industrial company members acted consistently with that choice. It was the specific timing and sequence of FSC's early efforts to gain support by focusing on foreign market pressure, followed by company decisions to support the FSC, that saw the associational system solidify its support for the FSC rather than work to create an industrial alternative.

We would also note that while the factors we identify did largely explain broad country or regional level forest company and non-industrial forest owner support for the FSC, we expect future research to explore why it was that *within* the same country or region, some forest companies and non-industrial forest owners departed from general trends. Explaining these differences may shed light on those factors specific to an individual firm that influence its evaluations over forest certification that may not be able to be explained by our broad level analysis.

Factor 3: The History of Forestry on the Public Policy Agenda

Hypothesis 6: Forest companies and non-industrial forest owners in a country or region with sustained and extensive environmental group and public dissatisfaction with forestry practices are more likely to be convinced to support the FSC than those in a country or region with less dissatisfaction.

Hypothesis 7: Forest companies and non-industrial forest owners in a country or region where access to state forestry agencies is shared with non-business interests are more likely to be convinced to support the FSC than those in a country or region where forest companies and non-industrial forest owners enjoy relatively close relations with state forestry agencies vis-à-vis non-business interests.

In each of the regions under review, the traditional public policy approach to forest management was identified as being key to understanding support for FSC-style certification, as it influenced whether forest companies and landowners viewed FSC certification as a threat or an opportunity. All of our cases showed support for our predicted relationship regarding sustained conflict and closed public policy networks.

The lack of business-government dominated public policy processes and the experimentation of a range of multi-stakeholder processes in BC during the early to mid-1990s meant industry was less threatened by the FSC multi-stakeholder, tripartite approach. The industry recognized that the closed processes dominant in the 1970s and 1980s would be difficult to reconstruct, even with the election of a more sympathetic administration in 2000. And the sustained scrutiny on BC forest practices both domestically and internationally meant that industry was more amenable to market solutions provided by the FSC approach.

Likewise, in Sweden increasing and sustained societal criticism of Swedish forestry practices, also from both domestic and international sources, meant that its industry was open to alternative solutions. And while lower level implementing networks were still closed, the increasing use of multi-stakeholder processes at the national level, with clear goals governing environmental stewardship, helped enhance support for the FSC as the way of addressing these goals. Similarly, in the UK increasing concern about domestic forestry practices helped the FSC in its efforts to convert forest owners to support the FSC. Relatively closed government-business networks mitigated against forest owner support, but once government decided to help facilitate certification discussions, these closed networks ended up, indirectly, supporting FSC efforts. This is because business interests entered into these discussions only because it was government, and *not* the FSC, with its highly disputed decision-making procedures, that was convening the process. And yet it was the FSC that was able to capitalize on this agreement by positioning itself as the dominant *certifier* of this negotiated standard.

The absence of such features on the German and the US public policy agendas worked to limit efforts to converting private forest owners to supporting the FSC. In Germany there was simply no discernible widespread society critique of domestic German forest practices and, hence, no strong rallying cry that FSC-style certification was needed to address a policy problem in this country. And unlike most other countries' domestic forest policy processes, Germany continues to maintain close relations between its state forestry agencies and its landowner clientele — at all levels of their policy process. In the German case the FSC multi-stakeholder approaches posed a radical departure in the way regulations would be made, one that found disfavor among most forest owners.

The role of the public policy process in the US was similar to Germany. National forest lands were, for the most part, off limits to FSC certification, which meant that core FSC environmental supporters would come to see the FSC as most relevant for privately owned commercial forestlands. Unlike the long tradition of sustained conflict on national forestlands, US private land regulation, for the most part, has not received high degrees of sustained and extensive scrutiny. In addition, forest industry and non-industrial private forest landowners are relatively more successful at influencing public policy networks at these levels. Both of these features worked against FSC efforts to gain support: in contrast to the educational and voluntary approaches encouraged by most state forest management agencies, companies and forest owners felt they had much to lose with FSC-style certification, in terms of both reduced access to the policy process and the perceived prescriptive and "stringent" approach of the FSC.

Direct and Intersecting Effects

Our empirical cases illustrate the direct role of our seven identified factors in shaping each country or region's specific logic. At the same time the cases reveal the importance of understanding what happens when active legitimacy achievement strategies begin and when some strategic choices end up having "path dependency" effects that influence and (partially) alter initial legitimacy achievement logics.

We believe that existing breakthroughs in path dependency literature (Pierson 2000; Hacker 2001) within comparative public policy studies may help shed light on these questions. Pierson and Hacker argue that path dependencies can occur and create "lock-in" effects (Pierson 1993) that constrain and direct future political struggles and policy choices. They argue that the concept of "increasing returns" is the key feature; this describes a situation in which a policy choice becomes more durable and increasingly difficult to change over time. Unlike punctuated equilibrium theory that focuses on big "exogenous" choices affecting stable policy subsystems (Baumgartner and Jones 1993), they argue that very small choices can have huge consequences. The key for Pierson, Hacker, and other path dependency scholars is to understand how the "timing" and "sequence" of choices governing the initial development of a policy can impact future choices and political struggles. Such concepts seem ideally suited for exploring further the role of intersecting effects between the identified factors above in influencing the emergence of non-state market-driven systems such as forest certification. This is because these systems are currently at their early stage of development, and choices made now are bound to have profound effects on the structure and function of these new systems and the manner in which these systems will (or will not) become institutionalized in the long run.

Certainly the choice by the FSC and its supporters to proceed without the support of non-industrial private forest landowners in Sweden (despite them being very close to offering support) made it decidedly more difficult for the FSC to capitalize on the relatively supportive Swedish environment and opened the door for the PEFC to emerge as a viable alternative in Europe (and globally). Similarly, recent choices in BC to impose arguably the strictest standards of any FSC process on industrial forest companies might prove to be appropriate for long-run environmental protection, but these choices may also limit what was relatively strong industrial support for FSC in the region. Without industry support in BC, it is highly unlikely that the system will ever institutionalize as a legitimate private governance system in this region, thus reducing future on-the-ground impacts of FSC-style certification and precluding any possibility of "continual improvement" in the standards once the system was fully institutionalized.

Recognition that both direct and intersecting effects occur and that our relatively static structural features intersect with more fluid "early" strategic choices led us to discover a remarkable systematic relationship that appears to govern our legitimacy achievement logics. We believe that our discovery and identification of this relationship will push forward future research aimed at understanding the overall legitimacy achievement logics as well as how structural and more fluid factors might influence this relationship over time.

Market Pressures and Receptiveness

A careful analysis of our cases revealed that evaluations on the part of forest companies and forest owners as to whether or not to support the FSC were largely the result of two factors: the degree of FSC market pressure transmitted to companies and landowners and the receptiveness of companies and landowners to the FSC. This distinction is important because the former is about factors that occur outside of industry and/or landowner internal evaluations, while the latter is an internal feature that, as Suchman would say, is subjectively given but can be objectively analyzed. When we represented our cases in this way it became clear that the degree of market pressure transmission and company and landowner receptiveness strongly correlate to whether the FSC is able to draw on converting or informing strategies, or whether it must undertake conforming efforts. Table 8.2 also identifies a strategic environment in which a "standoff" exists — where market incentives are high but receptiveness to these pressures is extremely low. And table 8.3 reviews whether each of our hypothesized factors influence market pressure, receptiveness, or both (see figure 8.1) and whether other factors we did not explicitly identify in chapters 1 and 2 are worthy of future exploration.

The logic presented in table 8.4 is that in those regions where there is a high degree of market pressure and a high degree of receptiveness to this pressure converting strategies are rendered relatively effective. This certainly describes the case of BC and (to a lesser extent) Sweden. Likewise when there are low market pressure and low receptiveness the FSC must conform if it hopes to gain support, and even here support is far from inevitable. This dynamic

Table 8.2 Impact of market pressure and receptiveness on FSC strategic environment

		Degree of FSC market pressure transmission	
		High	Low
Degree of receptiveness to	High	Convert	Inform
FSC market pressure	Low	Standoff	Conform

illustrates what transpired in the US and Germany. And in the UK, where market pressure and receptiveness were neither high nor low but rather "mid," we see a middle ground situation in which support for the FSC is conditional upon highly strategic and informed choices.

Table 8.2 also hypothesizes that there could be cases of low market transmission but high receptiveness — in which case such an environment would appear to promote the use of an informing strategy. While we found no general case that described this situation exactly, we would note that it seems to fit the limited number of small landowners and forest companies in the US that chose to support the FSC in the absence of market pressure (Hayward and Vertinsky 1999). These landowners arguably supported the FSC for "moral" reasons — which raises the importance of understanding better whether table 8.2 sheds light on non-pragmatic evaluations (a subject we return to below). Likewise table 8.2 speculates that a standoff occurs when there is high market pressure but a low degree of responsiveness to the FSC. In this case, conflicting market and moral pressures would potentially lead to a high degree of tension. And while we did not observe this situation in our cases, the evaluations of non-industrial forest owners in Germany, Sweden, the UK, and the US toward the FSC certainly fit in the "low receptiveness" box; if they eventually come under stronger market pressure we may very well see this subset of our research population fitting under such a logic.

The relationship identified in this matrix is important because it places our attention on understanding the array of structural (static) and dynamic features that may shape either the transmission of market pressure, the receptiveness to this pressure, or both. Reflecting on the historical narratives of our cases, we hypothesize now about the effects of three additional factors — two of which are fairly static (one structural, one ideational) and one that is

Table 8.3 *Factors that facilitate high market pressure and receptiveness*

Features	Hypothesis	Specific factors	Market pressure	Receptiveness
Original Hypotheses				
Place in global economy	1, 2	Dependence on foreign markets	✓	✓
Structure of domestic forest sector	3	Large, concentrated industrial forest companies	✓	✓
	4	Unfragmented non-industrial forest ownership		✓
	5	Diffuse or non-existent associational system		✓
Public policy approaches	6	Sustained and extensive public dissatisfaction with forestry practices	✓	✓
	7	Forest companies and non-industrial forest land owners share access to state forestry agencies with other societal interests		✓
New Hypotheses				
Supply chain structure	8	Cohesive supply chain	✓	
Culture of forest owner non-industrial private forest owner independence	9	Absence of non-industrial private forestlands		✓
Early innovator status	10	Absence of early FSC competitor		✓

Table 8.4 Empirical fit of our five cases with market pressure and receptiveness features

		Degree of FSC market pressure transmission	
		High	Low
Degree of	High	BC	
receptiveness		Sweden	
to FSC market			UK
			US
pressure	Low		Germany

dynamic — that enhance our matrix and an understanding the relationship between direct and indirect effects.

HYPOTHESIZING INFLUENCE OF THREE ADDITIONAL FACTORS

The Structure of the Supply Chain

Hypothesis 3 identified the explanatory importance of forest company concentration, in terms of the degree to which these companies are vertically and horizontally integrated. However, our historical narrative, which revealed the importance of choices made by customers further down the supply chain, leads us to assert that future research should place more careful attention to the way that the structure of the supply chain influences the strength of economic pressures that face forest companies and forest owners. Since certification does require either formal or informal tracking of products, and because pro-FSC pressure is often transmitted by companies who were themselves targeted by environmental groups, all parts of the chain are important. Understanding the influence of monopsony and oligopoly interests and "fragmented bottlenecks" on the certification decisions made at all steps along the supply chain appears to be essential. Our cases, along with the work of others, show how large companies are easy targets for environmental group campaigns (see especially Sasser 2002, 2003). We need to understand if and how the transmission of market pressure is influenced by the *location* of large companies along the supply chain, as well as the ramifications of multiple concentrated points.

In the US case, for instance, concentration exists in the retail sector (e.g., Home Depot and Lowe's) and the primary manufacturing sector, (e.g., International Paper and Weyerhaeuser). Systematically analyzing under what conditions concentrated interests occupying these and other locations along the supply chain influence efforts by the FSC and alternative programs to gain support appears to be a crucial future phase of research. Certainly existing strategic efforts will necessarily have to direct increased attention to more companies along the supply chain — many of whom currently have little or no knowledge of certification (Auld, Cashore, and Newsom 2003; Vlosky and Ozanne 1998, 1997). While much more theoretical research needs to be done to develop specific hypotheses in this vein we offer a general one to facilitate this process:

Hypothesis 8: Forest companies and landowners in a country or region with a high degree of concentrated ownership all along a product's supply chain are more likely to be convinced to support the FSC than those in a region where product supply chains are segmented and represented by diffuse ownerships.

Independence of Forest Owners

An emerging theme from most of our cases, and one we appear to have underestimated in our original hypotheses, was the way the "independence" value held by a region's non-industrial private forest owners might strongly influence legitimacy achievement strategies for these ownership types. For example, in Germany, the UK, and the US, where non-industrial private land-ownership has a long and established history, we observed evidence that land-owner opposition to the FSC was not based solely on the nature of the FSC standards, but rather on the *groups that were backing* the FSC. In Germany, numerous forest owners and landowner association officials stated that if state forestry agencies promulgated standards identical to those of the FSC, non-industrial private forest owners would likely not oppose them. In the UK, the development of the UKWAS created an agreed-to standard for sustainable forest management that both private forest landowners and FSC supporters were, on the whole, willing to endorse. Yet, the primary UK landowner association, the Timber Growers' Association, still created a PEFC national initiative that would offer an alternative to the FSC, even though it would be certifying to the same standard. And in Sweden all sides agree that the substantive differences between the FSC and the PEFC Swedish standards are relatively minimal.

Given that we witnessed such feeling across our cases, we propose that ownership of private forest land — and associated common values — is a key explanatory factor. Recognition of this is important, because it raises the extent to which *ideology may trump market incentives* (Murray and Abt 2001), or "tip the scales" against support for the FSC. We return to this theme below when addressing the role of "core audiences" for each program, but offer a hypothesis that we believe will be useful for future research.[1]

While we did assert in the introduction that this "independence" feature was a reason for hypothesizing that the more fragmented and prevalent non-industrial forest ownership is the more conforming strategies will be required, it may be that this feature warrants consideration on its own. At the least, even if it is covered by our existing hypotheses, more work should be done to understand why it matters and has such an impact on non-industrial private forest owner evaluations and why it does not appear to have the same effect on profit-maximizing industrial activity. This should be done so that the "cultur-ally" important traits or rationale for our rather static "fragmentation of non-industrial landownership" variable are better understood.

Hypothesis 9: Non-industrial private forest owners are less likely to be con-vinced to support the FSC than industrial forest companies because they place a higher value on "independence," rendering certification programs in which they feel they do not play a key role in standard development less likely to gain support.

The Role of the Competition: Early Innovators and Early Entry of FSC Competitors

We consciously chose to focus our hypotheses on institutional or struc-tural features that did not change during the course of our examination, which allowed us to identify their effects in influencing agent-based efforts to obtain forest company and forest owner support. However, what did emerge from our cases is that not only did strategic agent-centered choices matter, but that the early entry of a competitor program was an important explanatory factor that influenced forest company and forest owner evaluations of the FSC. We noted that the presence (US) or absence (Sweden, UK) of an early certification alternative significantly influenced how the ultimate support for the FSC was shaped. This strongly indicates that our structural factors were not, by them-selves, the only influence on facilitating (or debilitating) strategic efforts of FSC supporters and that a highly *variable* factor also mattered.[2] This leads us to offer the final of our three new hypotheses.

Hypothesis 10: Forest companies and landowners in a country or region with-out an FSC "competitor program" are more likely to be convinced to support the FSC than those in a region where a competitor program has emerged.

The Role of Biophysical and Ecological Characteristics of
the Region or Country

While our story has carefully detailed structural features that influence
efforts by the FSC and its supporters to gain forest owner and forest company
support, it is important to note that the biophysical and ecological characteristics of a region may also influence the receptiveness of forest companies and
forest owners to FSC-style certification. We have already noted the indirect
effects of such a feature when exploring the history of public policy controver-
sies in each region (hypotheses 6 and 7), but there are also other ways this
feature might influence evaluations. There are two ways in which this might be
the case. First, regions of the world that still have significant amounts of old-
growth forests, such as BC, may find it more difficult to become certified under
FSC rules vis-à-vis countries such as Sweden and Germany who long ago
harvested their old growth forests. This is because the FSC places special
attention on these forest types, with special rules required for "high conservation value forests." Hence, all else being equal, the type of forest a company is
harvesting may influence its choices to support the FSC. Second, some critics
have contended that support for or against the FSC could be influenced by
how "amenable" a region's forest type is to the lower impact forestry generally
promoted by the FSC. For example, we might expect support to be low in a
region where regenerating species are shade intolerant if the FSC standards
forbade clearcutting, while such standards could be viewed favorably by those
operating in shade tolerant spruce-fir forests of Sweden, Germany, and the
northeastern US. However, much more research needs to be done to understand whether and how FSC standards adequately address differences in forest
types and whether these differences have influenced forest company and forest
owner evaluations.[3]

Forest Certification and Its Lessons for the Emergence of Non-State Market-Driven Governance Systems

In addition to providing a more complete understanding of how the
debates and struggles over forest certification are emerging, our cases also
provide important insights for thinking broadly about the nature of non-state
market-driven governance and its potential ability to address forest sustainability. This second section of the conclusion explores the implications of a
governance system that turns to the market's supply chain for its (pragmatic)
authority, and what this means for long-term legitimacy and the role of traditional public policy approaches to environmental problem solving. The last
section directly addresses the impacts of forest certification in addressing the
key problems facing global forest management.

Implications of a Market-Driven Supply Chain Approach to Environmental Problem Solving

A key lesson that emerges from our story regarding the use of non-state market-driven governance systems is that forest companies and forest owners must perceive a direct or indirect economic benefit. This rather intuitive statement, widely supported by our cases, has enormous consequences. A non-state market-driven system cannot create standards that are so high that the potential benefits in support are outweighed by the costs of implementation. The trick for supporters of these systems who wish to create the highest of standards, then, is to work to improve the market incentives (either through increased boycott sticks that companies fear and often loathe or through increased market benefits carrots associated with undertaking forest certification) or to wait until such time as the system is institutionalized across such a large segment of the sector that any strengthening of rules would not put supporters at a competitive disadvantage vis-à-vis non-supporters. What is clear from the case of forest certification is that the latter has simply not happened yet (if it ever will) and that the perceived and real market benefits are such that standard development must be sensitive to "cost versus benefit" evaluations companies must necessarily make when deciding whether to support these systems. The case of BC reveals choices made at the regional standard-setting level that were more concerned with setting the highest possible standard than with the realities of what a supply chain system — which must rely on positive evaluations of forest companies and forest owners — can actually do (a topic we return to in the next section). This is an important point. The BC case illustrates that even when most factors facilitate FSC efforts to gain forest company support, *such an environment cannot reverse, or trump, the logic upon which non-state market-driven systems gain their authority:* the ability to demonstrate an economic incentive to support their system. (We explore in the next section how this logic stands in contrast to what traditional public policy approaches are able to do.) It may seem obvious, but profit maximizing forest companies and forest landowners have to believe that accepting the FSC will allow them to stay in business. Whether this is a drawback of certification, or its main strength, partly depends on how one perceives the depth of the environmental crisis facing global forestry and whether one believes that the focus on forest certification should initially be on institution building, rather than specific standards.

A second lesson, as indicated above, is that any analysis of supply chain systems must take into account industry programs that emerge. When US forest companies evaluated the often lengthy FSC regional standards-setting process, uncertainty over the future of the FSC and a fear of losing decision-making power led them to develop their own alternative to the FSC. Significant research underway on the role of business self-regulation (Haufler 2001; Cutler, Haufler, and Porter 1999a; Prakash 1999, 2001; Kollman and Prakash 2001; Moran 2002; Gunningham and Sinclair 2002; Webb 2002) has illustrated that the benefits of these systems is that, unlike traditional public policy processes or environmental group initiated programs, industry is less critical and adversarial when the procedures and approaches and rules of conduct have emanated from their own industry associations (Gunningham, Grabosky, and Sinclair 1998). At the same time, the industry self-regulation approaches must also fit within a market-driven logic, and hence upward movement of their rules will also be restricted by the need for economic incentives.

That these two competing governance systems gain their authority from the market's supply chain also has fundamental implications for the nature of the debate that occurs between them and the arguments and assertions that are made. We explained in chapter 1 that because of the two conceptions of forest certification (which would arguably equally apply in other emerging non-state market-driven sectors), the more stringent and wide in scope program is initially less likely to be viewed favorably by those companies who must implement the rules[4] than the more flexible, industry-initiated programs that emphasize the development of procedures. This means that conception one (more stringent, wide in scope) and conception two (flexible, procedural) non-state market-driven systems have different interests at different parts of the supply chain. At the level of those whose firms must actually implement the rules, conception one promoters have an interest in explaining why their system is not that different from conception two, while conception two supporters have an interest in explaining what makes their program different from conception one. We certainly saw much evidence of this in practice, with the FSC supporters emphasizing to companies and landowners the relative ease with which their program could be pursued, while the FSC competitors took care to explain exactly what was difficult about the FSC and how their programs better respected landowner autonomy in decision making.

This logic was reversed at that part of supply chain where companies did not have to implement the program but rather made commitments to purchase products whose production was governed by the certification standards. Here, we saw the FSC and its supporters, in all of our cases, using an array of strategies to assert heavily to customers of forest products that the FSC was superior to the other programs in addressing sustainable forestry management. And conversely, we saw the FSC competitors at this stage arguing that they were, in terms of environmental impact and rules, not all that different from the FSC. They made these claims in a variety of venues, ranging from meetings with retailers to conferences on cross-certification program collaboration. And when they found they had difficulty convincing retailers of this, they would initiate changes in their programs to buttress their arguments. These features appear to illustrate a universal "law" of non-state market-driven systems (which we defined in chapter 1 as gaining authority from "supply chains") that warrants future research.

Equally important for our research is to understand what has not (yet) happened in supply chain dynamics over global forest certification. The ultimate consumer — the individual who walks into a forest products retailer — has not played any significant role in how support is granted by forest companies and landowners. Retailers have made choices in response to direct action campaigns against their companies by environmental groups, but this action is based only on indirect societal support (environmental groups tend to enjoy high levels of trust and support from the general public). Whether individuals as consumers will ever play a role in supply chain dynamics appears to be an open question. If enough retailers and home builders decide to support certification, then it could be passed on to consumers as a *fait accomplit* — in which case consumers will have no role at all. On the other hand, it may be that if the retail market distinguishes between certified and non-certified products, then consumers will have a choice. However, most retailers do not appear to be moving in the direction that would give consumers a choice. Ikea, B&Q, and Home Depot, for example, are more interested in using the fact that they support and sell certified forest products as a means to enhance their own brand than to inform consumers about different forest certification labels and options. If this trend continues, then support for forest certification and its future will be determined by "business to business" operations along the supply chain, rather than ultimate consumers. Recognition of this is important because it may be that some forms of "political consumerism" (Micheletti 2003) fit outside traditional understandings in which the individual is thought to matter most as a consumer — it may be that they matter more as supporters of environmental groups.

Recognition of this leads to our call to understand better (hypothesis 8) how different supply chains within the forest sector and across sectors, such as in tourism, mining, fisheries, food production and coffee, may influence the political struggles over non-state market-driven authority. We suspect that in cases such as coffee, there may be no perceived need for a competitor program to emerge since those most threatened by certification programs are coffee traders found in the middle of the supply chain, rather than other coffee growers. It may also be that consumers will actually matter more in supply chain struggles in fisheries for the simple reason that humans physically *consume* the product and are thus more understandably concerned about the health benefits that might be associated with purchasing fish from certified sources — much akin to the certified organic movement now.

Legitimacy Granting and the Durability of Non-State Market-Driven Governance Systems

We explained in chapter 1 that forest company and forest owner support for the FSC was located within Suchman's pragmatic category and that the key issue for us centered on understanding better the legitimacy achievement logic and efforts that helped understand and explain the emergence of non-state market-driven governance in the forest sector. At the same time our empirical cases did shed light on understanding how those granting moral legitimacy strongly influence this debate. Our review also allows us to identify how cognitive legitimacy might be achieved.

Moral Legitimacy and Core Audiences

Our empirical cases revealed that both the FSC and FSC competitor programs each have a "core audience" that appears to grant moral legitimacy to the program. Many of the world's leading environmental groups as well as domestic environmental groups tend to grant the FSC moral legitimacy because the FSC's approach to sustainable forest management fits with their own organizational goals and values. This evidence arguably fits best with Sabatier and Jenkins-Smith's (1993) work on "deep core beliefs" in influencing the development of advocacy coalitions. While our historical narrative does not support Elliott's findings (Elliott 2000; Elliott and Schlaepfer 2003) that forest certification in Sweden led to a shift in "deep core" beliefs on the part of forest companies, nor that a new "advocacy coalition" developed as a result of this (Cashore 2003), our data do support the need to understand better how "deep core" beliefs figure in supply chain dynamics. Certainly our cases reveal strong evidence that those organizations that granted moral legitimacy served as an important check on the strategic choices that could be made by the FSC and its supporters in their attempt to woo industry and forest owners. And recognition of this is important because we found that the role of the moral audience varied depending on whether choices were made at the regional, national, or international levels.

The BC and Swedish cases illustrate these points. BC, which saw the strongest "converting" logic of any of our cases, was characterized at the regional standards setting level by a determined "core audience" group that shaped and developed high standards — standards that these groups believed were necessarily required to address sustainable forestry management in BC. Forest industry officials were taken aback and sought to influence other levels of FSC decision-making policy. They thus retained their overall strategy of attempting to influence the FSC from within and clearly "lost" the first round to the FSC moral legitimacy granting core audience in BC. The BC case also speaks to understanding better the divisions within the core audience — as some supporters come to see themselves as strategists for the institutionalization of the FSC, and others come to see themselves as "keepers" of the original intention. In the BC case the World Wildlife Fund came to support and promote industry buy-in to the program, but the FSC Canada board, dominated by non-industry interests, voted to accept the disputed regional standards, leaving the FSC International as the last arena in which BC companies could seek redress. And it was here where FSC strategists, from the WWF to senior FSC officials, worked to address industry concerns, all the while walking a "legitimacy" tightrope in the hopes of maintaining industry pragmatic legitimacy and the moral legitimacy of the BC environmental groups.[5]

In Sweden, the FSC's institutional structures worked in the opposite direction for those working to maintain industry's pragmatic support. Here, those groups in Sweden long involved in the process recognized the need to conform to percentage-in-percentage-out demands of industrial forest companies (where companies could sell the same percentage of lumber they harvested as certified, even if it was not from the actual forest that was certified), with the international FSC, while sympathetic, arguing that such a move would have dangerous implications elsewhere.

These cases illustrate how the support of core audiences who granted the FSC moral legitimacy can act to limit the options available to the FSC's strategists who are intent on first institutionalizing support for the FSC. This check was so important because FSC strategists were aware that they could not lose the bulk of their moral legitimacy granting core audience. There were two reasons for this. First, if FSC's moral legitimacy granting audience withdrew its support the FSC would lose its raison d'être — which was to offer a version of global sustainable forestry management that differed from what environmental groups were asserting to be domestic and global "logging charters" coming out of domestic and international governmental processes. Second, and related, it was support from environmental groups that provided a strong incentive for many companies to even consider the FSC — since these companies would then be approved by a system supported by their strongest critics. Future research should delve further into how and when circumstances might permit FSC and its strategists to act autonomously from its core audience, and the conditions under which this is not possible.

Indeed, evidence from our cases reveals that the FSC can indeed lose small amounts of "moral legitimacy," especially in cases where perceived radical groups fail to grant it. Hence, the FSC did not appear significantly hampered by the failure of Greenpeace Sweden to support the FSC there, and as Elliott (Elliott and Schlaepfer 2001; Elliott 2000) has argued, such opposition may have sent a signal to international markets about the "middle ground" approach of the FSC. Finally, we note the special role that US foundations have played as the FSC's key financial backers and identify for future research better understandings into the long-term implications of this financial dependence.[6]

Cognitive Legitimacy

While there is a clear relationship between the granting of moral legitimacy by environmental groups and the ability of the FSC to gain pragmatic legitimacy from forest companies and landowners, the role of cognitive legitimacy in non-state market-driven governance remains poorly understood. According to Suchman, cognitive legitimacy can be gained in two ways — either through "routinization" or by establishing "new allegiant constituencies" by mirroring the organization after other organizations that already possess cognitive legitimacy from the audiences from which the FSC is seeking support. Whether or not current levels of pragmatic legitimacy could become "cognitive" through routinization is an open question — and would obviously require the market incentives to be strong enough, and to be maintained over a long enough period of time, to see if this route might work for non-state market-driven systems. The other approach seems to offer a faster route to achieving cognitive legitimacy — which is important since cognitive legitimacy, as Suchman argues, is the most durable form of legitimacy. There are indications that FSC officials understand the need to take active efforts in achieving "cognitive" legitimacy through the "mirroring" path. FSC international officials have begun efforts to mirror their auditing procedures with international standards widely accepted by international businesses, arguing that the International Organization for Standardization[7] (Wenban-Smith and Elliot 1998) environmental management systems' approaches are compatible with the FSC. In addition, the FSC is currently restructuring itself in a way that closely resembles a decentralized United Nations system governing policy development.[8] Clearly more work has to be done to understand the impacts of such efforts in a non-state market-driven system and whether cognitive legitimacy is still the most durable in a system that, in order to survive in the long run, must demonstrate some kind of indirect or direct economic benefit to profit maximizing companies.

It is also important to understand at what stage, if ever, these systems might gain such a degree of cognitive legitimacy and would be covered by virtually all of the world's producers that economic evaluations would no longer be important — a situation that would exist when everyone would come to accept the program as appropriate and hence the idea of escaping the rules would not be an option. Our cases reveal that we are, of course, nowhere near such support, and it is unclear if non-state market-driven systems would ever get there. We do note that Teubner's work on autopoiesis and the role of private legal systems in gaining "autonomy" (Teubner 1993a and b, 1997) and Lawson and Cashore's (2001a) initial application of this work to forest certification provide important fodder for future research and reflection.

Relationship of Traditional Governmental Public Policy Approaches and Non-State Market-Driven Governance Systems

Two key questions emerge from an analysis of forest certification. The first comes from students of public policy, international private authority, and policy instruments who wonder whether non-state market-driven governance will pose a challenge to traditional public policy processes rooted in Westphalian sovereign authority or whether they will complement the state, acting in ways that the state is unable. The second comes from forest certification practitioners who are concerned about the state "taking over" or dominating the current private arena — and potentially undermining the system that, they believe, has a much better chance of addressing global sustainable forest management (detailed in the section below). Returning to our distinctions made in the introductory chapters, we note that there are a number of governmental functions that would enhance non-state market-driven legitimacy (such as governmental procurement policies, financial support to a certification program, and so on) and others that would detract from the non-state market-driven phenomenon (the most crude case would be a situation in which governments used their Westphalian authority to take over certification rule development and made private systems illegal).

Certainly our research revealed important ways in which governments directly or indirectly enhance the legitimacy of non-state market-driven forest certification systems. In the UK, for instance, it was the action of the Forestry Commission that tipped the balance in favor of the FSC by acting in a consensus-building role. Its decision to not create a public policy version of forest certification, which would have challenged non-state market-driven authority (Cashore 2002), but instead to facilitate rules governing private certification was instrumental in providing FSC strategic support. Likewise, in

Sweden the government's requirement that forest industry and forest owners promote ecological goals as much as economic ones, without explaining or determining how this be done, had the indirect effect of increasing industry support for the FSC as a way of accomplishing these goals. And in BC, recent efforts to reduce the province's forest practices regulations by introducing a "results based code" (Hamilton 2000a) appear to be enhancing credibility and support for non-state systems. Not only do these procedures leave environmental groups increasingly dissatisfied with BC's public policy approaches, but the government itself has asserted that, because of certification, "heavy government oversight" is no longer needed since certification represents "proof" that " industry has the knowledge and the incentive to do its job properly."

While more research must be conducted, what is clear is that environmental groups are clearly favoring and focusing much of their attention on forest certification relative to many domestic and international forest policy spheres. Indeed, the attention placed on international forestry governance by leading environmental groups at the 2002 Johannesburg environmental summit was negligible. Conversely, considerable attention and interest was placed on the FSC's 2002 General Assembly in Oaxaca, Mexico. Few environmental groups are asserting that traditional public policy approaches be ignored, but rather that, in many cases, FSC-style private authority is at present providing what they believe to be more opportunities for access to and influence over sustainable forestry management standards. In this sense the idea of "forum shopping" (Busch 1999a and b) shows promise for future research into understanding the calculations and motivations behind strategic choices about where to focus efforts. Certainly within the US, forum shopping meant that those concerned about forest preservation in the United States were more focused on federal forest policy and federal lands, where opportunities for access were much greater, than on the difficult-to-access state level (Cashore 1997). And as we show in the US and other cases, it was in response to dissatisfaction in the public policy process that led to support for the FSC.

Our cases demonstrated clearly a strong reciprocal relationship between the FSC and public policy processes in the emergence and support for the FSC and market-based legitimacy achievement campaigns. Whether forest certification as a non-state market-driven system ultimately challenges state-centered authority or simply provides another arena in which to mediate political struggles over forestry and other issues partly depends on the degree of institutionalization that the FSC and/or FSC competitors experience in the coming years. While we cannot answer this question, our research has shed light on how this process can be expected to occur and identified important avenues for future research.

Certification and the Promotion of
Global Sustainable Forest Management

Our analysis has revealed that forest certification is a unique and innovative governance system that is composed of an array of evolving procedures and policies designed to address and encourage sustainable forestry management globally.[9] We have clearly documented that because so much change is occurring as the FSC competes with industry and landowner-initiated programs for policy authority, it is difficult to predict, with any certainty, what the on the ground impacts of forest certification will be. What we have been able to do is to understand better the process through which these struggles occur and the way in which standards addressing forest management are being devised. What then, can we say about forest certification and its potential to address global concerns about sustainable forestry?

First, it is a mistake to use a comparison of existing standards, or existing lands certified under the FSC, CSA, PEFC, or SFI, to predict the long-term ability of forest certification to promote sustainable forest management. Because certification is in the process of institutionalizing, we cannot assume that those companies certified now will be the same kind of companies who get certified in the future (figure 8.2). Indeed, it may be that the "green" companies are the ones who chose FSC certification early on, while the less "green" companies are the ones the FSC or other certification systems will gain support from if, and when, they become institutionalized. Indeed, these tables tell us nothing about companies who have chosen to wait to get certified until the FSC's standards are developed for their region or country.

Second, since the FSC is only concerned about developing rules governing sustainable forest management, it is largely unable to address issues of deforestation where land is being removed from the forestland base (e.g., owing to urban sprawl in North America and conversion to agriculture in the tropics). And to the extent that certification makes harvesting more expensive, economists have argued that it may actually have a perverse "substitution" effect and create more demand for non-wood products, which might reduce economic incentives to keep land forested (Binkley 2000; Sedjo and Swallow 1999; Swallow and Sedjo 2000). Others have argued that sometimes it is unregulated non-industrial private forest owners who use the most low-impact practices, but given the reporting requirements and assessment costs of certification, these landowners may have a harder time receiving recognition for such practices from the FSC than those companies practicing higher impact, broader scale harvesting (Murray and Abt 2001).

These rather sobering reflections must be considered with the other poten-

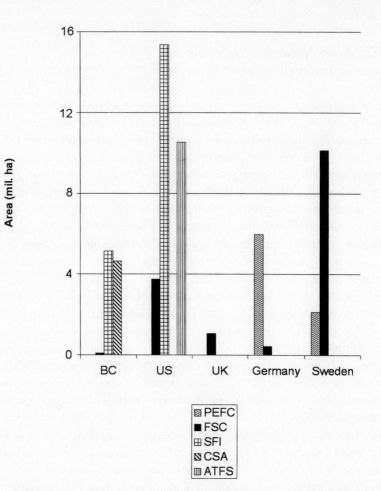

Figure 8.2. Area certified by the FSC and competitor programs as of the middle or end of 2002. Competitor programs include the PEFC in Germany, Sweden, and the UK; SFI third party in the US; and the CSA in Canada. SFI third-party certification is also available in Canada. As of June 1, 2002, five companies in BC were SFI third-party certified (5,130,000 ha certified). ATFS stands for the American Tree Farm System, a certification program specifically designed for US non-industrial private forest owners. It currently operates in 48 states and includes nearly 66,000 certified landowners. In Sweden and Germany, the areas certified under the PEFC were done on a regional basis; in Sweden 15,951 landowners participated and in Germany 3,256 participated. Sources: FSC data represent certified areas as of September 30, 2002, for Germany, the UK, Sweden, and the US; for BC, the data include areas certified as of June 1, 2002 (source: http://www.sfms.com/pdfs/cscbco.pdf and http://www.fscoax.org/html/5-3-3 .html). The ATFS data are for areas in the program as of end of 2002 (http://www.treefarmsystem.org/); CSA data include areas certified as of June 1, 2002 (http://www.sfms.com/pdfs/cscbco.pdf); SFI data include areas certified as of October 3, 2002 (http://www.afandpa.org/forestry/sfi/ SFI%20Certification%20Data%20(10-1-02).pdf); and PEFC data include area certified as of the end of 2002 (http://www.pefc.cz/i_register/STATISTICS1.ASP?COUNTRY=Germany& COUNTRY_CODE=0 4 and http://www.pefc.cz/i_register/STATISTICS1.ASP?COUNTRY =Sweden&COUNTRY_CODE=05).

tially enormous impacts of the FSC and other forest certification systems. The first is that they represent a grand new experiment in developing rules and procedures in ways quite foreign to traditional public policy approaches. The extent to which the FSC (and other certification programs) has raised new and innovative ideas about the ways collaborative solutions might be found appears to be an important contribution by itself. The norms of accountability and transparency have been strongly promoted by the FSC, and FSC competitors have largely adopted these norms themselves. While difficult to measure, this apparent effect on the ideational framework in which forestry regulations are developed could be expected to have significant impacts.

And despite much of the debate over forest certification occurring in the North, it appears that environmental groups and much of the world's industry agree that forest certification is more needed in the South, where poor governmental institutions and limited resources have led to rampant illegal logging and high impact forestry. It is ironic that Hubert Kwisthout's idea, which originated over his concern about finding sustainable tropical sources, might now, over a decade later, finally be entering the tropics in a coherent and systematic way.

What is clear is that, measured by an array of indicators, the world's forests are under stress. They are home to an increasing number of threatened and endangered species (in both developed and developing countries), the number of intact watersheds is diminishing, natural forests are disappearing, and forest ecosystem structure and function are under considerable threat. The choices that are made in addressing these important problems in the next decades will say much about our generation's determination to systematically and thoughtfully address environmental destruction. The ability of forest certification to be a part of this solution is a question that needs to be carefully researched and analyzed, so that those in a position to make strategic choices in this regard are able to make the most well-informed and environmentally sensitive choices. It was, after all, for this reason that Hubert Kwisthout first became concerned — and why he developed an idea that has transformed the world of global environmental governance.

Appendix 1

Comparison of Initial and Revised Standards of the FSC and
Its Competitor Programs on Six Specific Facets of Forest Management

Tables A1 and A2 provide a snapshot of how the four main certification programs address six facets of forest management (plantations, chemicals, clear-cuts, exotics, reserves, and genetically modified organisms). Table A1 details the initial attention given to these issues by the programs. The PEFC is not applicable to this table, since by the time it emerged the other programs were already undergoing changes to their initial standards. Table A2 details how all the programs address these issues in their revised standards. In the case of the PEFC, we detail the program's initial standards. These tables facilitate a comparison of the programs both across the substantive policy issues as well as through time to determine how much they have changed and how different the standards still are.

The information comes from the following sources. FSC, the 1996 and 2000 versions of the FSC Principles and Criteria (Forest Stewardship Council 1996, 2000c); SFI, the 1999 and 2002–2004 versions of the SFI Standard (AF&PA 2002a, 2000g); CSA, the 1996 Sustainable Forest Management System: Specification Document (CAN/CSA-Z809-96) and the 2002 Sustainable Forest Management: Requirements and Guidance Document (February Draft — CAN/CSA-Z809-2002) (Canadian Standards Association 1996, 2002); PEFC, the 1998 Pan European Criteria, Indicators, and Operational Level Guidelines (PEFC International 2001a).

Table A1. Initial attention given to six facets of forest management by the standards of the FSC and its competitor programs

	Programs FSC	CSA	SFI	PEFC
Plantations	Specific details limiting: (1) Representation on landscape; (2) Date of establishment; (3) Characteristics of management blocks (i.e., promote species, structural, and genetic diversity)	No specific standard Plantations not defined or regulated.	No specific standard Plantations not defined or regulated.	N/A[a]
Chemicals	Require minimizing use (prefer IPM[b] approach). Require documentation, strict monitoring, and control. Ban certain chemicals (e.g., WHO type 1a and 1b).	No specific policy beyond following government regulations.	Specify prudent use. Government regulations set standard (i.e., require adherence to all label requirements, laws, and regulations pertaining to chemical use).	N/A
Clearcuts	No specific policy in P&C (early regional standards restrict size and location).	No specific policy beyond following government regulations.	Average of 120 ac, with exceptions for forest health emergencies and natural catastrophes.	N/A
Genetically modified organisms	Prohibited.	No specific policy beyond following government regulations.	No specific policy beyond following government regulations.	N/A
Exotics	Careful use under strict monitoring to avoid adverse environmental impacts.	No specific policy beyond following government regulations.	No specific policy beyond following government regulations.	N/A

Table A1. (continued)

Programs			
FSC	CSA	SFI	PEFC

	FSC	CSA	SFI	PEFC
Reserves	Require conservation zones to protect rare, threatened, and endangered species. Representative samples of ecosystems on landscape mapped and protected. Prohibits harvesting in old growth forests.	No specific policy requiring set asides or reserves.	Require identification and management of sites with ecological, geological, or historical significance. Manager has discretion on how best to manage these sites.	

[a] The Pan European Forest Certification initiative was not underway until 1999; thus, its initial program is not reviewed in this table.

[b] IPM stands for integrated pest management, an approach to pest control that seeks to minimize chemical use through the use of alternative prevention and biological control techniques.

Table A2. Revised attention given to six facets of forest management by the standards of the FSC and its competitor programs

	Programs FSC	CSA	SFI	PEFC
Plantations	Specific details limiting: (1) Representation on landscape; (2) Date of establishment. (3) Specific characteristics of management blocks (e.g., require species, structural, and genetic diversity).	No specific policy. Plantations not defined or regulated.	No specific policy. Plantations not defined or regulated.	No specific policy. Plantations not defined or regulated.
Chemicals	Require minimizing use (prefer IPM[b] approach). Require documentation, strict monitoring, and control. Ban certain chemicals (e.g., WHO type 1a and 1b).	No specific policy beyond government regulations.	Require minimizing use given management objectives (promote IPM where feasible).	Guided[a] to avoid inappropriate chemical use with potential harm to water quality. Guided to minimize pesticide and herbicide use.
Clearcuts	Restrict size and location (varies among national/regional standards).	No specific policy beyond following government regulations.	Average of 120 ac; exceptions for forest health emergencies and natural catastrophes.	No specific policy (varies among national initiatives).
Genetically modified organisms	Prohibited.	Guided to address their use through consultation with public advisory group.	Require adherence to government regulations and international protocols. Use governed by scientifically sound methods.	No specific policy (varies among national initiatives).

Table A2. (continued)

	Programs FSC	CSA	SFI	PEFC
Exotics	Permitted, but not promoted. Require careful monitoring to avoid adverse environmental impacts.	No specific policy beyond following government regulations.	Minimize use. Research documentation available that indicates exotics pose minimal risk.	Guided to prefer native species and local provenances where available.
Reserves	Require conservation zones to protect rare, threatened, and endangered species. Representative samples of ecosystems on landscape mapped and protected. Require maintaining and enhancing attributes of high conservation value forests.	Respect government protected areas. Determine existence of under protected ecosystems (at the landscape level) in defined forest area and ensure their protection.	Require identification and management of sites with ecological, geological, or historical significance. Manager has discretion on how best to manage these sites.	Guided to inventory and map ecologically important forests ecosystems. Guided to protect and when necessary restore these important forests.

[a] "Guided" is used since the direction on how to implement the Pan European Criteria and Indicators is contained in the Pan European Operational Level Guidelines. The Pan European Criteria and Indicators are descriptive in nature and contain limited normative language.

[b] IPM stands for integrated pest management, an approach to pest control that seeks to minimize chemical use through the use of alternative prevention and biological control techniques.

Appendix 2

Europe

	Organization	Location of Interview	Date
1	Bundesministerium für Verbraucher-schutz, Ernährung und Landwirtschaft (Federal Ministry for Consumer Protection, Food and Agriculture)	Bonn, Germany	March 20, 2001
2	Arbeitsgemeinschaft Naturgemäße Waldwirtschaft member (Working Group for Natural Silviculture)	Germany	March 21, 2001
3	Deutscher Forstwirtschaftsrat (German Forestry Association)	Rheinbach, Germany	March 21, 2001
4	OBI-Bau- und Heimwerkermärkte GmbH & Co. KG (Obi Do-It-Yourself Center)	Wermelskirchen, Germany	March 22, 2001
5	Ministerium für Umwelt und Forsten Rheinland-Pfalz (Ministry of Environment and Forests Rheinland-Pfalz)	Mainz, Germany	March 23, 2001

Research Interviews

Europe (*continued*)

	Organization	Location of Interview	Date
6	Städte und Gemeindebund Rheinland-Pfalz (Association of Municipalities in Rheinland-Pfalz)	Mainz, Germany	March 23, 2001
7	Forest Stewardship Council, German Working Group	Freiburg, Germany	March 27, 2001
8	Gesamtverband Holzhandel (BD Holz-VDH)	Wiesbaden, Germany	March 29, 2001
9	Bundestag (Parliament)	Berlin, Germany	April 2, 2001
10	Holzabsatzfond	Bonn, Germany	April 4, 2001
11	Pan European Forest Certification Deutschland Sekretariat (PEFC Germany Secretariat)	Stuttgart, Germany	April 5, 2001
12	Ministerium Ländlicher Raum Baden-Württemberg (Ministry of Rural Space Baden-Württemberg)	Stuttgart, Germany	April 5, 2001
13	Verband Deutscher Papierfabriken (Association of German Paper Producers)	Hösbach, Germany	April 6, 2001
14	SCA Holzeinkauf GmbH	Hösbach, Germany	April 6, 2001
15	Bayerische Waldbesitzverband member (Bavarian Landowner Association)	Bavaria, Germany	April 20, 2001
16	Friends of the Earth	London, England	April 20, 2001
17	World Wide Fund for Nature	Godalming, England	April 20, 2001
18	Forest Stewardship Council, UK	Llanidloes, Powys, Wales	April 24, 2001
19	WoodMark Soil Association	Bristol, England	April 24, 2001
20	Canadian High Commission, London	London, England	April 25, 2001
21	Association of Professional Foresters	Belford, Northumberland, England	April 25, 2001
22	Woodland Trust	Auchterarder, Perthshire, England	April 26, 2001
23	Timbmet Ltd.	Oxford, England	April 30, 2001
24	SGS Qualifor	Oxford, England	April 30, 2001
25	Oxford Forestry Institute	Oxford, England	April 30, 2001
26	Ministerium für Umwelt, Raumordnung und Landwirtschaft des Landes Nordrhein-Westfalen	Düsseldorf, Germany	April 30, 2001
27	Anglia WoodNet	Eastern England	May 1, 2001
28	English Nature	Peterborough, England	May 1, 2001
29	Timber Growers Association, Eastern England	Eastern England	May 1, 2001
30	BM Trada	Edinburgh, Scotland	May 3, 2001
31	UK Forest Products Association	Stirling, Scotland	May 3, 2001

Research Interviews

Europe (*continued*)

	Organization	Location of Interview	Date
32	Haindl	Augsburg, Germany	May 4, 2001
33	Timber Growers Association	Edinburgh, Scotland	May 4, 2001
34	Forestry Commission for Great Britain, Northern Research Station	Edinburgh, Scotland	May 4, 2001
35	Forestry Commission for Great Britain, Practice and Policy Division	Edinburgh, Scotland	May 4, 2001
36	Forestry Commission for Great Britain (Former official)	Edinburgh, Scotland	May 7, 2001
37	Forestry Commission for Great Britain, Practice and Policy Division	Edinburgh, Scotland	May 8, 2001
38	Nexfor	Cowie Stirlingshire, England	May 8, 2001
39	Forest Industries Development Council	Edinburgh, Scotland	May 8, 2001
40	Forestry Consultant	Stirling, Scotland	May 9, 2001
41	Scottish Woodlands Ltd.	Edinburgh, Scotland	May 9, 2001
42	Wood Panelboard Industry Federation	Grantham, England	May 11, 2001
43	World Wide Fund for Nature	Matlock, Derbyshire, England	May 11, 2001
44	Industriegewerkschaft Bauen-Agrar-Umwelt (IG BAU)	Frankfurt, Germany	May 31, 2001
45	World Wide Fund for Nature Germany	Frankfurt, Germany	June 1, 2001
46	Greenpeace Germany	Hamburg, Germany	June 6, 2001
47	holz.conZert	Hamburg, Germany	June 7, 2001
48	B&Q	London, England	July 2, 2001
49	Timber Trade Federation	London, England	July 2, 2001
50	International Institute for Environment and Development	London, England	July 2, 2001
51	BBC Magazine	London, England	July 3, 2001
52	Fern	Morten on Marsh, England	July 3, 2001
53	Beacon Certified Timber	Edinburgh, Scotland	July 5, 2001
54	Royal Society for the Protection of Birds	Edinburgh, Scotland	July 5, 2001
55	Private forest owner, Sweden	Geneva, Switzerland	October 18, 2002
56	Swedish Forest Owner's Association	Geneva, Switzerland	October 18, 2002
57	World Wide Fund for Nature, Forests for Life Campaign	Geneva, Switzerland	October 19, 2002
58	Pan European Forest Certification Council, Secretariat	Luxembourg, Luxembourg	October 22, 2002
59	Confederation of European Forest Owners	Geneva, Switzerland	October 22, 2002

Research Interviews

Europe (*continued*)

	Organization	Location of Interview	Date
60	Forest Stewardship Council Europe	Oaxaca, OAX, Mexico	November 24, 2002
61	Church of Sweden	Oaxaca, OAX, Mexico	November 24, 2002
62	World Wide Fund for Nature, European Forest and Trade Network	Oaxaca, OAX, Mexico	November 24, 2002
63	Sveaskog	Oaxaca, OAX, Mexico	November 24, 2002
64	Korsnas AB	Oaxaca, OAX, Mexico	November 24, 2002
65	IKEA	Oaxaca, OAX, Mexico	November 25, 2002
66	Forest Stewardship Council, Swedish National Initiative	Oaxaca, OAX, Mexico	November 25, 2002
67	Environmental Consultant	Oaxaca, OAX, Mexico	November 26, 2002
68	Forest Stewardship Council, Secretariat	Oaxaca, OAX, Mexico	November 27, 2002
69	Forest Stewardship Council, Swedish National Initiative	Uppsula, Sweden	August 19, 2003
70	Church of Sweden	Västerås, Sweden	August 20, 2003
71	Swedish Society for Nature Conservation	Stockholm, Sweden	August 20, 2003
72	World Wide Fund for Nature Sweden (Former official)	Stockholm, Sweden	August 20, 2003
73	Swedish Environmental Protection Agency	Stockholm, Sweden	August 21, 2003
74	Swedish Environmental Protection Agency	Stockholm, Sweden	August 22, 2003
75	World Wide Fund for Nature Sweden	Solna, Sweden	August 22, 2003
76	StoraEnso	Stockholm, Sweden	August 22, 2003

Research Interviews

Europe *North America*

	Organization	Location of Interview	Date
1	Sierra Legal Defence Fund	Vancouver, BC, Canada	August, 1999
2	Weyerhaeuser Company	WA, USA	August 31, 1999
3	MacMillan Bloedel Ltd.	Vancouver, BC, Canada	September, 1999
4	Washington Environmental Council	Seattle, WA, USA	September, 1999
5	Northwest Natural Resource Group	Seattle, WA	September 1999
6	Rainforest Action Network	San Francisco, CA, USA	September 17, 1999
7	Business for Social Responsibility	San Francisco, CA, USA	September 17, 1999

Research Interviews

North America (*continued*)

	Organization	Location of Interview	Date
8	Natural Resources Defense Council	San Francisco, CA, USA	September 17, 1999
9	Canadian Forest Service	Ottawa, ON, Canada	January, 2000
10	Canadian Department of Foreign Affairs and International Trade, Office of Forestry and Environment (Former official)	Ottawa, ON, Canada	January, 2000
11	Maine Forest Products Council	Augusta, ME, USA	January, 17 2000
12	Canadian Pulp and Paper Association	Montreal, QC, Canada	January 24, 2000
13	Izaak Walton League of America	Washington, D.C., USA	February 4, 2000
14	American Forest Foundation	Washington, D.C., USA	February 4, 2000
15	American Forest & Paper Association	Washington, D.C., USA	February 4, 2000
16	Library of Congress Congressional Research Service	Washington, D.C., USA	February 4, 2000
17	Forest Stewardship Council, US National Initiative	Washington, D.C., USA	February 4, 2000
18	Weyerhaeuser Company	WA, USA	June 2002
19	Forest Stewardship Council, Canadian National Initiative	Toronto, ON, Canada	July 3, 2000
20	Champion International	Jacksonville, FL, USA	July 6, 2000
21	Georgia-Pacific	Atlanta, GA, USA	July 5, 2000
22	US Fish and Wildlife Service	Atlanta, GA, USA	July 5, 2000
23	US Department of Agriculture, Forest Service, Southeast Division	Atlanta, GA, USA	July 5, 2000
24	Forest Management Trust	Gainesville, FL, USA	July 6, 2000
25	Florida Defenders of the Environment	Gainesville, FL, USA	July 7, 2000
26	SmartWood	Richmond, VT, USA	July 8, 2000
27	World Resources Institute	Washington, D.C., USA	July 11, 2000
28	The Home Depot	Atlanta, GA, USA	July 14, 2000
29	Forest Stewardship Council, Canadian National Initiative	Toronto, ON, Canada	July 24, 2000
30	International Paper	Auburn, AL, USA	July 25, 2000
31	Forest Stewardship Council, Southeast Working Group	Auburn, AL, USA	July 25, 2000
32	Bloedel Forest Products	Montgomery, AL, USA	July 27, 2000
33	Alabama Forestry Commission	Montgomery, AL, USA	July 27, 2000
34	Private landowner	Auburn, AL, USA	July 28, 2000
35	Industry Canada	Ottawa, ON, Canada	July 31, 2000
36	Canadian Pulp and Paper Association	Montreal, QC, Canada	August 1, 2000
37	World Wildlife Fund Canada, head office	Toronto, ON, Canada	August 3, 2000
38	Sierra Club, Atlantic Canada	Bedeck, NS, Canada	August 3, 2000

Research Interviews

North America (*continued*)

	Organization	Location of Interview	Date
39	New Brunswick Landowners Association	Truro, NS, Canada	August 3, 2000
40	New Brunswick Forest Products Association	Fredericton, NB, Canada	August 4, 2000
41	Falls Brook Centre	Fredericton, NB, Canada	August 4, 2000
42	Forest Stewardship Council, BC Working Group	Nelson, BC, Canada	August 8, 2000
43	Silva Forest Foundation	Slocan Park, BC, Canada	August 8, 2000
44	Hancock Timber Resource Group	Boston, MA, USA	August 10, 2000
45	Natural Resources Defense Council	Washington, D.C., USA	August 11, 2000
46	British Columbia, Ministry of Forests Small Business Forest Enterprise Programme	Victoria, BC, Canada	August 15, 2000
47	British Columbia, Private Forest Land Owners Association	Victoria, BC, Canada	August 15, 2000
48	Ecoforestry Institute	Victoria, BC, Canada	August 16, 2000
49	British Columbia, Ministry of Forests	Victoria, BC, Canada	August 16, 2000
50	British Columbia, Forest Practices Board	Victoria, BC, Canada	August 17, 2000
51	British Columbia Government Green Economy Secretariat	Victoria, BC, Canada	August 17, 2000
52	Iisaak Forest Resources	Ucluelet, BC, Canada	August 25, 2000
53	US Department of Agriculture, Forest Service, State and Private Forestry	Vancouver, BC, Canada	August 29, 2000
54	Silva Forest Foundation	Vancouver, BC, Canada	August 29, 2000
55	Council of Forest Industries	Vancouver, BC, Canada	September 1, 2000
56	PriceWaterhouseCoopers	Vancouver, BC, Canada	September 7, 2000
57	EcoTrust Canada	Vancouver, BC, Canada	September 7, 2000
58	British Columbia Government International Relations & Trade Branch	Vancouver, BC, Canada	September 8, 2000
59	Industrial Wood and Allied Workers of Canada	Vancouver, BC, Canada	September 9, 2000
60	US Department of Agriculture, Forest Service, PNW Region, Cooperative Programs	Portland, OR, USA	September 12, 2000
61	Collins Pine	Portland, OR, USA	September 12, 2000
62	EcoTrust	Portland, OR, USA	September 13, 2000
63	Political Science Faculty, University of Victoria	Victoria, BC, Canada	September 13, 2000
64	The Wilderness Society	Seattle, WA, USA	September 14, 2000
65	Certified Forest Products Council	Portland, OR, USA	September 14, 2000

Research Interviews

North America (*continued*)

	Organization	Location of Interview	Date
66	Northwest Natural Resource Group	Portland, OR, USA	September 14, 2000
67	Washington Forest Protection Association	Seattle, WA, USA	September 15, 2000
68	David Suzuki Foundation	Vancouver, BC, Canada	September 18, 2000
69	EcoTrust Canada	Vancouver, BC, Canada	September 18, 2000
70	Lignum	Vancouver, BC, Canada	September 18, 2000
71	EcoTrust Canada	Vancouver, BC, Canada	September 18, 2000
72	Citizens Bank	Vancouver, BC, Canada	September 19, 2000
73	Forest Alliance of British Columbia	Vancouver, BC, Canada	September 19, 2000
74	Weyerhaeuser Company	Vancouver, BC, Canada	September 20, 2000
75	Western Forest Products	Vancouver, BC, Canada	September 20, 2000
76	Western Canada Wilderness Committee	Vancouver, BC, Canada	September 20, 2000
77	EcoTrust Canada	Vancouver, BC, Canada	September 27, 2000
78	The Vancouver Sun	Vancouver, BC, Canada	September 29, 2000
79	Consultant FSC std team	Nanaimo, BC, Canada	October 1, 2000
80	World Wildlife Fund	Vancouver, BC, Canada	October 2, 2000
81	West Coast Environmental Law	Vancouver, BC, Canada	October 3, 2000
82	Greenpeace	Vancouver, BC, Canada	October 5, 2000
83	Yakima Nation	WA, USA	October 10, 2000
84	SGS	Vancouver, BC, Canada	October 10, 2000
85	World Wildlife Fund	Vancouver, BC, Canada	October 10, 2000
86	Plum Creek Timber Company	USA	October 13, 2000
87	Izaak Walton League of America	Washington, D.C., USA	November 16, 2000
88	US Department of the Interior, Bureau of Land Management	Washington, D.C., USA	November 16, 2000
89	Pinchot Institute	Washington, D.C., USA	November 16, 2000
90	National Woodland Owners Association	Washington, D.C., USA	November 17, 2000
91	Forest Stewards Guild	Washington, D.C., USA	November 18, 2000
92	Forest Stewardship Council, US Southwest Working Group	Washington, D.C., USA	November 18, 2000
93	American Forest Foundation	Washington, D.C., USA	November 19, 2000
94	International Association of Fish and Wildlife Game Agencies	Washington, D.C., USA	November 19, 2000
95	US Department of Agriculture, Forest Service	Washington, D.C., USA	November 19, 2000
96	World Resources Institute	Washington, D.C., USA	November 20, 2000
97	The Wilderness Society	Seattle, WA, USA	November 21, 2000
98	American Lands Alliance	USA	November 30, 2000
99	Scientific Certification Systems	USA	December 19, 2000

Research Interviews

North America (*continued*)

	Organization	Location of Interview	Date
100	Forestry Faculty, University of Maine	Orono, ME, USA	January 8, 2001
101	Seven Islands Land Company	Bangor, ME, USA	January 8, 2001
102	Mid-Maine Forestry	Warren, ME, USA	January 9, 2001
103	Global Forest Policy Project	USA	January 9, 2001
104	Irland Associates	Winthrop, ME, USA	January 9, 2001
105	Two Trees Forestry	Augusta, ME, USA	January 10, 2001
106	AE Sampson and Sons	Warren, ME, USA	January 10, 2001
107	Natural Resource Council of Maine	Augusta, ME, USA	January 11, 2001
108	Small Woodland Owners Association of Maine	Augusta, ME, USA	January 11, 2001
109	Forest Ecology Network (Maine)	Augusta, ME, USA	January 11, 2001
110	James W. Sewall Co.	Old Town, ME, USA	January 12, 2001
111	Forestry Faculty, University of Maine	Orono, ME, USA	January 12, 2001
112	JD Irving Ltd.	Saint Johns, NB, Canada	January 15, 2001
113	Irving Woodlands	Ashland, ME, USA	January 15, 2001
114	Forest Stewardship Council Secretariat	Oaxaca, OAX, Mexico	November 10, 2001
115	Forest Stewardship Council Secretariat	Oaxaca, OAX, Mexico	November 11, 2001
116	Forest Stewardship Council, US National Initiative	Washington, D.C., USA	November 17, 2001
117	Weyerhaeuser Company	Vancouver, BC, Canada	June, 2002
118	British Columbia, Ministry of Forests	Victoria, BC, Canada	August 2002
119	Morseby Consultants	Victoria, BC, Canada	August, 2002
120	Sierra Club of BC	Victoria, BC, Canada	August, 2002
121	Lignum	Vancouver, BC, Canada	November 12, 2002
122	Natural Resources Defense Council	Washington, D.C., USA	May 15, 2003
123	American Forest & Paper Association	New Haven, CT, USA	October 17, 2003

Notes

Chapter 1: The Emergence of Non-State Market-Driven Authority

1. See Pierson (2000, 1993).

2. This distinction draws on that first made by Doern et al. (1996).

3. Personal interview senior official, World Wildlife Fund, Geneva, October 19, 2002, and Elliott (2000).

4. The Rainforest Alliance created its own "SmartWood" sustainable forestry certification program in 1989 and would become an integral player in the creation of the FSC. SmartWood now serves as an accredited FSC auditor.

5. The "stewardship" part of the name can be traced to Robert Simioni, who was an early champion of forest certification in the US, and the "council" part can be traced back to Hubert Kwisthout (personal interview, Hubert Kwisthout, November 8, 2002).

6. Originally a two-chamber format was created: a social and environmental interests chamber with 75 percent of the votes and an economic chamber with 25 percent of the votes. There are currently three equal chambers among these groups, with one-third of the votes each. Each chamber is further divided equally between North and South.

7. We use the term "non-industrial forest owner" to refer to all types of forest land-ownership outside of industrial forest companies, including lands owned by governments, timber investment companies, and social institutions such as churches. The term "non-industrial private forest" refers to the more limited category of non-industrial non-investment, and non-government owned forest land.

8. Our intent here is not to offer a systematic analysis of current rules, which, as we show in our empirical cases, are constantly changing as each program attempts to gain

legitimacy. For detailed comparisons see Meridian Institute (2001a), Fletcher, Adams, and Radosevich (2001), and Ozinga (2001).

9. In Denmark, for example, the government subsidized farmers' extra expenses during the first two years after conversion to organic production, and subsidies were paid for producing and disseminating information about organic concepts and products to farmers, consumers, and agents in the food market.

10. The standards to which farms must comply were developed by the National Organic Standards Board (NOSB), which was appointed by the USDA at the behest of the mainstream organic farming community in 1990 after it experienced a number of problems in the marketing of organic products (USDA 1995). Although the standards-setting process involved much (often contentious) stakeholder input and the final standards are still optional for farmers, the government control of the standard-setting process in this sector goes beyond the bounds of non-state market-driven.

11. The BC Salmon Marketing Board explained that, "In September 2000, the State of Alaska was successful in getting its entire salmon fishery certified under MSC principles and criteria, giving it a tremendous competitive advantage in some important markets. Major buyers of Canadian salmon are telling us that we must get certified or they will stop buying from us (BC Salmon Marketing Council 2002)."

12. Likewise, in the case of the MSC, sustainable fisheries are promoted by offering market demand that can be accessed by companies adhering to their rules through a chain-of-custody provision (Simpson 2001).

13. Multiple types of coffee certification currently exist, but most fall under one of three types: "organic," in which no pesticides are used, "fair trade," in which coffee producers are guaranteed a higher price per pound and more favorable terms of trade than usual, and "shade-grown," which guarantees that the beans are derived from a coffee strain that can be grown in the understory of native species and does not require monocropping. Combinations of these exist; the Smithsonian "Bird Friendly" certification requires both organic and shade-tolerant conditions (Sasser 2002). A move is currently underway to consolidate some of these approaches into a single label.

14. See http://www.msc.org.

15. Interestingly, these demands from buyers of MSC certified products do not appear to be the result of high-publicity pressure campaigns done by NGOs, as is the case in the market for certified forest products and coffee.

16. Similar verification procedures exist under other non-state market-driven systems, such as the case of socially and environmentally responsible coffee production, in which producers are audited to ensure they are following the program's rules and the label is given to firms that sell this certified coffee (Transfair USA 2000). Here, the desire to be seen as a good corporate citizen is linked to a market advantage — Starbucks and Peets can sell their coffee as socially responsible, allowing them to maintain or increase market share and perhaps to charge a price premium compared to what other coffee retailers are able to charge (Seattle Post-Intelligence Staff 2000; Transfair USA 2000a, b).

17. Lipschutz and Fogel (2002) diverge from Cutler, Haufler, and Porter because they do invoke a wider definition of private governance that would encompass the non-state market-driven category analyzed here, but they have not located non-state market-driven governance as a unique category operating under quite different logics. Legal analyses (Meidinger 1997) have also made this distinction.

Chapter 2: The Research Design

1. Cashore (2002) modified this third category, which Suchman identified as "selection strategies." Since converting and conforming usually have a "selection" aspect to them, Cashore felt this third category did not capture well a distinct achievement strategy. Much of what Suchman's "selection" category fits with this new "informing" label.

2. While we retain the same definition, we have altered this category's label from "manipulation" to "converting" for two reasons. First, "converting" better captured what is taking place with respect to FSC efforts to alter initial landowner and forest company evaluations of the FSC. Second, a number of reviewers felt that the term "manipulation" was pejorative, even when defined analytically.

3. The definition of pragmatic legitimacy falls outside existing political science international relations work on legitimacy (for a review see Bernstein 2001), which tends to treat legitimacy as entailing a morally or cognitively derived "logic of appropriateness" and which stands in contrast to rational self interest support for governance structures (March and Olsen 1998).

4. We also recognize that environmental groups and other social organizations also act in what could be referred to as self-interested calculations. For example, the SFI's current efforts in the US to change its program and wean itself from the American Forest and Paper Association have led some conservation groups to grant the SFI pragmatic approval, if only because of these groups' desire to keep the idea of certification alive in the US context. For a classic treatment of narrow self-interest see Hardin (1968). As Sethi and Somanathan (1996:766) explain, under this "tragedy of the commons" treatment, "the assumption [is] that human behavior is driven by a particular, narrowly defined conception of self-interest: the degree of resource exploitation undertaken by each individual is assumed to be that at which marginal private material gains are brought into equality with the marginal costs of the extractive effort."

5. This point has been made by reviewers of this manuscript, including Steven Bernstein and Kathryn Harrison as well as Charlene Zietsma.

6. Our decision to treat all forest ownership types in one dependent variable conflates differences within the dependent variable. The chapters to follow reveal that private forest landowners are more difficult to persuade to support the FSC than are industrial forest landowners, which means that the effects of our explanatory variables partly differ according to forest ownership type. The decision to conflate these factors was done on purpose and for methodological reasons. If we had separated out the dependent variable according to landownership types, we would not have been able to compare general patterns across all forest ownership regimes. And because the various landownership types interact to affect how feasible it is for the FSC and its supporters to use converting strategies, such an approach would have created more methodological hurdles than it would have resolved. Instead, we carefully delineate these differences in our cases, explore the interactive nature of these differences empirically, and address the implications of these differences in our conclusion.

7. Popular sovereignty is distinct from efforts to challenge Westphalian sovereign authority since Westphalian sovereign authority is about the ability of domestic governments to make their own choices. As Bernstein and Cashore (2000) explain, direct "infiltration" of the political system actually respects Westphalian sovereign authority

since it seeks to alter domestic politics. On the other hand, efforts to change governmental policies directly from outside through markets or international rules do directly challenge sovereign Westphalian authority if they are focused on governmental policy changes. However, this is not the case in our study since certification bypasses sovereign Westphalian authority.

8. In addition, even if public policies are available to challenge the demands of customers, the limited political resources available to small domestic producers would make it more difficult for them to make full use of these opportunities.

9. For clarity, we measure these features in our cases by exploring the amount of total and commercial forestland owned by industrial forest companies vis-à-vis other non-industrial ownerships and the extent to which a relatively small number of forest companies dominate landownership and primary production.

10. Howlett and Ramesh (1995) note the distinction between the "societal policy agenda" and the "institutional policy agenda" where problems are actively considered by governmental policy makers. The societal policy agenda is clearly important as well, since it is the source of many actors' involvement in the policy process. However the logic of this hypothesis is that certification is more likely to be pursued when there is a perceived failure of governmental actors to adequately address the policy problem; thus, we focus on the "institutional policy agenda."

11. For a detailed analysis of this approach, see Skocpol (1995).

12. A key research text within sociology (Babbie 2001) supports the argument that qualitative and quantitative techniques are inherently different in design and method.

13. Examples abound. For a classic treatment, see Hall (1986), and for recent work Pierson (2000). Comparative studies can also occur within countries across different sectors; Jacob Hacker explores divergent responses in health care and pension policies within the US (2001).

14. Much of the data for our story came from in-person research interviews, which are listed in appendix 2. Given the primary importance of these interviews to our historical narrative, we have decided to make specific reference to these interviews in the text only when a direct reference was required, such as when quoting an official.

15. A well-known example of selecting on the dependent variable is Michael Porter's (1990) analysis of the rise of the Asian Tigers, which were all chosen because they fit a single value of the dependent variable (countries with a competitive advantage across a range of industries). As King, Keohane, and Verba (1994) point out, "Without a control group of nations (this is, with this explanatory variable set to other values), he cannot determine whether the conditions associated with success are not also associated with failure" (134).

16. The reader will note that four of our cases, Sweden, the US, the UK, and Germany are country wide, whereas BC focuses on a single Canadian province. The reason for this choice was that we felt it appropriate to identify territories in terms of the arena in which contests were taking place. We had originally chosen three US sub-regions, but quickly came to the realization that the competition was a national one, with regional dynamics playing a contextual role. However, we came to recognize that BC certification politics were equivalent to those in every other country we explored—the interests, actors, and institutions involved in certification in BC were unique and distinct from every other

province—rendering our choice to focus on BC as the appropriate comparison to our other case.

17. Given our conscious choice to engage in what McKeown (1999:188) calls "interactive processing" or "moving back and forth between theory formulation and empirical investigation," it could be asserted that we should expect such a process to lead to "fit" between identified causal processes (our hypotheses) and our empirical cases. While this is a valid critique, two points are in order. First, if our hypotheses "fit" our empirical story, then it is evidence that we did a good job of interactive processing. Second, our interactive processing largely centered on four of our five cases, as we chose not to explicitly incorporate the Swedish case until later in the research process. When we applied our argument and hypotheses to the Swedish case, we found that very little of argument had to be changed. While far from a pure "test" of our model, it does explain why we believe the "fitness" of the Swedish case helps illustrate the validity of our argument.

18. See Van Evera (1997:30) for a discussion of "smoking gun" and "hoop" tests important in qualitative comparative case study research.

19. Our identified factors that facilitate FSC efforts to gain support could be labeled either "intervening" or "independent." They are intervening in the sense that we focus on agent-based efforts to pursue support from forest companies and non-industrial forest owners; these factors mediate these pressures and shape variations in the dependent variable. They are independent in that they directly explain variations in the dependent variable (though these effects can only be activated by the existence of "antecedent" agent-based efforts to gain support). For simplicity, we have chosen to refer to them as independent variables, though it is the clear relationship we identify, not the specific label, that is important for our analysis.

20. There is debate in the literature about whether our approach can be considered to "test" hypotheses, in addition to raising them. McKeown (1999) argues that the debate has misdirected the most important task of social science research, which is not about whether one fits a pure definition of testing, but whether a researcher develops the most appropriate model for explaining empirical phenomenon—the best approach for researchers wishing to assist future research and knowledge building. McKeown argues that those focused on designing the perfect "test" often fail in "the goal of finding the model of a causal mechanism that best accounts for observations" (185). In order to escape the "test" debate, McKeown offers that qualitative case study research replace "test" with the "identification of causal processes" (185).

Chapter 3: British Columbia, Canada

1. The Canada-US Softwood Lumber Agreement exacerbated the effects of the Asian collapse because a strict export quota system to the US market meant that BC's hardest hit coastal companies could not turn to the US market for relief.

2. In 1996, the value of BC exports to Japan totaled C$6.4 billion. This decreased modestly to C$6.0 billion in 1997, but plunged in 1998 to C$4.5 billion; a drop that took Japanese imports from 22.4 percent to 17.5 percent of BC's total (BC Stats 2001).

3. For an in-depth discussion of the influence of the annual allowable cut and the tenure system see Cashore et al. (2001, chaps. 2 and 5).

4. This information comes from personal interviews with forest company officials in BC and with German publishers (see appendix 2).

5. The province carries out formal audits of BC forest companies through the BC Forest Practices Board, which has made one component of FSC certification — third party audits — a familiar task for BC companies.

6. The forest industry contributes to provincial revenues through stumpage, royalties, corporate income taxes, sales taxes, and municipal property taxes; between 1984 and 1993 the Ministry of Forests estimates that this contribution averaged 8.3 percent of total provincial revenues (BC Ministry of Forests 1994).

7. Unlike the highly litigious policy environment in the US that institutionalized environmental group access at the federal level, the BC government has more authority to decide the nature of its policy-making processes. Power in BC is concentrated both in a centralized Westminster model cabinet setting but also in the Ministry of Forests, which for the most part successfully fended off efforts by the Ministry of Environment, Lands, and Parks to gain an equal role in forest management in the mid-1990s (Wilson 1998).

8. BC members of the coalition included BC Pulp and Paper Association, Council of Forest Industries, and Interior Lumber Manufacturers' Association (Canadian Sustainable Forestry Certification Coalition 2000).

9. This group included the Confederation of Canadian Unions, the Pulp, Paper and Woodworkers of Canada Union, the Union of B.C. Indian Chiefs, the Canadian Environmental Law Association, Greenpeace Canada, and a number of others.

10. This is pointedly made by Stanbury (2000), but it also came up during personal interviews with numerous industry officials (see appendix 2).

11. Personal interviews with forest industry officials (see appendix 2).

12. While the very core goals of the FSC meant that it could not recognize competing programs, the reverse was true regarding forest company and landowner support for the FSC. That is, it did not matter for the FSC whether forest companies that agreed to operate under its rules also operated under other certification systems rules. What did matter was that retail companies only supported the FSC — the combination of retailers demanding FSC and companies agreeing to abide by its rules was what it sought — whereas the FSC competitors such as CSA sought the reverse — they wanted retailers to include the CSA in their certification procurement policies and did not want its company supporters to support the FSC.

13. Personal interviews, forest sector officials (see appendix 2).

14. Francis Sullivan with the WWF clarified the threat being posed by the WWF 95 group when he was quoted saying: "Canadian forest companies won't be able to sell to 24 of their biggest UK customers next year if they can't prove their products come from sustainably managed forests. The firms have aligned with the Forest Stewardship Council, which was established last year to accredit organizations around the world so they can "eco-label" products. That will reassure consumers wood and wood products come from known, well-managed sources (*Vancouver Sun*, April 9, 1994, H4, cited in Stanbury 2000:94)."

15. See Stanbury (2000) for an excellent account of the environmental NGO campaigns occurring throughout the 1990s.

16. Ninety-five percent of forests logged in BC in 1997 were in old growth forests (McKinnon cited in Greenpeace 1997). Primary forests are considered areas where industrial logging has not yet occurred. Depending on the definition of old growth, primary forests may not qualify. If 120 years was used as an age threshold to delineate the area of old growth forests, then approximately 43 percent of the province's forests would qualify. The remaining 57 percent of the forested landbase is split into young forests resulting from natural disturbance, 41 percent, and harvesting activities, 16 percent (Ministry of Forests British Columbia 2001).

17. Personal interview, official, Canadian High Commission, London, England, April 25, 2001.

18. Stanbury (2000:01) offers an example from an ad the Forest Alliance published in the UK-based *Daily Telegraph* that stated, "Greenpeace is not telling you the truth about the state of British Columbia's forests, or what really goes on here. It is time for facts, not half-truths and innuendo."

19. The "Stumpy" tour is one notable example. Greenpeace UK took a 400-year-old Western red cedar stump on tour in Europe to raise general public and customer awareness about the types of trees being harvested in BC. BC forest companies sent representatives to the UK to mend the damage done to BC's reputation (Greenpeace UK 1994).

20. Personal interviews, official, Western Canada Wilderness Committee, Vancouver, Canada, September 20, 2000, and official, Greenpeace, Vancouver, Canada, October 5, 2000.

21. Personal interview, senior official, *British Broadcasting Corporation Magazine,* London, England, July 3, 2001.

22. They contracted the Soil Growers Association, a UK-based FSC-accredited certifier, to perform a pre-assessment and develop an interim checklist for FSC certification in BC, as at the time there was no endorsed FSC-BC standard.

23. Personal interviews, senior official, Forest Alliance of British Columbia, Vancouver, Canada, September 19, 2000, and senior official, British Columbia Council of Forest Industries, Vancouver, Canada, September 1, 2000.

24. Personal interview, official from BC forest industry (see appendix 2).

25. Personal interview, official, Forest Stewardship Council British Columbia working group, Nelson, BC, Canada, August 8, 2000.

26. This was important because the FSC auditor indicated that without this commitment, it would have denied the certification.

27. Largely owing to the lack of FSC regional standards and company decisions to wait until they were complete, the vast majority of certified land in the province was under CSA approval. As of August 2001, 8,148 ha of BC forests were FSC certified (Certified Forest Products Council 2001b). The total amount of forest certified with the CSA is over 4 million hectares (Canadian Sustainable Forestry Certification Coalition 2001)

28. Some of these buyers also criticized CPPA for being too aggressive in promoting the CSA.

29. The principle now states: "Management activities in high conservation value forests shall maintain or enhance the attributes which define such forests. Decisions regard-

ing high conservation value forests shall always be considered in the context of the precautionary approach" (Forest Stewardship Council 1999b).

30. While there were a number of potential issues of "conflict" between the FSC P&C (principles and criteria) and BC's public forest policies (Haddock 2000), these were open to adaptation, whereas BC could not get around the issue of old growth forests. They were a physical reality that the BC companies and government had to contend with.

31. With the exception of rules governing US National Forests, BC's forest practices code riparian harvesting rules are roughly equal to or more stringent than riparian zone rules governing private forestland management in the US (Cashore and Auld 2003).

32. An internal FSC report, referring to tables in the draft standards, echoed these issues. "Table 1 specifies the thresholds for each category of stream/wetland/lakeshore. *These thresholds are consistently higher than those required by the Forest Practices Code.* This issue is not a significant one. Table 4 specifies the minimum budgets to be deployed at the Riparian Assessment Unit level. Utilization of this approach may result in buffer zones higher or lower than required by the Forest Practices Code and would require justification. *The significant issue is that this approach, while innovative and creative is untested at large operational scales and creates uncertainty in terms of potential costs (implementation and impact on timber supply), and overall effectiveness of these measures"* (FSC Canada 2002, italics added).

33. The opposing view was raised in a number of personal interviews with environmental group officials (see appendix 2).

34. The other economic member of the steering committee was a small woodlot owner.

35. Personal interview, official, Sierra Club of British Columbia, Vancouver, BC, Canada, August 2002.

36. This logic is illustrated nicely by the case of Scott Papers UK, which suspended a C$5.5 million contract with MacMillan Bloedel in 1994 because of the controversy surrounding MacMillan Bloedel's forest practices, while at the same time Scott Papers Canada published a full-paged ad in a BC newspaper with the headline "Boycotts are not solutions" and criticized Greenpeace for threatening jobs and communities on Vancouver Island and undermining BC's domestic land use decision-making processes (Stanbury 2000).

Chapter 4: The United States

1. The US produced 430 million cubic meters (27 percent of the global total) of industrial roundwood in 1999 (FAO 2001).

2. In 1999 US consumption of sawnwood exceeded 150 million cubic meters, paper and paperboard products approached 100 million metric tons, and wood pulp fell just short of 60 million metric tons (FAO 2001).

3. In nominal terms, the US was the world's largest importer of sawnwood, wood-based panels, wood pulp, and paper and paperboard products by volume 38 percent, 23 percent, 17 percent, and 16 percent, respectively (FAO 2001). Even though most of its production stays within the US domestic market, the sheer size of the US forest sector has resulted in it being among the world's leading forest product exporters. In 1999 the US exported over 6 million cubic meters of sawnwood (5 percent world total) and nearly 5.5

million metric tons of wood pulp (15 percent world total), giving it a rank of 5th and 2nd globally for these respective products. The majority of its forest product exports go to Canada and Mexico (see figures 4.1 and 4.2), but its other major market, Asia, has been in decline following the collapse of that region's economy.

4. Concentration of production along the supply chain varies depending on the specific forest product being manufactured. Georgia-Pacific, for instance, ranks first in its estimated share of the total North American tissue and structural panel production capacity, 40 percent and 17 percent, respectively (Georgia-Pacific 2001).

5. For example, International Paper, the world's largest forest products company, owning and managing approximately 4.2 million hectares, only harvests enough timber to meet 28 percent of its wood fiber requirements (International Paper 2002). Boise, Louisiana-Pacific, Bowater, and Potlatch, respectively, obtain 42 percent, 11 percent, 18 percent, and 50 percent of their fiber needs from their own lands (Boise Cascade 2002; Bowater 2002; Louisiana-Pacific 2002; Potlatch 2002).

6. Commercially productive "Forest land that is producing or is capable of producing crops of industrial wood and not withdrawn from timber utilization by statue or administrative regulation. (Note: Areas qualifying as timberland are capable of producing in excess of 1.4 cubic meters per acre per year of industrial wood in natural stands. Currently inaccessible and inoperable areas are included.)" (Smith and Sheffield 2000).

7. The AF&PA was created by merging three associations, the National Forest Products Association, the American Paper Institute, and the American Forest Council (Cashore 1997).

8. The Endangered Species Act places a much greater role on federal forest landowners than on private forests landowners (Cashore 2001a).

9. Personal interview, company official, Champion International, Jacksonville, FL, July 6, 2000.

10. Personal communication, Scott Wallinger, March 27, 2003.

11. Personal interview, company official, Champion International, Jacksonville, FL, July 6, 2000.

12. Personal interview, national environmental group, September 17, 1999.

13. SmartWood's choice to undertake regional standard setting bodies can be traced to their concern that rules be sensitive to very diverse forest types within North America.

14. Such an approach veered dramatically from the AF&PA approach that views standard setting and auditing as necessarily separate activities. The ability of FSC verifiers to also influence standards was deemed an unacceptable practice by most industry observers.

15. Personal interview, official, Oregon Forest Industries Council, Salem, OR, September 13, 2000.

16. As one official noted, "If FSC is marginalized to have one thing over here and one thing over there it's not doing landscape conservation. It's not done squat . . . the point that we are trying to drive home is landscape level conservation, and the only way that we are going to get there is if we have a lot of people sign up."

17. Research has documented that willingness to pay varies with both consumer and product characteristics. Higher levels of willingness to pay are attributed to consumers who are female, well educated, and in high income brackets and who support the Demo-

cratic Party and live on either the East or West Coast or in urban rather than rural areas (Ozanne and Vlosky 1997; Forsyth, Haley, and Kozak 1999). High price items generally are ascribed lower percentage premiums.

18. The CFPC merged the Good Wood Alliance and the North American buyer group started by Environmental Advantage (World Wildlife Fund 1998; Lyke 1996).

19. In the US Southeast, the FSC established an extensive stakeholder identification process in which 3,600 individuals were identified and invited to three regional workshops, from which many of the representatives forming the regional working group were elected.

20. One AF&PA member did express an interest in being involved in the FSC Southeast US working group, but withdrew the offer after speaking with more senior company officials.

21. Personal interview, official, Natural Resource Defense Council, San Francisco.

22. Personal interview, AF&PA, October 17, 2003.

23. Personal interview, member of the Forest Stewardship Council, Pacific Coast regional standard working group (see appendix 2).

24. Personal interview, member of the Forest Stewardship Council, Pacific Coast regional standards working group (see appendix 2).

25. 11.3 million hectares of the national forests (48 percent of the region's timberland) and 2.4 million hectares of other public forests (10 percent of the region's timberlands) (United States Forest Service 2000).

26. This point came up in personal interviews with social members of the Forest Stewardship Council Pacific Coast standards working group (see appendix 2). Trinity County, California, typifies this problem, as nearly 80 percent of the county land area is owned by the federal government, much of it being National Forests (Danks 1996). High levels of unemployment and a tendency for the value derived from forest production to be realized by interests located outside counties such as Trinity made the initial conception of community-based eco-forestry appealing. Not allowing certification of National Forests, while not hugely important in terms of volume of wood produced, did have grave significance to many local stakeholders in the West who saw the FSC as a vehicle for local empowerment and economic growth and diversification.

27. The FSC P&C (principles and criteria) stated that, "Plantations established in areas converted from natural forests after November 1994 normally shall not qualify for certification. Certification may be allowed in circumstances where sufficient evidence is submitted to the certification body that the manager/owner is not responsible directly or indirectly for such conversion" (Forest Stewardship Council 1999b). The working group chose to remove this criterion (10.9), stating that "[this] criterion does not 'raise the bar' for stewardship; it is simply exclusionary. It arbitrarily reduces the number of people who could ever participate in the program and, therefore, serves to marginalize the impact of this program. The FSC should reconsider what it is trying to accomplish with this criterion, and should, at least, change the date to that point in time when FSC standards for the region are endorsed and published" (Forest Stewardship Council, U.S. Southeast Working Group 1999).

28. Personal interview, official, Natural Resource Defense Council, San Francisco.

29. This program allowed non-AF&PA members to take part in the SFI program (by

the end of 2001, seventy-seven licensees had enrolled, adding 9.3 million hectares to the program's total) (AF&PA 2002b).

30. It would take almost four years to develop the 1999 draft standard in the Pacific Coast region, which heightened some groups' disappointment in the process and led many observers to feel that the FSC was not working on "business time."

31. Personal interview, senior official, American Forest and Paper Association, Washington, DC, February 4, 2000.

32. By the 1970s Tree Farm was broadened to promote "Multiple Use Forest Management" and now focuses on promoting its conception of "sustainable forestry."

33. An AF&PA press release described the decision of Fraser Papers, a company with holdings in both the US and Canada, to certify their Canadian lands under the SFI: "The SFI program has taken a monumental leap forward . . . With Fraser's announcement, today the SFI program has solidified its position as an international forestry standard" (AF&PA 2000f).

34. There were, however, instances where informing strategies ran counter to their intentions, especially when the strategy involved publicizing the SFI program's environmental "achievements." SFI proponents have actively advertised awards the program has won from environmentally oriented organizations (e.g., Renew America and the President's Council on Sustainable Development [AF&PA 1999]). However, in one instance, the AF&PA ran an advertisement announcing the SFI program had won an Association Advance America award (American Society of Association Executives 1999; AF&PA 1999), an award that had been granted without some members of the American Society of Association Executives knowing — most notably, certain environmental groups. When the AF&PA advertisement was released it included the names of these environmental groups, implying they were endorsing the SFI program. This provoked a strong reaction as groups such as the Sierra Club denounced the award and castigated the SFI program (Sierra Club 1999). As one Sierra Club spokesperson stated, "If they were truly friends of the environment, they wouldn't be sending highly paid lobbyists to Washington to weaken forest protection laws. They do not deserve an award, nor do they deserve patronage from customers — they simply deserve scorn (Sierra Club 1999)."

35. Personal interview, senior official, World Wildlife Fund, Washington, DC, February 3, 2000.

36. Personal interview, senior official, American Forest and Paper Association, Washington, DC, February 4, 2000.

37. TREES stands for Training, Research, Education, Extension, and Systems.

38. State and municipal owners of forestlands control 7 percent of the total US timberland base and less than 6 percent of the commercial harvest (United States Forest Service 2000).

39. A landowner choice to pursue both FSC and the FSC competitor does more for FSC legitimacy efforts than the competitor, since the competitor has an interest in being an alternative to the FSC, rather than being certified in conjunction with the FSC. Likewise, since the FSC legitimacy achievement strategies are focused on having their rules adopted, it does not matter, for legitimacy achievement purposes, whether landowners supporting the FSC also choose to certify with another program as well.

40. The report revealed that a forest tract managed by the USDA Forest Service for the

Department of Energy in South Carolina was seeking quotes to have their lands assessed against the FSC P&C (Mater et al. 1999). Public lands certification—particularly Forest Service lands—is troubling to particular environmental groups such as the Sierra Club and is viewed with concern given the potential precedent it could set.

41. Personal interview, official, The Wilderness Society, Seattle, Washington, September 14, 2000.

Chapter 5: The United Kingdom

1. In the 1980s support for the forest sector waned, as environmental and financial concern over the impacts and rationale for industrial afforestation reached the public agenda (Bainbridge et al. 1987; Turner 1992, 1991; Ghazi 1994a).

2. Only 15 percent of the UK's 48.3 million cubic meters of consumed timber are harvested domestically (Forestry Industry Council of Great Britain 2000). In 1996 the UK ranked second in the world, behind Japan, in terms of net value of forest products imported. In 1999, in the solid wood product market, imports were required to meet 52 percent of wood-based panel demand and 75 percent of sawnwood demand, and for pulp and paper markets imports met 78 percent of wood pulp demand and 60 percent of paper and paperboard demand.

3. 7.108 million cubic meters (planks, beams, joist, boards, rafters, scantlings, laths, boxboard, "lumber," etc.).

4. 2.847 million cubic meters (veneer sheets, plywood, particle board, and fiberboard).

5. Brazil, Indonesia, and Malaysia account for over 50 percent of the total volume of plywood imported into the UK (FAO 2002). Official statistics reveal that they produced 21 percent of the total value of all UK wood and cork manufactured imports in 1995 (OECD 2001), though others argue that this estimate is low because of the amount of illegally harvested forest products that comes from these countries (Ghazi 1994b).

6. The exception is Tilhill Economic Forestry owned by Shotton Paper Company, a subsidiary of UPM-Kymmene (Tilhill Economic Forestry 2002).

7. In 1997 domestic production of medium density fiberboard, particleboard/chipboard, and oriented strand board controlled 66 percent, 63 percent, and 65 percent of British market, respectively (Forestry Industry Council of Great Britain 2000).

8. FICGB represents constituents from the growing community (e.g., the Timber Growers Association), forest management community (e.g., the Association of Professional Foresters, Institute of Chartered Foresters, and Scottish Woodlands), timber merchant and wood product processing (e.g., the Wood Panel Industry Federation and Iggesund Panelboard), and education, training, research, and learned societies (e.g., the Scottish Council Development and Industry and Aberdeen University Department of Forestry) (Forestry Industry Council of Great Britain 2000).

9. Immediately after its creation in 1987, the FICGB released a report entitled *Beyond 2000: The Forestry Industry of Great Britain,* and two years later (1989) it published *The Impact of Forestry in the United Kingdom.* These publications set out the challenges facing the forest sector and responded to the findings of the NAO report arguing that social, economic, and environmental benefits conferred by forest management activities had been ignored by the NAO's conclusions (Forest Industries Development Council 2001).

10. On the ecological side, forest conversion to non-forested land uses and fragmentation were threatening ancient woodlands, and afforestation that created forest plantations in ecologically sensitive flow lands of Caithness and Sutherland in the North of Scotland placed pressure on endangered bird species reliant on open habitat (Bainbridge et al. 1987; Foottit 1989). Walking and hiking associations (known as ramblers in the UK) voiced concern over access to lands under threat of privatization (Mead 1993).

11. Until 1999 UK forest regulations were developed by the Forestry Commission, which was responsible to the Forestry Ministers of Great Britain, and the Forest Service for the Department of Agriculture for Northern Ireland (Forestry Authority 1998). After 1999, constitutional reform devolved responsibility over Scottish and Welch forests previously held in Westminster to the Scottish Ministers and the National Assembly of Wales. The Forestry Commission remains in existence, but its board of commissioners now receives direction from the three separate governments (United Kingdom 2001).

12. The Forestry Commission has regulatory authority over tree felling in almost all circumstances, on both private and public lands. The exceptions include trees in gardens and public open spaces, trees under 8 cm and 15 cm for coppice at diameter 1.3 m above the ground, and trees that are obstructing developments that have planning permission. Also, owners have the right to fell, on a quarterly basis, up to 5 cubic meters of wood without any permission (Forestry Commission, United Kingdom 1998; Tickell 2000).

13. While other bodies, such as local authorities and statutory conservation agencies, play minor roles in regulating forest practices, they did not figure prominently as factors influencing the direction in which certification developed in the UK.

14. During the late 1980s the domestic forest sector struggled to justify its operations on a financial and ecological basis due in large part to the increased scrutiny of the National Auditing Organisation and large conservation organizations such as the Royal Society for the Protection of Birds. In 1993, this attention culminated with the threat of privatization. Both environmental groups and the forest sector opposed privatization (Mead 1993; Christie-Miller 1995). The government left the commission intact, but transformed the Forest Enterprise into a "next step agency" with a greater business-like focus (Pringle 1994).

15. Groups like Friends of the Earth did not give up on their boycott campaigns, targeting B&Q even after it had committed to FSC certification (A. Knight 2000).

16. Alan Knight, who would become an integral part of the development of the FSC internationally and domestically in the UK, explained the origins of B&Q's interest in forest certification: "What we recognized was that the [market] pressure at that time was pretty low [but] that it was going to increase and we had to give more intelligent and sensible answers to what we were doing. . . . We weren't sure if it was a six-month project just to do a good bit of PR or something bigger. I then started off asking the same old awful questions to our suppliers, which we had over a hundred. Only to find that many couldn't answer themselves, they themselves didn't know where their timber was coming from. And we had this bizarre statistic that over half our suppliers couldn't tell us where their timber was coming from, from the country level. But 90 percent of them reassured us they were from well-managed forests in the same sort of survey. More than that over 25 percent of our suppliers were putting some form of claim or reassurance of their products about forestry. . . . I then went to the trade association particularly the British

Timber Association who went out of their way to say that the environmental groups have got it all wrong and that the timber is perfectly all right. And in fact you should buy more timber because the more timber you buy the more forests you protect and by the way can we have 25,000 pounds for membership so that we can sort of pay our campaign. Then I went to people like Friends of the Earth who basically said all tropical timber is bad, you shouldn't buy it immediately, you should boycott it, and actually we are planning to really shaft you if you don't stop buying it. So then you're sort of stuck." (personal interview, Alan Knight, London, July 2, 2001).

17. The ITTO comprises twenty-three tropical producer countries and twenty-seven tropical consumer countries. Its key objective, as Humphreys notes, is "the expansion and diversification of international trade in tropical timber" (Humphreys 1996b:221).

18. Bernstein and Cashore (2001) cite Ross (1996), who found that the ITTO committed only US$46 million from 1989 to 1991 in implementation of Target 2000.

19. Environmental groups originally proposed that certification be undertaken through the auspices of the ITTO, but groups like the Friends of the Earth and the World Wildlife Fund came to the conclusion that a process outside and independent of government would be more likely to achieve sustainable forest management goals (P. Knight 1996; Elliott 1999).

20. In a 1994 report commissioned by the WWF, clear criticism was directed at UK timber importers, represented by the Timber Trade Federation (TTF) (Read 1994). The report noted, "Only 84 TTF members could identify the countries of origin of their suppliers. No more than 52 claimed that they could 'trace some of their sources in more detail' and as few as four claimed all their timber could be traced to this level of accuracy" (7). This was a central concern of the environmental groups such as the WWF, as they and others lacked political avenues to influence the practices occurring in these supplier nations (Cabarle et al. 1995; Dudley, Jeanrenaud, and Sullivan 1995).

21. Many of these member companies were simultaneously involved in plans to create the FSC (A. Knight 2000; Viana et al. 1996).

22. Elliot Morley, member of the British Parliament, asserted, on June 3, 1992, that "I welcome the opportunity to make some comments while the Rio summit is under way and to encourage the Government to press for some much needed policies in an important event of this kind. However, the Government must recognize that every member state at the conference must put its own house in order and they must ensure that each of us has a responsibility for dealing with environmental issues in our own country; rather than simply pointing the finger at the more spectacular cases of environmental destruction—especially those in the developing countries, such as the destruction of the rainforests in Latin America" (Morley 1992).

23. Initially, due to limited interest in Northern Ireland the FSC set up a working group to develop standards for Great Britain. Later, this changed because of the participation of the representatives from Northern Ireland participating in the development of the United Kingdom Woodland Assurance Scheme.

24. As a reporter at the time noted, the WWF had a clear vision such that "once the market [had] been primed it [would] favour those products bearing an eco-label and exclude anything that [did] not. Good producers will win, bad will lose" (P. Knight 1996).

25. Timbmet, who joined the WWF 95+ Group in late 1999 (Timbmet Group 2000),

had been subject to campaign pressure during the mid 1990s due to its role as an importer of wood products from tropical and temperate forests (personal interview, company official, Timbmet, Oxford, England, April 30, 2001).

26. Convincing companies of future increases in FSC market demand was a key part of the WWF's efforts (personal interview, senior official, World Wildlife Fund, Godalming, England, April 20, 2001). Initially the organization focused on convincing retailers to support the FSC, which in part offered incentives for timber brokers and merchants to get involved. The WWF also published information on the level of retail demand in order to elicit the cooperation of companies further up the supply chain. Crossley (1996) reported that in 1995, the WWF released a report estimating that the potential market for FSC wood in Europe was approximately US$3 billion. Such information played an important role in convincing companies that, while the market was not yet developed, market advantage would emerge eventually (World Wildlife Fund United Kingdom 2001b). The reputation of the WWF was so strong that some of these businesses were simply content to do what WWF told them to do (personal interview, company official, Beacon Certified Timber, Edinburgh, July 5, 2001; Webster 2000).

27. Peter Wilson expressed this view when he wrote the following about the first WWF "Forests for Life" conference held in the summer of 1996: "I am not sure what the average WWF supporter would have made of last month's affair in Brussels, but despite the title, the seminar is basically akin to a well orchestrated Nazi rally save that instead of the Third Reich it is the Mexico based Forest Stewardship Council that is being promoted. The problem from WWF/FSC's point of view (for it is difficult to determine quite where WWF finishes and FSC starts) is that the audience is not quite as well schooled as it might be" (1996:9).

28. As detailed in chapter 1, initially, the FSC's social and environmental chambers shared 75 percent of the voting rights, leaving economic interests with only a 25 percent stake (Moffat 1998; Upton and Bass 1996). Voting rights have been a cause of controversy for both economic and social audiences (P. Knight 1996). In 1993, Greenpeace withdrew from the FSC in protest when economic audiences were given voting rights. They were reported to say: "The FSC will be a naïve venture, and at worse, it will legitimize efforts by business to acquire no more than a green veneer" (P. Knight 1993).

29. This view was repeated during personal interviews with officials from the Association of Professional Foresters (London, April 25, 2001) and Timber Growers Association (Edinburgh, May 4, 2001). See also Christie-Miller 1994; Timber Grower 1994a.

30. This program was developed and released just after the release of the Soil Association Woodmark program, which provoked some to conjecture that it had been designed purely to undermine the FSC-GB national initiative. While many agree that the FICGB Woodmark scheme had this in mind, observers reported that it was a pure coincidence that the two programs shared the same name (personal interview, participant in development of FICGB Woodmark).

31. BM Trada offered certification services that assessed the implementation of this scheme (Certification Information Service 2001).

32. One exception was Scottish Woodlands. It took part in the FSC-GB process by helping to field test the standards during their early development stage (personal interview, company official, Scottish Woodlands, Edinburgh, May 9, 2001).

33. Recognizing the role of government as a landowner and facilitator, UK FSC officials diverged from the international body by encouraging governmental participation. In fact, as early as 1995 they requested the Forestry Commission to chair their standards setting process, but private sector opposition at the time led them to decline this invitation. In the end, the Forestry Commission of Great Britain became an "observer-member" of the FSC-UK working group.

34. Other businesses and organizations that participated in the UK FSC standards setting process included: Boots the Chemists, Charles Bentley and Son, Chindwell Co., David Craig, Do It All, Ecological Trading Company, Environmental Investigation Agency, European American Industries, F. W. Mason and Sons, Fauna and Flora International, Fern, Forest Management Foundation, Friends of the Earth, Friends of the Earth Scotland, Habitat UK, Jac by the Stowl, John Dickinson Stationary, Overseas Development Institute, Premium Timber Products, Reforest the Earth, SGS Qualifor, Shireclose Houseware, Soil Association, Woodland Trust, and WWF-UK.

35. Information on the exact quantity of wood consumed by local authorities was not available, however most observers agreed that they were an important consumer of wood given their role in municipal development projects.

36. Personal interview, official, World Wide Fund for Nature, Matlock, Derbyshire, England, May 11, 2001.

37. The program began as a joint effort between the Soil Association and the WWF, but later became the sole responsibility of the WWF.

38. The WWF also recommended that each local Authority have their procurement policies reviewed by legal staff to avoid legal conflicts (World Wildlife Fund United Kingdom 2001b).

39. Personal interview, company official, B&Q, London, July 2, 2001.

40. Personal interview, company official, B&Q, London, July 2, 2001.

41. Personal interviews with officials from the Wood Panelboard Industry Federation, Grantham, England, April 25, 2001; the UK Forest Products Association, Stirling, Scotland, May 3, 2001; Nexfor, Cowie, Stirlingshire, England, May 8, 2001; and the Forest Industries Development Council, Edinburgh, May 8, 2001.

42. The strong British pound (see, e.g., Sutherland 2000), falling timber prices, and low-cost wood coming from competitor countries (Bills 2001) all contributed to concerns on the part of UK forest landowners that the FSC commitments from foreign competitors could hasten domestic market erosion.

43. The first FSC policy was released in October 1997. It required solid wood products carrying the FSC label to be made from 100 percent certified wood and manufactured goods from chip and fiber products to include a minimum of 70 percent FSC certified wood by dry weight (Forest Stewardship Council 2000b).

44. The private sector, and in particular the Timber Growers Association, saw the Forestry Commission as an ally in their efforts to rebuild public support for British forestry; and thus publicly opposed the proposed privatization of the Forest Enterprise (see, e.g., Ghazi 1994a).

45. One report published by the Royal Society for the Protection of Birds noted that "with the advent of Indicative Forestry Strategies, improvements in consultation pro-

cedures and changes to the Woodland Grant Scheme, much of the unproductive confrontation has left the debate between foresters and conservationists. Increasingly, the common ground between the two sides is being recognized and extensive consensus on the future development of forestry is becoming a possibility" (Marshall 1996).

46. As one report released by the Environmental Investigation Agency and Telepak Indonesia noted, "Even if you could track an illegally cut tree from a National Park in one of these countries to a port in a timber consuming country, and supply conclusive evidence that it was illegally cut, none of the consuming countries have legislation in place that would allow their enforcement authorities to seize the shipment" (Currey 2001).

47. While the FSC had originally intended to develop standards for all of the UK, the lack of participation from stakeholders in the Northern Irish Republic made them choose to limit the standard's scope to Great Britain (Scrase et al. 1998).

48. In addition to comparing the standards, the report assessed the changes needed to make the UK Forestry Standard auditable. The recommendations mainly focused on the need for clarifying vague and ambiguous wording, the existence of poorly specified requirements, and the general structural problems associated with the Standards Notes representing individual, but overlapping, guidance documents for different woodland types (SGS Forestry 1997).

49. Mike Garforth, who represented the commission in the national FSC working group and subsequently facilitated the UK Audit Protocol process, had developed a keen knowledge of the participants in each process, which helped to generate a friendly atmosphere surrounding these discussions (personal interview, officials, Forestry Commission, Edinburgh; Webster 2000).

50. Various officials involved in development of UKWAS made this point during personal interviews (see appendix 2).

51. Some observers noted that key concessions were made by all participants; most notably, Friends of the Earth were reported to have dropped their demands that chemical use be prohibited, which clearly would have made the standards appear impractical to many landowners and managers (personal interviews with officials involved in developing the UKWAS; see appendix 2).

52. Some insiders noted a rift between WWF market-based strategies and the approach of the FSC in the UK, which focused on the FSC as a "service provider" (personal communication, former FSC official March 15, 2003). FSC officials in the UK did not want to be part of the rift between the WWF and TGA, which they felt hindered their efforts to gain broader domestic support.

53. The chemical issue was handed off for further discussion by the UKWAS Steering Group (UKWAS Steering Group 2000). Division surrounding pesticide use was also an issue during the development of the FSC-GB standard, with a group called "Reforest the Earth" opposing what it felt were too liberal provisions, and Friends of the Earth (England, Wales, and Northern Ireland) all abstaining on this policy until a working group on pesticides reported its findings (Scrase et al. 1998).

54. Personal interviews, officials participating in the UKWAS process (see appendix 2); Forestry Commission, Great Britain, 1999a and b; Goodall 2000.

55. Since the 1970s, the Soil Association has been a key certifier of organic agriculture:

during periods of the 1990s it accounted for up to 70 percent of the certifications occurring in the UK (Wenban-Smith 2001). Its public profile benefited the FSC in its early attempts to gain public recognition for its activities.

56. The government's release of this policy traces back to pressure from environmental groups (see, e.g., Sollander 1999), and to international commitments and discussions on illegal logging held by members of the G8 (Foulkes 2001).

57. Personal interviews, officials with Nexfor, Cowie, Stirlingshire, England, May 8, 2001; and Wood Panelboard Industry Federation, Grantham, England, May 11, 2001.

58. It reduced the thresholds for solid wood products to a minimum of 70 percent FSC certified wood and not more than 30 percent non-certified sources. For chip and fiber products and components the threshold was lowered to a minimum of 30 percent FSC virgin fiber and 17.5 percent FSC of the total fiber measured by dry weight (Forest Stewardship Council 2000b). The negotiations involved the FSC taking account of the views voiced by an informal technical committee composed of interested European stakeholders, including officials from a number of firms with manufacturing facilities in the UK (Forest Stewardship Council 1999e). Unlike the debates over management standards, these policy deliberations were not consensus oriented; they were reported to have involved "tough bargaining" over the issues at hand (personal interviews, company official, Nexfor, Cowie, Stirlingshire, England, May 8, 2001).

59. As early as 1997, this arose as an issue for the supply side, which felt that, one, the FSC's policies on recycled wood contradicted the organization's interest in conserving forests and, two, the policy was impractical for processors who drew on multiple wood fiber sources.

60. The products included oriented strand board, medium density fiberboard, chipboard, plywood, hardwood, particle board, and construction grade timber, all of which were made available in the UK market in certified form due in large part to the change in percentage based claims threshold (World Wildlife Fund United Kingdom 2000).

61. As one beneficiary of free certification noted: "It is questionable whether we would have gone through the process had we had to pay the market rate for assessment and audit of the woodland" (Adlard 2001).

62. Following its involvement in the development of UKWAS and partly while the process was ongoing, the FSC working group reworked its own standard, taking account of some of the lessons learned from the broader consultative process. Subsequently, it took to gaining the endorsement of stakeholders in the Northern Irish Republic, which enabled it to submit a FSC-UK draft standard to the FSC secretariat and board for formal endorsement, hence the change in name and scope of the FSC national standard from GB to the UK (Scrase et al. 1998).

63. While Forest Check has been accredited by the UK Accreditation Service (UKAS) to certify forest management to the UKWAS standard, they do not provide a label. At the time, the UKWAS had no on-product label independent of the FSC's: operators audited by an FSC-accredited certifier could use the label, others could not.

64. Alan Knight recently argued in a speech to a forest trade conference in Atlanta in April 2002 that the FSC and its competitors need to stop their battles and focus on the original sustainability goals for which they were intended.

Chapter 6: Germany

1. The federal government manages approximately 2 percent of Germany's forests; this amount is included in state lands.

2. Fragmentation occurs when formerly continuous forest is broken up into smaller parcels due to conversion of forest to other uses.

3. In addition, it is estimated that there are an additional 1 million Germans who own less than 1 hectare of forestland (Volz and Bieling 1998).

4. In the 1980s, the *Waldsterben* (forest dieback), a widespread thinning of tree tops eventually leading to tree death in many cases, was the issue that began the ongoing discussion about the effects of air pollution of forest ecosystems (Schraml and Winkel 1999). Although this condition is thought to be related to the predominance of even-aged monocultures in Germany and was given widespread media attention in the eighties, it does not hold the same imagery and direct connection to forest practices that large-scale clearcuts in Canada or slash-and-burn forestry in tropical countries did.

5. Interestingly, this pulp and paper sector, which imports a high proportion of its products, has been the segment that was most active in bringing the discussion of forest certification to Germany and making concrete demands from its suppliers both domestic and abroad, giving indirect support to hypothesis 1.

6. Personal interview, Bavarian Landowner Association member, Bavaria, Germany, April 20, 2001.

7. These data come from several interviews conducted with a number of officials involved in forest certification in Germany (see appendix 2) and from Anonymous 2001e.

8. The Naturland program would go on to be recognized as meeting FSC standards in 1998, allowing all Naturland-certified forest products to bear the FSC label (http://www.naturland.de).

9. Personal interview, Bavarian Landowner Association member, Bavaria, Germany, April 20, 2001.

10. This critique is made even more threatening to state forest agencies as their financial and personnel resources have been reduced in recent years (Schraml and Winkel 1999).

11. Personal interview, German Forestry Association, Rheinbach, Germany, March 21, 2001.

12. Personal interview, March 2001.

13. The reference to traitor was made on a number of occasions during research interviews conducted for the German case (appendix 2).

14. Many companies had stopped importing tropical wood products in the late 1980s due to environmental group and public pressure (e.g., Praktiker 2001; Anonymous 2001c).

15. Personal interview, Bavarian Landowner Association member, Bavaria, Germany, April 20, 2001.

16. Personal interview, Mainz, Germany, March 23, 2001.

17. The Naturland program existed in Germany in addition to the FSC but was not seen as a major threat to forest owners for reasons discussed earlier.

18. Personal interview, Mainz, Germany, March 23, 2001.

19. The Helsinki Process, which began in 1990, developed the general guidelines for the sustainable management of forests in Europe. The process has sought to identify measurable criteria and indicators for the evaluation of how European countries have progressed in their efforts to follow the principles of sustainable forest management and conservation of the biological diversity of European forests. As part of this process, from June 2 to 4, 1998, ministers responsible for forests in European countries gathered in Lisbon, Portugal, for the Third Ministerial Conference on the Protection of Forests in Europe; the outcome of the conference was two declarations and two resolutions regarding sustainable forestry in Europe.

20. While funding was eventually also given to the FSC (e.g., World Wildlife Fund Germany 2000b), the amount allocated to each program was based on that program's certified area, giving the PEFC an advantage. The *Holzabsatzfond* has since declared its neutrality vis-à-vis certification and does not give any funding to certification programs.

21. Personal interview, March 2001.

22. Personal interview, Germany, March 2001.

23. The logic being that profit maximizing firms operating in a competitive market will choose, everything else being equal, the less rigorous, or PEFC standard. As a result, and on the other hand, the FSC's success in gaining forest manager support hinges on it being the only certification program accepted by the purchasers of certified wood.

24. Personal interview, Germany, March 2001.

25. This view was expressed from a number of personal interviews listed in appendix 2 as well as in Anonymous 2001b.

26. Personal interview, Working Group for Natural Silviculture member, Germany, March 21, 2001.

27. Personal interview, Working Group for Natural Silviculture member, Germany, March 21, 2001.

28. There are approximately 10.7 million hectares of forestland in Germany (Hofmann et al. 2000).

29. Personal interview, Working Group for Natural Silviculture member, Germany, March 21, 2001.

Chapter 7: Sweden

1. Industrial companies own less than 20 percent of sawmills, with independently owned companies producing the vast majority of sawn products.

2. The bulk of the data from this paragraph comes from Leif Öster, Information department, Sveaskog, as reported by telephone to Tage Klingberg, University of Gävle, Sweden, Friday, March 28, 2003. Other data sources include an in-person interview with a Sveaskog official in Oaxaca, Mexico, November 2003, and from van Kooten, Wilson, and Vertinksy (1999:170).

3. This features was true for pulp and paper and sawmill production. It is true for pulp and paper because it relies on chips to produce paper, which necessarily mixes FSC and non-FSC sources (chips are often residuals from sawmills, so unless the sawmill produced

100 percent certified FSC wood, it is even more difficult). It is also true for sawmill production because of a "swapping" system that exists in Sweden in which trees harvested from one company's sites will be traded with trees harvested from another company or landowner's site that is closer to the company's mill. This system greatly reduces transportation costs, but also makes it more difficult for an FSC-certified company to harvest its own FSC-certified sources—as it would require significantly increased costs and inefficiencies (not to mention being worse for the environment owing to energy costs of transportation).

4. The limited amount of old growth forests were certainly subject to considerable scrutiny (Swedish Society for Nature Conservation 1996). Our point here is that they were not as important to forest conservation debates as in regions such as BC or the US Pacific Northwest.

5. van Kooten et al. (1999:170) reports that regional forestry boards tend to focus on technical issues, are more oriented toward forest owner clientele than the general public, and prefer to use "persuasion and education" over "reliance on regulatory instruments."

6. The Swedish government invoked a Forestry Commission in 1993 with which to conduct this process (Elliott 2000).

7. We are grateful to Tage Klingberg for this clarification.

8. AssiDomän was a founding member of the FSC, in part because it perceived an opportunity for economic advantage (personal interviews, AssiDomän/Sveaskog, Oaxaca, Mexico, November 2002). However, such membership did not translate into immediate support for the FSC domestically for two reasons. First, agreeing to support specific standards to which it must comply is different than supporting in principle, responsible forest management—which was all that was required to be a founding member of the FSC. Second, the Swedish forestry association had a longstanding history of speaking with one voice when it came to domestic forestry issues. AssiDomän would have to convince other members of the sector before it could offer its support.

9. Personal interview, first secretary of US FSC national initiative, Oaxaca, Mexico, November 25, 2002. See also van Kooten et al. (1999:173).

10. Personal interviews and Moller (1996) and Terstad (1999).

11. Personal interviews, senior officials, Swedish Forest Owners Association, Geneva Switzerland, October 18, 2002.

12. Personal interviews, senior officials, Swedish Forest Owners Association, Geneva Switzerland, October 18, 2002.

13. Landowner officials involved in the standards discussion identified four factors as key to their withdrawal: (1) the Sami issue; (2) the setting of the "altitude line," where there could be no forest management, in which it was asserted that the standards set a line lower than what scientists had prescribed; (3) the kind and number of trees dictated to be left in the final cut; and (4) sentiments about governance, arguing that there was no due process and that it was dominated by urban elites from the outside telling forest owners how to run their business (personal interviews, senior officials, Swedish Forest Owners Association, Geneva Switzerland, October 18, 2002).

14. Telephone interview with Forest Trade Network Official, January 24, 2003.

15. Personal interview, Korsnäs official, Oaxaca, Mexico, November 24, 2002.

16. The solid wood threshold was reduced to 70 percent; the chip, fiber, and component threshold to 30 percent FSC wood for virgin fiber; and to 17.5 percent FSC wood for all fiber (Forest Stewardship Council 2000b).

17. Sodra immediately pursued EMAS for its pulp mills following the landowner withdrawal from the FSC system (Elliott 2000).

18. Personal interview, PEFC official, Luxembourg, October 22, 2002.

19. As revealed in chapter 5, the PEFC in the UK accepts the UKWAS standards, which are also the same as those for the FSC.

20. Billboards with actor Pierce Brosnan, "007," advertised the FSC logo in the UK and Germany at the end of December 1999.

21. The FSC's headquarters moved to Bonn, Germany, in January 2003.

22. Personal interviews, FSC officials, Oaxaca, Mexico, November 25–27, 2002. Korsnäs officials have explained that if the FSC does not change its policy, it would be easier for them to simply support the PEFC (since the PEFC recognizes the FSC) than to no longer participate in the Swedish industry wood swapping system (personal interview, Korsnäs official, Oaxaca, Mexico, November 25, 2002). Korsnäs officials have also noted that they are facing a conundrum between ISO rules, which indicated they must be efficient in their operations, and FSC chain-of-custody tracking rules (the swapping system is efficient and environmentally friendly as it reduces transportation and hence carbon dioxide emissions).

23. Officials on all sides in the UK, and particularly those who participated from the Forestry Commission, noted that had the PEFC existed before the talks on the UKWAS standard commenced, the choices made by the commission and others would likely have been significantly different (personal interviews). They explained that the FSC would have been unable to obtain the dominant position that it did.

24. Personal interviews, FSC officials, Oaxaca Mexico, November 25–27, 2002.

25. Other supporters of the FSC include WWF Sweden, the Swedish Society for Nature Conservation, Swedish Forest Industries Association, Taiga Rescue Network, Swedish Association for the Protection of the Environment, Swedish Wood Industry Workers Union, Swedish Union of Forest Workers, National Union of the Swedish Sami People, Swedish Youth Association for Environmental Studies and Conservation, Friends of the Earth, Swedish Ornithological Society, Forestry Society (Skogssallskapet), Swedish Association for Hunting and Wildlife Management (SAHWM), Swedish National Property Board (Elliott 2000).

Chapter 8: Competing for Legitimacy

1. This is not to argue that a select number of small NIPF owners are not ideologically disposed to supporting the FSC because it fits with their own values. Indeed, research by Hayward and Vertinsky (1999) found that initial FSC supporters fell into this category. But these were in the minority and did not result in any significant levels of support for the FSC.

2. Here, we make a distinction between a program that existed for years, such as the American Tree Farm system or even the SFI in its original incarnation as a "code of conduct" program, to the instances in which these programs became launched and identified by their supporters as certification programs.

3. Our thanks to an anonymous reviewer for making this point.

4. We recognize that not all companies will search for lower standards. There are two reasons why this is so. The first is that companies who receive market benefits from being certified to high standards have an interest in excluding companies who do not practice the same high standards, since it could erode any competitive market advantage they enjoy (Vogel 1995, 2001). Second, while these companies would necessarily want those non-participating companies to have rules as high as possible (rendering their green premium relatively less expensive), they likewise have an interest in having joiners meet the standards they are meeting—otherwise they might not be able to compete with joiners.

5. Very similar dynamics occurred in the Canadian Maritimes regional standards setting process where the large industry player, JD Irving, deemed them inappropriate for industrial forestry and returned its certificate to the FSC (Lawson and Cashore 2001b).

6. In 2000 and 2001, respectively, donations accounted for 85 percent (US$1,611,443) and 91 percent (US$3,118,596) of the FSC total annual income (Forest Stewardship Council 2001).

7. Cashore (2002) has argued that the FSC competitors CSA and SFI, by mirroring themselves after ISO 14001 environmental management standards, gained indirect cognitive legitimacy from an array of business and governmental organizations.

8. Personal interviews.

9. Gunningham and colleagues (Gunningham and Sinclair 2002; Gunningham, Grabosky, and Sinclair 1998) refer to this as part of a broader "baskets" approach to policy development.

Glossary of Terms

Term	Definition
Buyer groups	A term referring to the groups set up around the world by the World Wildlife Fund for companies, mostly at the demand side of the value chain, that wanted to support FSC certification.
Cognitive legitimacy	A form of legitimacy that rests on either the "comprehensibility" or the "taken-for-grantedness" of a program in view of an audience. An audience will support a program and the services that program provides when the program is easily understood, given the audience's social context, or when the program must exist in order for the audience's world to properly function.
Conditions for non-state market driven	Role of the market: Products being regulated are demanded by purchasers further down the supply chain. Role of the state: Sovereign authority is not used to directly require adherence to rules. Role of stakeholders and broader civil society: Authority granted through an internal evaluative process. Enforcement: Compliance must be verified.
Conforming strategies	A category of strategies whereby a program changes its procedures or standards in order to gain the support of an external audience (i.e., the program changes while the audience remains the same).

Term	Definition
Converting strategies	A category of strategies whereby a program changes the evaluations of an external audience without modifying program procedures and standards (i.e., the audience changes its evaluations, but the program remains the same).
Demand side	A term referring to businesses that generally do not process products but are involved in the sale of goods to end consumers and/or institutional buyers.
Dependent variable	The presence or absence of pragmatic legitimacy granted to the FSC by forest companies and forest landowners.
Direct effects	The effects attributable to a factor under the condition that "everything else is held constant" (i.e., independent effects).
Evaluations	The process through which a program's audience member determines whether or not to give the program any form of support.
External audiences or audiences	Includes two groups: an "immediate audience" that is made up of stakeholders with a political or material interest in a program's procedures and standards, and a "general audience" or "civil society" that does not have direct interests in a program's procedures and standards.
Factor	A structural characteristic of a region or country that directly and/ or indirectly influences the success of legitimacy achievement strategies.
Forest companies	Businesses that own and/or manage forestlands from which they extract timber for commercial use.
Forest landowners	A category including government owners and individual private non-industrial owners, often referred to as NIPFs.
FSC competitor	Any certification program developed by company and landowner interests that were set up in part to offer an alternative to the FSC.
Indirect effects	Effects that occur as a factor is mediated by the values taken on by other factors (i.e., intersection effects or interactive effects).
Industrial roundwood	An aggregate forest product category defined by the FAO to include "sawlogs or veneer logs, pulpwood, other industrial roundwood and, in the case of trade, also chips and particles and wood residues" (FAO 2001).
Informing strategies	A category of strategies whereby a program does not change; it only contacts external audiences that are already predisposed to evaluate the program positively (i.e., the audiences are informed of the program and immediately offer their support).
Legitimacy	"A generalized perception or assumption that the actions of an entity are desirable, proper, or appropriate within some socially constructed system of norms, values, beliefs, and definitions" (Suchman 1995).
Legitimacy achievement strategies	Active efforts on the part of a certification program and its supporters to gain support from other stakeholders.

Term	Definition
Moral legitimacy	A form of legitimacy that rests on the pro-social evaluations of an audience as to what is the "right thing to do." An audience will support a program and the services that program provides when its support benefits the audience's perceptions of broad societal goals.
Non-state market-driven (NSMD) governance	An institution of governance, where rule-making authority is derived from evaluations of enterprises along a market supply chain. State sovereign authority is not granted (or ceded) by the state to these new systems, nor is it used to enforce compliance.
Paper plus paperboard	Defined by the FAO to include "newsprint, printing and writing paper, other paper and paperboard" (FAO 2001).
Particleboard	Defined by the FAO to include "a sheet material made of ligno-cellulosic material (chips, flakes, splinters, etc . . .) combined with binder and exposed to heat, pressure, humidity or catalyst to make a solid product" (FAO 2001).
Plywood	Defined by the FAO to include "plywood, veneer plywood, core plywood including veneered wood, blockboard, laminboard and battenboard. Other plywood such as cellular board and composite plywood. Veneer plywood produced with two or more bonded veneer sheets, grain of alternative sheets crossed generally at right angles. Core plywood has core (centre layer) thicker than plies, is solid and consists of narrow strips, blocks or boards of wood that may or may not be glued together (includes products with low quality core covered with thin veneer attached with glue under high pressure). Core of cellular board not solid, and composite plywood has non-wood or non-solid wood core" (FAO 2001).
Pragmatic legitimacy	A form of legitimacy that rests on the self-interest evaluations of an audience. An audience will support a program and the services that program provides when its support benefits the audience's material interests.
Sawnwood	Defined by the FAO to include "sawnwood, unplaned, planed, grooved, tongued, etc., sawn lengthwise, or produced by a profile-chipping process (e.g., planks, beams, joists, boards, rafters, scantlings, laths, boxboards, 'lumber,' sleepers, etc.) and planed wood which may also be finger jointed, tongued or grooved, chamfered, rabbeted, V-jointed, beaded, etc. Wood flooring is excluded. With few exceptions sawnwood exceeds 5 mm. in thickness" (FAO 2001).
Supply side	A term referring to businesses, government agencies (when they own the means of production), and private individuals involved in the production of raw materials that are generally used as inputs in further manufacturing or value-added processes.
Wood-based panels	Defined by the FAO to include "veneer sheets, plywood, particleboard, and fibreboard compressed (building board)" (FAO 2001).

Term	Definition
Wood pulp	Defined by the FAO to include "mechanical, semi-chemical, chemical and dissolving wood pulp" (FAO 2001).

References

Abusow, Kathy. 1997. *Towards Sustainable Forest Management Certification with Canada's National Sustainable Forest Management System Standards.* Montreal, QC: Canadian Sustainable Forestry Certification Coalition.

Adlard, Philip. 2001. UKWAS for small woods. *Woodland Owner: The Newsletter of the TGA Woodland Initiative,* Spring, 4–5.

Alabama Forestry Commission. 1993. *Alabama's Best Management Practices for Forestry.* Montgomery: Alabama Forestry Commission.

Alden, Edward. 1998. MacMillan Bloedel bows to the pressure from Greenpeace: The company is to stop clear-cut logging in British Columbia. *Financial Times,* June 19, 28.

American Forest and Paper Association (AF&PA). 1999. *SFI: 4th Annual Progress Report.* Washington, DC: American Forest and Paper Association.

———. 2000a. *AFF's American Tree Farm System and the AF&PA's Sustainable Forestry Initiative (SFI)sm Program Collaborate to Expand the Practice of Sustainable Forestry.* Washington, DC: American Forest and Paper Association.

———. 2000b. *Application of the Sustainable Forestry Initiative (SFI)SM Standard in Canada.* Washington, DC: American Forest and Paper Association.

———. 2000c. Fraser Third Party Certifies all its Forestlands: Lands in Canada First Outside U.S. to Be Independently Audited under SFI. Washington, DC: American Forest and Paper Association's SFI Program.

———. 2000d. Maine Garners Second Annual SFIsm State Implementation Committee Award. Washington, D.C.: American Forest and Paper Association.

———. 2000e. *Multi-Stakeholder Sustainable Forestry Board to Manage SFIsm Program.* Washington, DC: American Forest and Paper Association.

——. 2000f. *SFISM AFF's American Tree Farm System® and AF&PA's Sustainable Forestry Initiative (SFI)SM Program Collaborate to Expand the Practice of Sustainable Forestry.* Washington, DC: American Forest and Paper Association.

——. 2000g. *Sustainable Forestry Initiative Standard: Principles and Objectives.* Washington, DC: American Forest and Paper Association.

——. 2001a. A resolution commending America's forest and paper products industry and the Sustainable Forestry Initiative summary statement. http://www.afandpa.org.

——. 2001b. Summary of key enhancements to the SFI program in 2000. http://www .afandpa.org/forestry/sfi_frame.html.

——. 2002a. *The 2002–2004 Edition Sustainable Forestry Initiative (SFI) Program.* http://www.afandpa.org/forestry/sfi/Standard_0204.pdf.

——. 2002b. *The Sustainable Forestry Initiative (SFI) Program 2002 7th Annual Progress Report.* External Review Panel. http://www.afandpa.org/forestry/sfi/AnnualRep7 _long.pdf.

——. 2002c. *Sustainable Forestry Initiative (SFI) Program Companies That Have Completed Third Party Certification.* February 12, 2002. http://www.afandpa.org/forestry/ sfi/SFI_comp_acreage21202.pdf

American Lands Alliance. 2000. *Perspectives on AF&PA's "Sustainable Forestry Initiative" and Forest Certification.* Portland, OR: American Lands Alliance.

American Society of Association Executives. 1999. Associations put muscle behind advocacy messages: AAA Awards hail heroes of association philanthropy. *Association Management* 51 (1): 29–31.

American Tree Farm System. 2000. *The American Tree Farm System: Voluntary Certification of Sustainable Forestry.* Washington, DC: American Tree Farm System.

Ankrah, Paul. 1994. For richer, for poorer. *Timber and the Environment: A Supplement to Timber Trade Journal* 29 (January).

Anonymous. 2000a. Almost half of Sweden's forests FSC-certified. Press release. WWF Newsroom, September 28. http://www.panda.org/news/press/news.cfm?id=2071.

——. 2000b. PEFC does not fulfill minimum credibility requirements. Appeal/petition. Finnish Nature League, April 8. http://www.luontoliitto.fi/forest/certification/ fmeappeal.htm.

——. 2001a. The forest gets certified. Swedish Society for Nature Conservation, September 6. http://www.snf.se/verksamhet/skog/skog-markning_english.htm.

——. 2001b. Holzwirtschaft will nur ein Zertifikat: Streit der Zertifizierung schädlich für Holzabsatz. *Holzzentralblatt,* April 4, 591.

——. 2001c. Lohnt sich das Engagement in der "Gruppe 98"? *Holzzentralblatt,* February 9, 254.

——. 2001d. PEFC stands on shaky ground, a new WWF report shows. WWF Newsroom, April 2. http://www.panda.org/forests4life/news/PEFC.cfm.

——. 2001e. Zertifizierung bietet Chancen fuer Kommunalwald. *Holzzentralblatt,* February 26.

Arbeitsgemeinschaft Deutscher Waldbesitzerverbaende. 1997. Bundeskongress fuer Fuehrungskraefte forstwirtschaftlicher Zusammenschluesse zum Thema "Fortentwicklung forstwirtschaftlicher Zusammenschluesse."

AssiDomän. 1996. AssiDomän joins the FSC. http://www.asdo.se/english/presscenter/ pressarchive.html.

Auld, Graeme. 2001. Explaining certification legitimacy: An examination of forest sector support for forest certification programs in the United States Pacific Coast, the United Kingdom, and British Columbia, Canada. MS thesis, School of Forestry and Wildlife Sciences, Auburn University, Auburn, AL.

Auld, Graeme, Benjamin Cashore, and Deanna Newsom. 2003. Perspectives on forest certification: A survey examining differences among the US forest sectors' views of their forest certification alternatives. In *Forest Policy for Private Forestry: Global and Regional Challenges,* edited by L. Teeter, B. Cashore, and D. Zhang. Wallingford, UK: CABI Publishing.

Aulen, Gustaf, and Stefan Bleckert. 2001. *The Stockdove.* Report supported by Association of South Forest Owners, World Wide Fund for Nature, Swedish Society for Nature Conservation, Forestry of Swedish Forest Industries Association.

Axelson, L. 1996. Demand and supply factors on the market for organic vegetables: The case of Sweden. *Acta Horticulturae* 429:367–375.

Babbie, Earl. 2001. *The Practice of Social Research.* Ninth ed. Belmont, CA: Wadsworth.

Bainbridge, I. P., D. W. Minns, S. D. Housden, and A. N. Lance. 1987. *Forestry in the Flows of Caithness and Sutherland.* Edinburgh: Royal Society for the Protection of Birds.

Balachandran, Vani, and Alister Henderson. 2001. *Sawmill Survey 2000: Report on the Survey of Sawmill Consumption and Production in Great Britain in 2000.* Edinburgh: Forestry Commission, Great Britain, Economics and Statistics Unit.

Baldrey, Keith. 1994. Premier told he's fighting a lost cause: Greenpeace leader calls German trip 'totally useless'. *Vancouver Sun,* January 31, 1994, A1.

Bancroft, Bryce, and Ken Zielke. 2002. Spatial analysis on AAC impact relating to Draft 3 of the BC FSC Standards on three selected Tree Farm Licenses. In *For Bill Bourgeois, FSC-BC Steering Committee, in cooperation with Canfor (IFS), Tembec, Weyerhaeuser (Timberline).* Vancouver: Symmetree Consulting Group.

Barclay, Reg. 1993. British Columbia: War of the words. *Timber and the Environment: A Supplement to Timber and Trade Journal,* January, 9–13.

Barker, John. 1996. Breakthrough for forest certification. *Forestry Chronicle* 72 (2): 133.

Barker, Mary Lou, and Dietrich Soyez. 1994. Think locally, act globally? The transnationalization of Canadian resource-use conflicts. *Environment* 36 (5 June): 12–20, 32, 35–36.

Baumgartner, Frank, and Bryan Jones. 1991. Agenda dynamics and policy subsystems. *Journal of Politics* 53.

——. 1993. *Agendas and Instability in American Politics.* Chicago: University of Chicago Press.

BC Ministry of Forests. 1994. *1994 Forest, Range and Recreation Resource Analysis.* Report. Victoria, BC.

——. 2001. Wilson supports co-operative support to certification. Press release. Victoria, BC. http://www.for.gov.bc.ca/pscripts/pab/newsrel/mofnews.asp?refnum=2001%3A043.

BC Salmon Marketing Council. 2002. Salmon Marketing Council seeks Marine Stewardship Council certification for BC salmon. http://www.bcsalmon.ca/bcmsc/news1.htm.

BC Stats. 1996. *British Columbia Still a Natural Resource Economy.* Feature report. Ministry of Finance and Corporate Relations, April. http://www.bcstats.gov.bc.ca/data/bus_stat/trade.htm.

———. 2001. Exports (BC origin) 1991–2000. Statistics. Ministry of Finance and Corporate Relations, February. http://www.bcstats.gov.bc.ca/data/bus_stat/trade.htm.

BC Wild. 1994. *Forest Practices in British Columbia: Not a World Class Act*. Vancouver: BC Wild, Earthlife Canada Foundation.

Becker, Michel. 1999. Marktperspektiven der Holzproduktzertifizierung in Deutschland. Paper read at Fachkongress "Verantwortung fuer den Wald ueber Generationen" des Deutschen Forstvereins e.V., September 24, Schwerin, Germany.

Berger, Suzanne. 1996. Introduction. In *National Diversity and Global Capitalism*, edited by S. Berger and R. Dore. Ithaca: Cornell University Press.

Berger, Suzanne, and Ronald Dore, eds. 1996. *National Diversity and Global Capitalism*. Ithaca: Cornell University Press.

Bernstein, Steven. 2001a. *The Compromise of Liberal Environmentalism*. New York: Columbia University Press.

———. 2001b. Legitimacy in global governance: Three conceptions. Paper read at annual meeting of the American Political Science Association, August 29–September 1, San Francisco.

Bernstein, Steven, and Benjamin Cashore. 2000. Globalization, four paths of internationalization and domestic policy change: The case of eco-forestry policy change in British Columbia, Canada. *Canadian Journal of Political Science* 33 (1): 67–99.

———. 2001. The international-domestic nexus: The effects of international trade and environmental politics on the Canadian forest sector. In *Canadian Forest Policy: Regimes, Policy Dynamics and Institutional Adaptations*, edited by M. Howlett. Toronto: University of Toronto Press.

Bernstein, Steven, Richard Ned Lebow, Janice Gross Stein, and Steven Weber. 2000. God gave physics the easy problems: Adapting social science to an unpredictable world. *European Journal of International Relations* 6 (1): 43–76.

Bills, David. 2001. Certification — The UK experience. Paper read at Forests in a Changing Landscape: 16th Commonwealth Forestry Conference, 18–25 April, Fremantle, Australia.

Binkley, Clark. 2000. Forestry in the new millenium: Creating a vision that fits. In *A Vision for the US Forest Service*, edited by R. A. Sedjo. Washington, DC: Resources for the Future.

Boise Cascade Corporation. 2002. *Work, Build, Create: Boise 2001 Annual Report*. Boise Cascade Corporation. http://www.shareholder.com/visitor/dynamicdoc/document.cfm/documentid=28.

Bond, Patti. 2000. Lowe's to stop selling products from endangered forests. *Atlanta Constitution*, August 8, D1.

Boström, Magnus. 2003. How State-Dependent Is a Non-State-Driven Rule-Making Project? The Case of Forest Certification in Sweden. *Journal of Environmental Policy and Planning* 5 (2):165–180.

Bourgeois, Bill. 2002. Letter to FSC Canada and FSC BC Regional Initiative. Vancouver: Economic Chamber Member, Forest Stewardship Council — British Columbia Steering Committee.

Bowater, Corporation. 2002. *Bowater 2001 Annual Report*. Bowater Corporation, March 14. http://media.corporate-ir.net?media_files/NYS/BOW/reports/BOW2001 ar.pdf.

British Columbia. 2000. *Advisory Council to Aid Forest Management Certification.* Vancouver, BC: Government of British Columbia.

British Columbia, Ministry of Forests. 1994. Petter promotes B.C.'s forest practices and land use initiatives on European trip. News release. Government of British Columbia, September 21. http://www.for.gov.bc.ca/pscripts/pab/newsrel/mofnews.asp?renum=1994%3A128.

———. 1995. *B.C.'s Forest Practices Code: A Living Process—1995.* Reference document. Ministry of Forests. http://www.for.gov.bc.ca/pab/publctns/fpcliv/process.htm.

———. 1998. Province delivers B.C.'s sustainable forest message in Holland. News release. Ministry of Forests.

———. 2002. *2001/02 Annual Report: A New Era Update.* Victoria, BC: Ministry of Forests.

Bruce, Robert A. 1998. The comparison of the FSC forest certification and ISO environmental management schemes and their impact on a small retail business. MBA thesis, University of Edinburgh Management School.

Bryant, Dirk, Daniel Nielsen, and Laura Tangley. 1997. *The Last Frontier Forests: Ecosystems and Economies on the Edge. What Is the Status of the World's Remaining Large, Natural Forest Ecosystems?* Washington, DC: World Resources Institute, Forest Frontiers Initiative.

Busch, Marc L. 1999a. Overlapping institutions and global commerce: The calculus of form shopping for dispute settlement in Canada–US trade. Paper read at the 1999 annual meeting of the International Studies Association, Washington, DC.

———. 1999b. *Trade Warriors: States, Firms, and Strategic Policy in High Technology.* Cambridge, UK: Cambridge University Press.

Buthe, Tim. 2002. Taking temporality seriously: Modeling history and the use of narratives as evidence. *American Political Science Review* 96 (3 September): 481–494.

Cabarle, Bruce, Robert J. Hrubes, Chris Elliot, and Timothy Synnott. 1995. Certification accreditation: The need for credible claims. *Journal of Forestry* 93 (4): 12–16.

Canadian Standards Association. 1996. *A Sustainable Forest Management System: Specification Document.* Etobicoke: Canadian Standards Association.

———. 2001a. *Certification Status in Canada.* http://www.sfms.com/decade.htm.

———. 2001b. CSA International launches forest products marking program. Press release. CSA, July 17, Toronto.

———. 2002. *Sustainable Forest Management: Requirements and Guidance.* Document CAN/CSA-Z809–2002. Mississauga, ON: Canadian Standards Association.

Canadian Sustainable Forestry Certification Coalition. 2000. Canadian forest management certification status report: Company info. *Bulletin: Sustainable Forestry* 6 (1): 4.

Carlton, Jim. 2000a. Against the grain: How Home Depot and activists joined to cut logging abuse—if a tree falls in the forest, the small, powerful FSC wants to have its say. *Wall Street Journal,* September 26.

———. 2000b. Home builders Centex and Kaufman agree not to buy endangered wood. *Wall Street Journal,* March 31.

Cashore, Benjamin. 1997. Governing forestry: Environmental group influence in British Columbia and the US Pacific Northwest. PhD diss., University of Toronto.

———. 1999. US Pacific Northwest. In *Forest Policy: International Case Studies,* edited by B. Wilson, K. V. Kooten, I. Vertinsky and L. Arthur. Oxon, UK: CABI Publications.

———. 2001. Comparing endangered species protection in Canada and the United States:

What are the lessons for future policy development? Auburn, AL: Auburn University Forest Policy Center. Working paper 118.

——. 2002. Legitimacy and the privatization of environmental governance: How non state market-driven (NSMD) governance systems gain rule making authority. *Governance* 15 (4): 503–529.

——. 2003. Perspectives on Forest Certification as a Policy Process: Reflections on Elliott and Schlaepfer's Use of the Advocacy Coalition Framework. In *Social and Political Dimensions of Forest Certification,* edited by E. Meidinger, C. Elliott, and G. Oesten. Remagen-Oberwinter: Forstbuch.

Cashore, Benjamin, and Graeme Auld. 2003. The British Columbia environmental forest policy record in comparative perspective. *Journal of Forestry,* December.

Cashore, Benjamin, Graeme Auld, and Deanna Newsom. 2003. Forest certification (eco-labeling) programs and their policy-making authority: Explaining divergence among North American and European case studies. *Forest Policy and Economics* 5 (3).

Cashore, Benjamin, George Hoberg, Michael Howlett, Jeremy Rayner, and Jeremy Wilson. 2001. *In Search of Sustainability: British Columbia Forest Policy in the 1990s.* Vancouver: University of British Columbia Press.

Cashore, Benjamin, and James Lawson. 2003. Private policy networks and sustainable forestry policy: Comparing forest certification experiences in the US Northeast and the Canadian Maritimes. *Canadian American Public Policy,* Spring.

Cashore, Benjamin, and Ilan Vertinsky. 2000. Policy networks and firm behaviours: Governance systems and firm responses to external demands for sustainable forest management. *Policy Sciences* 33 (March): 1–30.

Cashore, Benjamin, Ilan Vertinsky, and Rachana Raizada. 2001. Firm responses to external pressures for sustainable forest management in British Columbia and the US Pacific Northwest. In *Sustaining the Pacific Coast Forests: Forging Truces in the War in the Woods,* edited by D. Salazar and D. Alper. Vancouver: University of British Columbia Press.

Certification Information Service, European Forest Institute. 2001. *Forest Certification Sourcebook,* edited by A. Cullum. Joensu, Finland: European Forest Institute.

Certified Forest Products Council. 1999. Certified Forest Products Council packet. Beaverton, OR: Certified Forest Products Council.

——. 2001a. *How the CFPC operates.* Certified Forest Product Council. http://www.certifiedwood.org/.

——. 2001b. Certified Forest Product Council Web site. http://www.certifiedwood.org.

Christie-Miller, Andrew. 1994. Comment and analysis: A make or break year. *Timber Grower* 130 (Spring): 10.

——. 1995. Comment and analysis. *Timber Grower* 134 (Spring): 10.

Clapp, Jennifer. 1998. The privatization of global environmental governance: ISO 14000 and the developing world. *Environmental Governance* 4 (3): 295–316.

Coady, Linda. 2000. Forest companies and environmentalists talking about how to resolve conflict in the marketplace. British Columbia Coast Forest Conservation Initiative. Press release. Vancouver: Weyerhaeuser Canada.

Coady, Linda, and Arlin Hackman. 2000. Protection/certification link a step to peace in the woods. *Logging and Sawmilling Journal:* 57.

Coastal Rainforest Coalition. 2000. Forest companies and environmental groups pursue unprecedented solutions initiative: Will jointly sponsor consultation and scientific and technical work on conservation-based ecosystem management for temperate rainforests on the north and central coast of BC. Press release. Vancouver: Coastal Rainforest Coalition.

Coleman, William D. 1987. Federalism and interest group organization. In *Federalism and the Role of the State,* edited by H. Bakvis and W. M. Chandler. Toronto: University of Toronto Press.

———. 1988. *Business and Politics: A Study of Collective Action.* Kingston: McGill-Queen's University Press.

Coleman, William D., and Grace Skogstad. 1990. Policy communities and policy networks: A structural approach. In *Policy Communities and Public Policy in Canada: A Structural Approach.* Mississauga, ON: Copp Clark Pitman.

Coleman, William, and Anthony Perl. 1999. Internationalized policy environments and policy network analysis. *Political Studies* 47:691–709.

Confederation of European Paper Industries. 2000. *Comparative Matrix of Forest Certification Schemes: Confederation of European Paper Industries.* Report.

Coulombe, Mary, and Marvin Brown. 1999. *The Society of American Foresters Task Force on Forest Management Certification Programs: The Society of American Foresters.* Report. Bethesda, Md.

Council of Forest Industries. 2000. *Fact Book.* http://www.cofi.org.

Council of Forest Industries of British Columbia. 1999. *British Columbia Forest Industry Fact Book — 1998.* Council of Forest Industries of British Columbia.

Crichton-Maitland, Mark. 1995. Comment and analysis. *Timber Grower* 137 (Winter 1995): 12.

———. 1996. Comment and analysis: Time to push the woodmark. *Timber Grower* 141 (Winter): 11.

———. 1997. Comment and analysis. *Timber Grower* 143 (Summer): 9–10.

Crossley, Rachel. 1996. A review of global forest management certification initiatives: Political and institutional aspects. Paper read at Conference on the Economic, Social and Political Issues in Certification of Forest Management, May 12–16, Malaysia.

Cubbage, Frederick W., Jay O'Laughlin, and Charles S. Bullock. 1993. *Forest Resource Policy.* New York: Wiley.

Currey, Dave. 2001. Timber trafficking: Illegal logging in Indonesia, South East Asia and international consumption of illegally sourced timber. Environmental Investigation Agency and Telepak Indonesia, September. http://www.eia-international.org/campaigns/forests/reports/timber/index.html.

Curtis, Malcolm. 1995. Battle rages over arbiter of wood product labelling. *Times Colonist,* October 19, B4.

Cutler, Claire, Virginia Haufler, and Tony Porter. 1999a. Private authority and international affairs. In *Private Authority in International Politics,* edited by C. Cutler, V. Haufler, and T. Porter. New York: SUNY Press.

———, eds. 1999b. *Private Authority in International Politics.* New York: SUNY Press.

Danks, Cecilia. 1996. *Developing Institutions for Community Forestry in Northern California.* Rural Development Forestry Network.

DIY. 1998. Hodkinson tells World Bank of B&Q's FSC plans. *DIY,* January 23.

Doern, G. Bruce, Leslie A. Pal, and Brian W. Tomlin. 1996. The internationalization of Canadian public policy. In *Border Crossings: The Internationalization of Canadian Public Policy*. Toronto: Oxford University Press.

Dogan, Mattei, and Dominique Pelassy. 1990. *How to Compare Nations: Strategies in Comparative Politics*. Chatham, NJ: Chatham House Publishers.

Dolcini, Marie. 1997. Task force advances 'no-cut'. *Planet*, June.

Domask, Joe. 2003. From boycotts to partnership: NGOs, the private sector, and the world's forests. In *Globalization and NGOs: Transforming Business, Governments, and Society,* edited by J. P. Doh and H. Teegen. New York: Praeger.

Dudley, N., J. P. Jeanrenaud, and F. Sullivan. 1995. *Bad Harvest? The Timber Trade and the Destruction of the World's Forests*. London: Earthscan Publications.

Eckerberg, K. 1987. *Environmental Protection in the Swedish Forestry Industry*. Department of Political Science, University of Umea, Sweden.

Editorial. 2001. The battle of the forest. *Dagens Nyheter*. May 6.

Ekelund, H., and C. G. Dahlin. 1997. *Development of the Swedish Forests and Forest Policy During the Last 100 Years*. Jönköping, Sweden: National Board of Forestry.

Eklkofer, Elke, and Michael Suda. 2000. Wie informieren sich Waldbesitzer. *AFZ/Der Wald* 20:1059–1060.

Ellefson, Paul, Anthony Cheng, and R. J. Moulton. 1995. *Regulation of Private Forestry Practices by State Governments*. St. Paul: University of Minnesota, Station Bulletin 605–1995. Minnesota Agricultural Experiment Station.

———. 1997. Regulatory programs and private forestry: State government actions to direct the use and management of forest ecosystems. *Society and Natural Resources* 10:195–209.

Elliott, Chris. 1999. Forest certification: Analysis from a policy network perspective. PhD diss., Departement de genie rural, Ecole Polytechnique Federale de Lausanne, Lausanne, Switzerland.

———. 2000. *Forest Certification: A Policy Network Perspective*. Centre for International Forestry Research (CIFOR: Bogar, Indonesia).

Elliott, Chris, and Rodolphe Schlaepfer. 2001. The advocacy coalition framework: Application to the policy process for the development of forest certification in Sweden. *Journal of European Public Policy* 8 (4): 642–661.

———. 2003. Global governance and forest certification: A fast track process for policy change. In *Social and Political Dimensions of Forest Certification,* edited by E. Meidinger, C. Elliott, and G. Oesten. Remagen-Oberwinter: Forstbuch.

Environmental News Service. 2000. Forest certification battle axes clash. Press release. April 19. http://www.ens.lycos.com/.

Esty, Daniel C. 1994. *Greening the GATT: Trade, Environment, and the Future*. Vol. 16. Washington, DC: Institute for International Economics.

———. 1998a. Environmental protection and international competitiveness: A conceptual framework. *Journal of World Trade* 32 (3): 5–46.

———. 1998b. Environmentalists and trade policy making. In *Constituent Interests and US Trade Policies,* edited by A. V. Deardorff and R. M. Stern. Ann Arbor: University of Michigan Press.

———. 2000. *NAFTA and the Environment: Seven Years Later*. Vol. 13. Washington, DC: Institute for International Economics.

European Forest Institute. 2001. *Country Reports—Germany 2001.* http://www.efi.fi/cis/english/creports/germany.html.

Fair Trade.org. 2001. An introduction: Fair trade organisation and fair trade assistance. http://www.fairtrade.org.

FAO. 2001. FAOSTAT forestry data. http://www.foa.org/forestry/foris/index.jsp?start_id=4029.

*Federal Register.*2000. Regulatory Impact Assessment for Proposed Rules Implementing the Organic Foods Production Act of 1990. 65 FR 49. 13632–13645.

Fletcher, Rick, Paul Adams, and Steve Radosevich. 2001. *Comparison of Two Forest Certification Systems and Oregon Legal Requirements.* Corvallis: Oregon State University College of Forestry.

Fletcher, Rick, and Eric Hansen. 1999. *Forest Certification Trends in North America and Europe, at New Zealand.* Oregon State Extension report.

Food Alliance. 2001. Guiding principles. http://www.thefoodalliance.org/guidingprinciples.htm.

Foottit, Clarie. 1989. Flow country afforestation. *Timber Grower* 111 (Spring): 16.

Forest Alliance. 1996. Canada a leader on forest management certification. *Forest and the People,* October.

Forest Alliance of British Columbia. 1997. Greenpeace boycott campaigns hits BC. News release. http://www.forest.org/frameset.cfm?pagerequest=mediacentre.

Forest Industries Development Council. 2001. *Forest Industries Development Council—The UK's Wood-Chain Partnership—Information Note.* Edinburgh: Forest Industry Development Council.

Forestry Industry Council of Great Britain. 2000. *A Reference for the Forestry Industry.* London: Forest Industry Council of Great Britain.

Forest Products Association of Canada. 2000. Sixty Percent of Canadian Forests to be Certified by 2003. Prince George, Canada.

Forest Stewardship Council. 1996. *FSC Principles and Criteria.* Forest Stewardship Council.

———. 1999a. *Forest Stewardship Council Regional Certification Standard for British Columbia.* Nelson, BC: Forest Stewardship Council.

———. 1999b. *FSC Principles and Criteria.* Forest Stewardship Council.

———. 1999c. *FSC Principles and Criteria.* Document 1.2. Forest Stewardship Council, February http://www.fscoax.org/principal.htm.

———. 1999d. *FSC Process Guidelines for Developing Regional Certification Standards.* FSC document 4.2. Forest Stewardship Council, January 19. http://www.fscoax.org/html/4–2.html.

———. 1999e. *Revised FSC Policy on Percentage Based Claims: Draft For Discussion.* Oaxaca, Mexico: Forest Stewardship Council.

———. 1999f. *Status of FSC-Endorsed Certification in the United States.* Washington, DC: FSC.

———. 2000a. *Certification Standards for Best Forestry Practices in the Maritime Forest Region.* Forest Stewardship Council.

———. 2000b. *FSC Policy of Percentage Based Claims.* Oaxaca, Mexico: FSC International.

——. 2000c. *FSC Principles and Criteria.* Document 1.2. February. http://www.fscoax .org/html/1–2.html.

——. 2001. *Forest Stewardship Council, A. C. Financial Statements.* Oaxaca, Mexico: Forest Stewardship Council.

——. 2003. *Regional Certification Standards for British Columbia, Canada: Final Accreditation Report.* Oaxaca, Mexico: Forest Stewardship Council.

Forest Stewardship Council Canada. 2002. *Action Report from FSC Canada Meeting of Board of Directors.* Edmonton, Alberta: Forest Stewardship Council, Canada.

Forest Stewardship Council Germany. 1998. Deutsche FSC-standards. *AFZ/DerWald* 21:1324–1326.

——. 2001a. Stellungnahme zum Abschlussbericht ueber Modellprojekt "Ostwestfalen-Lippe." Forest Stewardship Council Germany. Press release.

——. 2001b. Unterschiede groesser als Gemeinsamkeiten. Paderborn. Press release.

——. 2002a. Keine Ausreden mehr fuer umweltfreundliche Beschaffung. Freiburg, Germany. Press release.

——. 2002b. *FSC-Zertifizierte Forstbetriebe in Deutschland.* October 1. http://www .fsc-deutschland.de/index1.htm.

Forest Stewardship Council United States. 2001. FSC-U.S. national indicators for forest stewardship. May 2. http://www.fscstandards.org/downloads/fsc_national_indicators .pdf.

Forest Stewardship Council, U.S. Southeast Working Group. 1999. *Forest Certification Standards for the Southern U.S.* Gainesville, FL: Forest Stewardship Council U.S.

ForestEthics. 2001. Employment opportunities: Job posting: Corporate advisor-traction consultant. Online job information. http://www.forestethics.org/html/eng/939-AA .shtml. Accessed March 20, 2001.

ForestEthics and the Dogwood Alliance. 2002. Staples corporate credibility gap: Company caught lying about destruction of old-growth forests in new report. San Francisco: ForestEthics, Dogwood Alliance.

Forestry Authority. 1998. *The UK Forestry Standard: The Government's Approach to Sustainable Forestry.* Edinburgh: Forestry Commission.

Forestry Commission. 1999a. Britain's public forests pass green test. News release. Forestry Commission Great Britain, November 29. http://www.forestry.uk.gov.

——. 1999b. Forest enterprise prepares for certification. News release. Forestry Commission Great Britain, April 9. http://www.forestry.gov.uk.

——. 1999c. Tony Blair welcomes world-first for UK forestry. News release. Great Britain Forestry Commission, June 3. http://www.forestry.gov.uk.co.

——. 2001. *Forestry Statistics 2001: A Compendium of Statistics about Woodland, Forestry and Primary Wood Processing in the UK.* Edinburgh: Great Britain Forestry Commission.

Forestry Commission, Great Britain. 1998. Forestry briefing. Edinburgh: Secretariat Division Forestry Commission.

Forestry Commission, United Kingdom. 1999. Tony Blair welcomes world-first for UK forestry. In *Issued on Behalf of the UKWAS Technical Working Group.* Edinburgh: United Kingdom Forestry Commission.

Forestry Source. 2000. Home builders give preference to certified wood: Announcements head off nationwide protests. *The Forestry Source,* May. http://www.safnet.org/archive/home500.htm.

Forsyth, Keith, David Haley, and Robert Kozak. 1999. Will consumers pay more for certified wood products? *Journal of Forestry* 2:18–22.

Foulkes, George. 2001. DFID backs sustainable forestry. *Focus: Timber Trade Federation Member Magazine* 1 (February): 4–5.

Fountains PLC. 2002. Environmental policy and UKWAS statement. Services/forestry—environmental policy. http://www.fountainsplc.com/services/forestry_9.htm.

Fridman, Jonas. 2000. Conservation of forest in Sweden: A strategic ecological analysis. *Biological Conservation* 96:95–103.

Frizzell, Alan, and Jon H. Pammett, eds. 1997. *Shades of Green: Environmental Attitudes in Canada and Around the World.* Ottawa: Carleton University Press.

FSC Canada. 2002a. *Action Report from FSC Canada Meeting of Board of Directors.* Edmonton, Alberta: FSC-Canada.

———. 2002b. *Staff Report on BC Standards.* Toronto. Internal document.

FSC Germany. 2002. Waldbesitzer auf dem Holzweg. Freiburg. Press release.

Gale, Fred, and Cheri Burda. 1997. The pitfalls and potential of eco-certification as a market incentive for sustainable forest management. In *The Wealth of Forests: Markets, Regulation and Sustainable Forestry,* edited by C. Tollefson. Vancouver: University of British Columbia Press.

Gale, Fred P. 1998. *The Tropical Timber Trade Regime.* International political economy series. New York: St. Martin's Press.

Gamlin, Linda. 1998. Sweden's factory forests: Sweden leads the world in the intensive exploitation of coniferous forests. Its environmental problems give a foretaste of what could happen to the boreal forest as a whole in the next century. *New Scientist,* 41–47.

Geddes, Barbara. 1990. How the cases you choose affect the answers you get: Selection bias in comparative politics. *Political Analysis* 2:131–150.

Georgia-Pacific Corporation. 2001. Georgia-Pacific 2000 operating + statistical information brandnew:G-P. http://www.gp.com/center/financials/stats.html.

Gereffi, Gary, Ronie Garcia-Johnson, and Erika Sasser. 2001. The NGO-industrial complex. *Foreign Policy,* July/August,: 56–65.

Ghazi, Polly. 1994a. Future of woodlands in John Major's hands. *Observer,* February 27, 1994, 6.

———. 1994b. UK imports hasten ruin of rainforest. *Observer,* June 5, 1994, 15.

Glastra, Rob, ed. 1999. *Cut and Run: Illegal Logging and Timber Trade in the Tropics.* Ottawa: International Development Research Centre.

Global Forest and Trade Network. 2000. *The Global Forest and Trade Network: Leaders for Responsible Forestry.* Geneva: World Wildlife Fund.

Global Witness. 1999. *Made in Vietnam, Cut in Cambodia: How the Garden Furniture Trade Is Destroying Rainforests.* London: Global Witness, in association with Friends of the Earth.

Good Wood Watch. 2001. Key attributes of a credible eco-certification standard. http://www.goodwoodwatch.org.

Goodall, Stuart. 2000. Forest certification and the UK woodland assurance scheme. *Quarterly Journal of Forestry* 94 (3): 239–244.

Grant, Wyn. 1989. *Government and Industry: A Comparative Analysis of the U.S., Canada and the U.K.* Aldershot, Hampshire, England: Edward Elgar Publishing Limited.

Greenpeace Canada. n.d. *Just did it! The Vernon model of ecologically responsible forestry: Canada's first eco-certified forest. Clearcut-free?* Vancouver: Greenpeace.

———. 1994. *Canadian Logging: Not a World Class Act.* Vancouver: Greenpeace Canada.

Greenpeace Canada, Greenpeace International, and Greenpeace San Francisco. 1997. *Broken Promises: The Truth About What's Happening to British Columbia's Forests.* Vancouver: Greenpeace Canada.

Greenpeace International. 1993. *Greenpeace Annual Report 1992–1993.* Greenpeace International.

Greenpeace UK. 1994. UK lead in saving rainforest. In *Campaign Report: Climate Time Bomb.* London: Greenpeace UK.

———. 1998. *Summer Campaign Report.* London: Greenpeace UK.

Guardian. 1998. Minister's mahogany table angers environmentalists. *Guardian,* August 27, 9.

Gunningham, Neil, Peter N. Grabosky, and Darren Sinclair. 1998. *Smart Regulation: Designing Environmental Policy.* Oxford and New York: Clarendon Press and Oxford University Press.

Gunningham, Neil, and Darren Sinclair. 2002. *Leaders and Laggards: Next-Generation Environmental Regulation.* Australia: Greenleaf Publishing.

Hacker, Jackob. 2001. *The Divided Welfare State.* New York: Cambridge University Press.

Haddock, Mark. 2000. *Certification Challenges for British Columbia.* Vancouver: WWF Canada.

Halbert, Cindy, and Kai Lee. 1990. The timber, fish, and wildlife agreement: implementing alternative dispute resolution in Washington State. *Northwest Environmental Journal* 6:139–175.

Haley, David, and Martin K. Luckert. 1998. Tenures as economic instruments for achieving objectives of public forest policy in British Columbia. In *The Wealth of Forests: Markets, Regulation and Sustainable Forestry,* edited by C. Tollefson. Vancouver: University of British Columbia Press.

Hall, Peter. 1986. *Governing the Economy: The Politics of State Intervention in Britain and France.* New York: Oxford University Press.

Hamilton, Gordon. 1998a. BC chops red tape to save forestry firms $300 million. *Vancouver Sun,* April 3, A1, A2.

———. 1998b. MacBlo decides to abandon pro-logging forest alliance. *Vancouver Sun,* August 18.

———. 1998c. New MacBlo boss axes five VPs. *Vancouver Sun,* January 28, A1.

———. 1999. Forestry Alliance pledges to join environmentalists in greening effort. *Vancouver Sun,* June 5.

———. 2000a. Campbell's forestry solution ranges wide: More access to timber, less regulation, end to FRBC included as Liberal leader unveils plan for BC industry. *Vancouver Sun,* January 14, D1.

———. 2000b. Coastal loggers seek eco-truce: 'Significant' industry proposal would halt much of the logging on the remote north and central coast, environmentalists says. *Vancouver Sun,* March 16.

Hannigan, Andrew J. 2000. Centex ends use of endangered wood. March 30. Press release.

Hansen, Eric. 1998. Certified forest products marketplace. In *Forest Products Annual Market Review, 1997–1998.* Geneva: United Nations Economic Commission for Europe.

Hansen, Eric, Keith Forsyth, and Heikki Juslin. 1999. *A Forest Certification Update for the ECE Region.* Geneva, Switzerland: United Nations Economic Commission for Europe and Food and Agriculture Organization of the United Nations, Timber Section.

Hansen, Eric, and Heikki Juslin. 1999. *The Status of Forest Certification in the ECE Region.* New York and Geneva: United Nations, Timber Section, Trade Division, UN Economic Commission for Europe.

Hardin, Garrett. 1968. The tragedy of the commons. *Science:* 1243–1248.

Harrison, Kathryn. 1999. Talking with the donkey: Cooperative approaches to environmental protection. *Journal of Industrial Ecology* 2 (3): 51–72.

Harrods faces rainforest protest. 1993. *Press Association,* March 5.

Hatzfeldt, Count Hermann, Oliver Tickell, Pieter Poldervaart, Donné Norbert Beyer, Jan Nässtrom, and WWF. 2001. *Keeping the Forests — Making the Money: Forest Owners Tell Their Own FSC Stories.* Dreieiche: WWF European Forest Team.

Haufler, Virginia. 2001. *A Public Role for the Private Sector: Industry Self-Regulation in a Global Economy.* Washington, DC: Carnegie Endowment for International Peace.

Hayward, Jeffrey. 1998. Certifying industrial forestry in B.C.: European market drives big timber to the FSC. *Understory* 8 (4).

Hayward, Jeffrey, and Ilan Vertinsky. 1999. What managers and owners think of certification. *Journal of Forestry* (2): 13–17.

Heidenheimer, Arnold J., Hugh Heclo, and Carolyn T. Adams. 1990. *Comparative Public Policy: The Politics of Social Choice in America, Europe and Japan.* 3rd ed. New York: St. Martin's Press.

Hessisches Ministerium für Umwelt Landwirtschaft und Forsten. 2000. Hessische Waldbesitzer können an PEFC-Zertifizierung für nachhaltige Waldpflege teilnehmen. Press release. Hessisches Ministerium für Umwelt, Landwirtschaft und Forsten, Wiesbaden, Germany. December 21.

Hoberg, George. 1993a. From logroll to logjam: Structure, strategy, and influence in the old growth forest conflict. Paper read at annual meeting of the American Political Science Association, Washington, DC.

———. 1993b. Regulating forestry: A comparison of institutions and policies in British Columbia and the US Pacific Northwest. Forest economics and policy analysis research unit, the University of British Columbia. Working paper.

———. 1997. From localism to legalism: The transformation of federal forest policy. In *Western Public Lands and Environmental Politics,* edited by C. Davis. Boulder, CO: Westview Press.

———. 2000. How the way we make policy governs the policy we choose. In *Sustaining the Pacific Coast Forests: Forging Truces in the War in the Woods,* edited by D. Alper and D. Salazar. Vancouver: University of British Columbia Press.

———. 2002. The British Columbia forest practices code: Formalization and its effects. In *Canadian Forest Policy: Regimes, Policy Dynamics, and Institutional Adaptations,* edited by M. Howlett. Toronto: University of Toronto Press.

Hoberg, George, and Edward Morawski. 1997. Policy change through sector intersection: Forest and aboriginal policy in Clayoquot Sound. *Canadian Public Administration* 40 (3): 387–414.

Hoeltermann, Anke. 2000. Ethische Grundlagen nachhaltigen forstlichen Handelns. Paper read at Forstwissenschaftliche Tagung 2000, October 11–15, Freiburg.

Hofmann, Frank, Jutta Kill, Roland Meder, Harald Plachter, and Karl-Rheinhard Volz. 2000. *Waldnutzung in Deutschland: Bestandsaufnahme, Handlungsbedarf und Maßnahmen zur Umsetzung des Leitbildes einer nachhaltigen Entwicklung.* Vol. 35, edited by D. R. v. S. f. Umweltfragen. Stuttgart: Metzler-Poeschel Stuttgart.

Hogben, Dave. 1998. Western seeks certification to satisfy European buyers. *Vancouver Sun,* June 5, F2.

Home Depot. 1999. The Home Depot launches environmental wood purchasing policy: Company promises to reduce wood sourced from endangered forests during next three years. News release. Home Depot, August 27. http://www.uswa329.org/1999/august99/aug27a.htm.

Hood, Christopher. 1986. *The Tools of Government.* Chatham: Chatham House Publishers.

Howlett, Michael. 1999. Complex network management and the paradox of modern governance: A taxonomy and model of procedural policy instrument choice. Paper read at annual meeting of the Canadian Political Science Association, June 6, University of Sherbrooke, PQ.

———. 2000. Managing the "Hollow State": Procedural Policy Instruments and Modern Governance. *Canadian Public Administration* 43 (4): 412–431.

Howlett, Michael, and M. Ramesh. 1995. *Studying Public Policy Cycles and Policy Subsystems.* Toronto: Oxford University Press.

Huette, Gero. 2000. Postitionsveraenderung wesentlicher Akteure seit 1990 im deutschen forst- und naturschutzfachlichen Nachhaltigkeitsdiskurs. Paper read at Forstwissenschaftliche Tagung 2000, October 11–15, Freiburg, Germany.

Hufbauer, Gary Clyde, Daniel C. Esty, Diana Orejas, Luis Rubio, and Jeffrey J. Schott. 2000. *NAFTA and the Environment: Seven Years Later.* Vol. 61, Policy Analyses in International Economics series. Washington DC: Institute for International Economics.

Humphreys, David. 1996a. The Global Politics of Forest Conservation Since the UNCED. *Environmental Politics* 5:231–256.

———. 1996b. Hegemonic ideology and the international tropical timber organization. In *The Environment and International Relations: Theories and Processes,* edited by John Vogler and Mark Imber. London: Routledge.

———. 1996c. NGOs and regime theory: the case of forest conservation. *Journal of Commonwealth and Comparative Politics* 33 (1): 90–115.

Humphries, Shoana. 1999. *Forest Certification Handbook: For the Southeastern United States.* Gainesville, FL: Forest Management Trust.

Hyde, Bill, and William B. Stuart. 1998. US South. In *Forest Policy: International Comparisons,* edited by L. Arthur. Oxon, UK: CAB International.

Independent Forestry. 2003. CoC Members list. Independent Forestry: Specialists in group certification schemes. http://www.independentforestry.co.uk/member/coc_members_%20list.htm.

International Paper. 2002. *2001 Annual Report.* http://media.corporate-ir.net/media_files/NYS/IP/reports/2001ar.pdf.

Island Press. 1998. *The Business of Sustainable Forestry: Case Studies.* Covelo, CA: Island Press.

IWA Canada. 1996. *The Forest Is the Future: I.W.A. Canada Forest Policy.* Vancouver: IWA Canada.

Janssen, Gerd. 1998. Vor- und Nachteile einer Zertifizierung einer Landesforstverwaltung (Advantages and disadvantages of certification of federal state forest administration offices). *Forst und Holz* 53 (14): 447–449.

Jenkins, Anna. 1999. *Description of the FSC UK Process That Has Led to the Modification of the Endorsed FSC GB Standard (October 1998).* Llanidloes, Wales: Forest Stewardship Council UK Working Group.

Jennings, P. Devereaux, and Paul A. Zandbergen. 1995. Ecologically sustainable organizations: An institutional approach. *Academy of Management Review* 20 (4): 1015–1052.

Johnson, Nels. 1993. Introduction to Part I, Sustain what? Exploring the objectives of sustainable forestry. In *Defining Sustainable Forestry,* edited by G. Aplet, N. Johnson, J. T. Olson, and V. A. Sample. Washington, DC: Island Press.

Jordan, David. 1997. Rival forestry certification groups clash. *Vancouver Sun,* December 9, 3.

———. 1999a. Forest alliance to join the forest stewardship council. *Business in Vancouver,* June 15, 3.

———. 1999b. Home Depot move pushes certification: Forestry firms, environmentalists get different message. *Business in Vancouver,* September 7.

Keck, Margaret E., and Kathryn Sikkink. 1998. *Activists Beyond Borders: Advocacy Networks in International Politics.* Ithaca: Cornell University Press.

Kiekens, Jean-Pierre. 1997. *Certification: International Trends and Forestry and Trade Implications.* Bruxelles: Environmental Strategies Europe.

King, Gary, Robert O. Keohane, and Sidney Verba. 1994. *Designing Social Inquiry: Scientific Inference in Qualitative Research.* Princeton: Princeton University Press.

Klingberg, Tage. 2002. *A European View of Forest Certification: Issues for Consideration.* Gävle: University of Gävle, Department of Business and Economics.

Klins, Ullrich. 2000. Die Zertifizierung von Wald und Holzprodukten in Deutschland: Eine Forstpolitische Analyse. Ph.D. diss., Forstwissenschaftliche Fakultaet, Technische Universitaet Muenchen, Muenchen.

Knight, Alan. 2000. *Seeing the Customer and the Trees: A Proposed Revised Buying Policy for B&Q for Beyond 2000 for Consultation.* Chandlers Ford, Eastleigh, Hampshire: B&Q.

Knight, Peter. 1993. Timber watchdog ready to bark. *Financial Times London Edition,* October 6, 12.

———. 1996. Forests for life: Eco-label—the mark of good management. *Observer,* September 29, 96.

Kohm, Kathryn A., and Jerry F. Franklin. 1997. *Creating a Forestry for the 21st Century: The Science of Ecosystem Management.* Washington, DC: Island Press.

Kolk, Ans. 1996. *Forests in International Environmental Politics: International Organisations, NGOs and the Brazilian Amazon.* Alexander Numankade, Netherlands: International Books.

Kollman, Kelly, and Aseem Prakash. 2001. Green by choice: Cross-national variations in firms' responses to EMS-based environmental regimes. *World Politics* 53 (3): 399–430.

Kuipers, Dean. 2001. Tax free speech. *LA Weekly,* July 27, 20.

Lapointe, Gerald. 1998. Sustainable forest management certification: The Canadian programme. *Forestry Chronicle* 74 (2): 227–230.

Lawson, James, and Benjamin Cashore. 2001. *Autopoiesis and Forest Certification.* Auburn, AL: Auburn University Forest Policy Center.

———. 2003. Firm choices on sustainable forestry forest certification: The case of JD Irving. In *Forest Policy for Private Forestry: Global and Regional Challenges,* edited by L. Teeter, B. Cashore, and D. Zhang. Wallingford, UK: CABI Publishing.

Liimatainen, Matti, and Sini Harkki. 2001. *Anything Goes? Report on PEFC Certified Finnish Forestry.* Helsinki: Greenpeace Nordic and the Finnish Nature League.

Lijphart, Arend. 1975. The comparable case strategy in comparative research. *Comparative Political Studies,* 158–177.

Lindahl, Karin Beland. 2001. *The Development, Standards and Procedures of the Forest Stewardship Council (FSC) and the Pan European Forest Certification Scheme (PEFC) in Sweden.* Jokkmokk, Sweden: FERN.

Lipschutz, Ronnie D. 2001. Why is there no international forestry law? An examination of international forestry regulation, both public and private. *UCLA Journal of Environmental Law and Policy* 19 (1): 153–180.

Lipschutz, Ronnie D., and Cathleen Fogel. 2002. The emergence of private authority in global governance. In *The Emergence of Private Authority in Global Governance,* edited by R. B. Hall and T. J. Biersteker. Cambridge, UK: Cambridge University Press.

Louisiana-Pacific, Corporation. 2002. *2001 Annual Report and 10-K.* http://media.corporate-ir.net/media_files/NYS/lpx/reports/LouisianaPacific10K.pdf.

Lush, Patricial. 1998. BC said to slash stumpage fees. *Globe and Mail,* Tuesday, January 27, B3.

Lyke, Julie. 1996. Forest product certification revisited: An update. *Journal of Forestry* 94 (10): 15–20.

MacCleery, Doug. 1999. *NFS Certification Issues.* Washington, DC: USDA Forest Service.

March, James G., and Johan P. Olsen. 1998. Institutional dynamics of international political orders. *International Organization* 52 (4): 943–969.

Marchak, M. Patricia, Scott L. Aycock, and Deborah M. Herbert. 1999. *Falldown: Forest Policy in British Columbia.* Vancouver, BC: David Suzuki Foundation and Ecotrust Canada.

Marshall, Nick. 1996. *Forestry Plan Scan '96: A Review of Indicative Forestry Strategies.* Sandy, Bedfordshire, UK: Royal Society for the Protection of Birds.

Martin, Glen. 2001. Attack on tax status of environment group: Conservatives ask IRS for new ruling. *San Francisco Chronicle,* June 21, A3.

Mater, C. 2002. Forest certification comparative analysis: FSC and SFI benchmark field results. Montreal. Presentation.

Mater, Catherine M., V. Alaric Sample, James R. Grace, and Gerald A. Rose. 1999. Third-party, performance-based certification: What public forestland managers should know. *Journal of Forestry* 97 (2): 6–12.

Matthew, Ed. 2001. Import of illegal tropical timber into UK. London: Friends of the Earth, Living World Campaign.

McAlexander, James, and Eric Hansen. 1998. *J Sainsbury plc and the Home Depot: Retailers' Impact on Sustainability — Case Study.* Washington, DC: Island Press.

McKeown, Timothy J. 1999. Case studies and the statistical worldview: Review of King, Keohane, and Verba's "Designing Social Inquiry: Scientific Inference in Qualitative Research." *International Organization* 53 (1): 161–191.

Mead, Chris. 1993. In the country: If you sell off the woods today . . . Chris Mead calls for urgent new measures to head off a potential crisis for forestry. *Daily Telegraph,* January 23, 2.

Meek, Chanda L. 2001. *Sustainable for Whom? A Discussion Paper on Certification and Communities in the Boreal Region: Case Studies from Canada and Sweden.* Jokkmokk, Sweden: Taiga Rescue Network and Boreal Footprint Project 2001.

Meidinger, Errol E. 1997. Look who's making the rules: International environmental standard setting by non-governmental organizations. *Human Ecology Review* 4 (1): 52–54.

———. 2000. *Incorporating Environmental Certification Systems in North American Legal Systems.* Buffalo: University of Buffalo.

Meridian Institute. 2001a. *Comparative Analysis of the Forest Stewardship Council and Sustainable Forestry Initiative Certification Programs.* Vol. 1, Introduction and Consensus Statement on Similarities and Differences Between the Two Programs. Washington, DC: Meridian Institute.

———. 2001b. *Comparative Analysis of the Forest Stewardship Council and Sustainable Forestry Initiative Certification Programs.* Vol. 2, Description of the Forest Stewardship Council Program. Washington, DC: Meridian Institute.

———. 2001c. *Comparative Analysis of the Forest Stewardship Council and Sustainable Forestry Initiative Certification Programs.* Vol. 3, Description of the Sustainable Forestry Initiative Program. Washington, DC: Meridian Institute.

Micheletti, Michele. 1999. Put your money where your mouth is! The virtues of shopping for democracy. Paper read at Nordic Political Science Association conference, October 28–29, Uppsala.

———. 2003. *The Politics Behind Products Using the Market as a Site for Ethics and Action.* New Brunswick, NJ: Palgrave.

Ministry of Forests, British Columbia. 2001. Old growth forests British Columbia, Canada. http://www.growingtogether.ca/facts/old_growth.pdf.

Mirbach, Martin von. 1997. Demanding good wood. *Alternatives Journal* 23 (Summer 1997).

Moffat, Andrea C. 1998. Forest certification: An examination of the compatibility of the Canadian Standards Association and Forest Stewardship Council Systems in the maritime region. Master's in Environmental Studies Dalhousie University, Halifax, Nova Scotia.

Moller, Lotta. 1996. AssiDomän's view of forestry certification. Summary of speech delivered by Lennart Ahlgren at the World Wildlife Fund seminar Forests for Life '96, June 12, Brussels. http://www.asdo.se/english/presscenter/pressarchive.html.

Moran, Ted. 2002. *Beyond Sweatshops, Foreign Direct Investment and Globalization in Developing Countries.* Washington, DC: Brookings Institution.

Morley, Elliot. 1992. House of Commons Hansard—June 3, 1992, column 902. Online archive. The United Kingdom Parliament. http://www.publications.parliament.uk/pa/cm199293/cmhansrd/1992-06-03/Debate-7.html.

Murray, Brian C., and Robert C. Abt. 2001. Estimating price compensation requirements for eco-certified forestry. *Ecological Economics* 36 (2001):149–163.

National Home Center News. 1998. B&Q pledges to work towards forest certification. *National Home Center News* 24 (2): 9.

National Research Council. 1998. *Forested Landscapes in Perspective: Prospects and Opportunities for Sustainable Management of America's Nonfederal Forests.* Washington, DC: National Academy Press.

Natural Resources Defense Council. 1993. *Saving the Ancient Forests of British Columbia.* New York: Natural Resources Defense Council.

——. 1997. *Forests for Tomorrow: Responsible Forest Stewardship Through Consumer Power.* New York: Natural Resources Defense Council.

Newsom, Deanna. 2001. Achieving legitimacy? Exploring competing forest certification programs' actions to gain forest manager support in the U.S. Southeast, Germany, and British Columbia, Canada. Auburn, AL: School of Forestry and Wildlife Sciences, Auburn University.

Newsom, Deanna, B. Cashore, G. Auld, and J. Granskog. 2003. Certification in the heart of Dixie: A survey of Alabama landowners. In *Forest Policy for Private Forestry: Global and Regional Challenges,* edited by L. Teeter, B. Cashore, and D. Zhang. Wallingford, UK: CABI Publishing.

Noah, Emily, and Benjamin Cashore. 2002. Revised discussion paper on forest certification. Paper read at the Forest Dialogue Meeting on Certification, Geneva, Switzerland.

OBI. 2001. *OBI Einstellung zur Holzzertifizierung.* http://www.obi.de/oekologie/fsc/index.html.

OECD. 2001. International trade by commodity statistics 1995/2000: Denmark, Korea, Mexico, Portugal, Turkey, Slovak republic, United Kingdom (volume 2001/4). Statistics. http://www.sourceOECD.org.

Oliver, Christine. 1991. Strategic responses to institutional processes. *Academy of Management Review* 16 (1): 145–179.

Olsson, Roger. 1995. FSC certification making progress (15). Press release. October 17. http://forests.org/archive/general/fscstar.html.

Ozanne, Lucie K, and Richard Vlosky. 1997. Willingness to pay for environmentally certified wood products: The consumer perspective. *Forest Products Journal* 47 (6): 1–8.

——. 1996. Wood products environmental certification: The United States perspective. *Forestry Chronicle* 72 (2): 157–165.

Ozinga, Saskia. 2001. *Behind the Logo: An Environmental and Social Assessment of Forest Certification Schemes.* Moreton-in-Marsh: Fern, based on case studies by WWF France, Taiga Consulting, Taiga Rescue Network, Natural Resource Defense Council (NRDC), Fern, Finish Natural League, and Greenpeace.

Paget, Gregg, and Barbara Morton. 1999. Forest certification and 'radical' forest stewardship in British Columbia, Canada: The influence of corporate environmental procurement. Paper read at Greening of Industry Network Conference, November, Kenan-Flagler Business School, University of North Carolina, Chapel Hill.

PEFC. 2000. Public invited to comment on Finnish, Norwegian and Swedish forest certification schemes. Press release. March 2. http://www.pefc.org/content.htm.

PEFC Germany. 1999a. Newsletter 2 Deutschland. Stuttgart: Pan European Forest Certification Program.

——. 1999b. WWF und NABU stellen ihre Kompetenz in Waldfragen infrage. Bonn: Pan European Forest Certification Program.

——. 2000a. Stellungnahme zum Abschlussbericht über die Begleitung und Begutachtung des Modellprojektes "Zertifizierung nachhaltiger Forstwirtschaft in Nordrhein-Westfalen im Raum Ostwestfalen-Lippe" von Prof. Dr. C. Thoroe. PEFC Deutschland e.V.

——. 2000b. Thurn und Taxis Wald Erhält PEFC Gütesiegel — Fürstin Gloria Nimmt Zertifikat für Nachhaltige Waldbewirtschaftung Entgegen. Munich, Germany: Pan European Forest Certification Program.

——. 2000c. Zertifizierungssystem für Deutsche Wälder Verabschiedet. Bonn: Pan European Forest Certification Program.

——. 2001a. Diskreditierung um Jeden Preis? PEFC-Netz zur Sicherstellung der Systemstabilitaet Funktioniert. Stuttgart: Pan European Forest Certification Program.

——. 2001b. NRW-Studie belegt Stärke des PEFC-Zertifikates. Paderborn: Pan European Forest Certification Program.

——. 2001c. PEFC-Zertifikat Bekommt Gute Noten aus Europa. Bestrebung zur Gegenseitigen Anerkennung der Zertifizierungssysteme. Brussels: Pan European Forest Certification Program.

——. 2002a. *PEFC erfuellt die VDZ-Anforderungen.* http://www.pefc.de.

——. 2002b. *Stand der Zertifizierung in Deutschland.* October 2. http://www.pefc. destand-zertifizierung/index.htm.

PEFC International. 2001a. Common elements and requirements of PEFC: The technical document. http://www.pefc.org/Ramme2.htm.

——. 2001b. Web site. http://www.pefc.org.

——. 2001c. Welcome to PEFC. http://www.pefc.org.

PEFC UK. 2001. *UK Certification Scheme for Sustainable Forest Management.* Report — submission to PEFC council. http://www.pefc.org/application/UK/Revised25JAn02 PEFCUKSchemeElectronicVersion.pdf.

Pierson, Paul. 1993. When effect becomes cause: Policy feedback and political change. *World Politics* 45 (4): 595–628.

——. 2000. Increasing returns, path dependence, and the study of politics. *American Political Science Review* 94 (2): 251–268.

Porter, Michael E. 1990. *The Competitive Advantage of Nations.* New York: Free Press.

Porter, Michael E., and Class van der Linde. 1995. Green and competitive: Ending the stalemate. *Harvard Business Review* (September–October): 120–138.

Potlatch Corporation. 2002. *Form 10-K Annual Report Pursuant to Section 13 or 15(d) of the Securities Exchange Act of 1934 — Potlatch Corporation.* http://www.potlatch corp.com/finance/graphics/01_10k.pdf.

Power, Michael. 1992. The politics of brand accounting in the UK. *European Accounting Review* 1:39–68.

Prakash, Aseem. 1999. A new-institutional perspective on ISO 14000 and responsible care. *Business Strategy and the Environment* 8:322–35.

——. 2000a. *Greening of the Firm: The Politics of Corporate Environmentalism.* Cambridge, UK: Cambridge University Press.

——. 2000b. Responsible care: An assessment. *Business and Society* 39 (2): 183–209.

Praktiker. 2001. PEFC-Label fuer Praktiker-Baumaerkte nicht Ueberzeugend. Press release. Kirkel.

Price, Will. 2000. *Forest Certification Pilot Projects Expanded on State and Tribal Lands.* Washington, DC: Pinchot Institute for Conservation.

PricewaterhouseCoopers. 1999. *B.C. Ministry of Forests Small Business Forest Enterprise Program (SBFEP) Certification Pre-Assessment.* Vancouver: Pricewater house-Coopers.

Pringle, Douglas. 1994. *The Forestry Commission: The First 75 Years.* Edinburgh: Forestry Commission, Public Relations Division.

Putnam, Robert D. 1988. Diplomacy and domestic politics: The logic of two-level games. *International Organization* 42 (3): 427–460.

Ragin, Charles C. 1987. *The Comparative Method: Moving Beyond Qualitative and Quantitative Strategies.* Berkeley: University of California Press.

Rainforest Action Network. 1999. *Fueling the Chainsaws: How Home Depot Is Driving Old Growth Forest Destruction.* San Francisco: Rainforest Action Network.

——. 2000. Nation's top homebuilders vow to end endangered wood use: Huge win for environmentalists as pressure brings dramatic turnabout. Press release. http://www.ran.org/news/newsitem.php?id=123&area=newsroom.

——. *Boise: An American disgrace.* n.d. http://www.ran.org/ran_campaigns/old_growth/bcresponse_printer.html.

Rainforest Alliance. 2002. SmartVoyager certification. http://www.rainforest-alliance.org/programs/sv/index.html.

Rametsteiner, Ewald. 1999. The attitude of European consumers toward forests and forestry. *Unasylva* 50 (196).

Rametsteiner, Ewald, Peter Schwarzbauer, Heikki Juslin, Jari Karna, Roger Cooper, John Samuel, Michael Becker, and Tobias Kuhn. 1998. *Potential Markets for Certified Forest Products in Europe.* Joensuu, Finland: European Forest Institute.

Read, Mike. 1994. *Truth or Trickery? Timber Labelling Past and Future.* Salisbury: Mike Read Associates and World Wide Fund for Nature.

Reinold, Martin. 1998. Die Naturland Zertifizierung von Wald. *Forst und Holz* 53 (14): 441–443.

Rettenmeier Holzindustrie Wilburgstetten. 2001. Rettenmeier erhaelt PEFC-Zertifikat. Wilburgstetten, Germany.

Reuters News Service. 1999. European body sets new forest certification scheme. Press release. http://www/planetark.org/dailynewsstory.cfm?newsid=4570.

Rickenbach, Mark, Rick Fletcher, and Eric Hansen. 2000. *An Introduction to Forest Certification*. Corvallis, OR: Oregon State University Extension Service.

Ripkin, Heiko. 1999. Die deutschen FSC-Standards zur Zertifizierung auf dem Pruefstand (A critical look at the draft of the German FSC standards as applied to the certification). *Forest und Holz* 54 (6): 170–174.

Risse-Kappen, Thomas, ed. 1995. *Bringing Transnational Relations Back In: Non-state Actors, Domestic Structures and International Institutions*. Cambridge, UK: Cambridge University Press.

Rosenbaum, Walter A. 1995. *Environmental Politics and Policy*. 3rd ed. Washington: CQ Press.

Ross, Michael. 1996. Conditionality and logging reform in the tropics. In *Institutions for Environmental Aid,* edited by R. O. Keohane and M. A. Levy. Boston: MIT Press.

Royal Society for the Protection of Birds. 1993. *Time for Pine: A Future for Caledonian Pinewoods*. Edinburgh: Royal Society for the Protection of Birds.

Royal Swedish Academy of Agriculture and Forestry. 2001. *The Swedish Forestry Model*. Stockholm.

Ruggie, John Gerard. 2002. Taking embedded liberalism global: The corporate connection. Paper read at Canadian Congress of the Social Sciences and Humanities, University of Toronto.

Sabatier, Paul A., and Hank C. Jenkins-Smith. 1993. The Advocacy Coalition Framework: Assessment, revisions, and implications for scholars and practitioners. In *Policy Change and Learning: An Advocacy Coalition Approach,* edited by P. A. Sabatier and H. C. Jenkins-Smith. Boulder, CO: Westview Press.

Sasser, Erika N. 2003. Gaining leverage: NGO influence on certification institutions in the forest products sector. In *Forest Policy for Private Forestry: Global and Regional Challenges,* edited by L. Teeter, B. Cashore, and D. Zhang. Wallingford, UK: CABI Publishing.

———. 2002. The certification solution: NGO promotion of private, voluntary self-regulation. Paper read at 74th Annual Meeting of the Canadian Political Science Association, May 29–31, Toronto.

Schmitter, Philippe C., and Wolfgang Streeck. 1981. *The Organization of Business Interests*. Berlin: Wissenschaftszentrum Berlin.

Schoon, Nicholas. 1992. Ecological wood product labels planned. *Independent,* March 16, 2.

Schraml, U., and H. Thode. 2000. Forstpolitik im nichtbaeuerlichen Kleinprivatwald — Einstellung und Verhalten der Waldbesitzer und Konsequenzen fuer die Forstpolitik der Laender. Paper read at Forstwissenschaftliche Tagung 2000, October 11–15, Freiburg, Germany.

Schraml, U., and G. Winkel. 1999. Germany. In *Forestry in Changing Societies in Europe,* edited by E. Rojas. Joensuu: Silva Network.

Scottish Woodlands. n.d. *Scottish Woodlands Group Certification Scheme*. Edinburgh: Scottish Woodlands.

———. 2002. Management. http://www.scottishwoodlands.co.uk/management.html.

Scrase, Hannah, John Palmer, Simon Pryor, and Anna Jenkins. 1998. *Forest Stewardship Council UK Working Group: Description of Process of Developing Draft FSC Standards for Great Britain.* Llanidloes, Wales: Forest Stewardship Council UK Working Group.

Seattle Post-Intelligence Staff. 2000. Starbucks, Tully's offer fair trade, organic coffee. *Seattle Post-Intelligence,* August 15.

Sedjo, Roger A., and Stephen K. Swallow. 1999. *Eco-Labeling and the Price Premium.* Washington: Resource for the Future.

Sethi, Rajiv, and E. Somanathan. 1996. The evolution of social norms in common property resource use. *American Economic Review* 86 (4): 766–788.

SGS Forestry. 1997. *The UK Forestry Standard: An Analysis for the Forestry Commission.* Oxford: SGS Forestry.

Shotton, Rachel. 2000. WWF/Soil Association local authorities project. In *Forest Footprint,* edited by WWF-UK. Matlock, Derbyshire, UK.

Sierra Club. 1999. Sierra Club warn of green scam by American Forest and Paper Association. Press release. http://lists.sierraclub.org/SCRIPTS/WA.EXE?A2=ind9905&L=ce-scnews-releases&P=R71.

Sierra Legal Defence Fund. 1996. *British Columbia's Clear Cut Code: Changing the Way We Manage Our Forests? Tough Enforcement.* Vancouver: Sierra Legal Defence Fund.

———. 1998. *Profits or Plunder: Mismanagement of BC's Forests.* Vancouver: Sierra Legal Defence Fund.

———. 1999. *British Columbia Forestry Report Card 1997–98.* Vancouver: Sierra Legal Defence Fund, Sierra Club of British Columbia, Silva Forest Foundation, B.C. Environmental Network, Forest Caucus.

Simpson, Scott. 2001. B.C. salmon industry seeks ecological certification: Industry's ability to meet global council's standards hangs on sustainability question. *Vancouver Sun,* Thursday, April 26, Internet posting.

Simula, Markku. 1998. *Timber Certification: Progress and Issues.* Yokohama: International Tropical Timber Organization.

Skocpol, Theda. 1995. Why I am an historical institutionalist. *Polity* 28 (1): 103–106.

Smartwood Program. 2000. *Forest Management Public Summary for Timfor Contractors Limited.* New York: Rainforest Alliance.

Smith, Charles. 1995. Smartwood: Environmentally sound building materials include the new, the old, and the recycled. *San Francisco Examiner,* January 18, ZA.

Smith, W. Brad, and Raymond M Sheffield. 2000. *A Brief Overview of the Forest Resources of the United States, 1997.* Washington, DC: United States Department of Agriculture, Forest Service.

Soil Association UK. n.d. *Soil Association Woodmark: Your Partners in Forest Certification.* Bristol, UK: Soil Association Certification.

Sollander, Erik. 1999. *European Forest Scorecards 2000.* Report. Gland, Switzerland: WWF-International.

Spalding, Steve. 2002. An assessment of the utility, usability and cost implications associated with the Forest Stewardship Council Regional Certification Standards for British

Columbia (Draft 3 April 22, 2002) when applied to large-scale forestry operations. In *For Bill Bourgeois, FSC-BC Steering Committee, In cooperation with Canfor (IFS), Tembec, Weyerhaeuser (Timberline)*. Vancouver.

Spathelf, Peter. 1997. Seminatural silviculture in Southwest Germany. *Forestry Chronicle* 73 (6): 715–722.

Speth, James Gustave. 2002. A new green regime: Attacking the root causes of global environmental deterioration. *Environment* 44 (7): 16–25.

Stanbury, William T. 2000. *Environmental Groups and the International Conflict Over the Forest of British Columbia 1990 to 2000*. Vancouver: SFU-UBC Centre for the Study of Government and Business.

Stanbury, William T., and Ilan Vertinsky. 1997. Boycotts in conflicts over forestry issues: The case of Clayoquot Sound. *Commonwealth Forestry Review* 76 (1).

Stanbury, William T., Ilan B. Vertinsky, and Bill Wilson. 1995. *The Challenge to Canadian Forest Products in Europe: Managing a Complex Environmental Issue*. Vancouver: Forest Economics and Policy Analysis Research Unit, University of British Columbia.

Statistisches Bundesamt. 2001. Holzmarktbericht 2000 fuer die Forstwirtschaft and Holzwirtschaft.

Suchman, Mark C. 1995. Managing legitimacy: Strategic and institutional approaches. *Academy of Management Review* 20 (3): 571–610.

Sutherland, Iain. 2000. Imports force UK prices down. *Timber Grower* 157 (Winter): 11.

Swallow, Stephen K., and Roger A. Sedjo. 2000. Eco-labeling consequences in general equilibrium: A graphical assessment. *Land Economics* 76 (1): 28–36.

Swedish Forest Industries Association. 1997. Agreement achieved on forest certification for Sweden. *Forestry Chronicle* 73 (6): 664–665.

Swedish Forest Industries Federation. 2000. *Europe Needs the Forest Industry*. Falkoping, Sweden: Swedish Forest Industries Federation Media Express.

——. 2001. *The Swedish Forest Industries 2000 Facts and Figures*. Falkoping, Sweden: Elanders Gummessons Tryckeri.

Swedish Society for Nature Conservation. 1996. Stop logging Sweden's last old-growth forests now! Appeal. http://norrbotten.snf.se/appeal_english.html.

Taiga Rescue Network. 2002. The paper chain. *Taiga News* (32).

Taylor, Russell E., and Gerry van Leeuwen. 2000. *Outlook for Hemlock Prices and Volumes in Japan and Western Red Cedar Prices and Volumes in the USA*. Vancouver: R. E. Taylor and Associates; produced for Coastal Forest and Lumber Association.

Terstad, Jan. 1999. Swedish experiences of incentives for the protection of nature. *Science of the Total Environment* 240 (May): 189–196.

Teubner, Gunther. 1993. *Law as an Autopoietic System*. Translated by A. Bankowska, R. Adler, and Zenon Bankowski (transl. ed.). Oxford: Blackwell.

——. 1993b. The 'state' of private networks: The emerging legal regime of polycorporatism in Germany. *Brigham Young University Law Review* 2:553–575.

——, ed. 1997. *Global Law without a State*. Aldershot: Dartmouth.

Thierme, Falko. 2001. FSC will seinen Marktanteil ausbauen. *Holzzentralblatt*, February 2, 201.

Thompson, Steve. 2001. Seeing the forest for the trees: Home Depot, Plum Creek and the effects of media spin. *Independent* 12 (5), February 1. http://www.missoulanews.com/News/News.asp?no=735.

Thoroe, C. 2000. Abschlussbericht ueber die Begleitung und Begutachtung des Modellprojektes "Zertifizerung nachhaltiger Forstwirtschaft in Nordrhein-Westfalen im Raum Ostwestfalen-Lippe." Hamburg: Institut fuer Oekonomie der Bundesforschungsanstalt fuer Forst- und Holzwirtschaft.

Tice, Carol. 1998. MacMillan Bloedel to end clearcutting. *National Home Center News* 24 (13): 17.

Tickell, Oliver. 1993. Direct action is the only way left: From Oxford to Skye, Britain's eco-activists are on the move. *Guardian,* May 28, 17.

——. 1996. Forests for life: Native woodlands: Time to turn over a new leaf. *Observer,* September 29, 94.

——. 2000. *Why the UK's Ancient Woodland Is Still Under Threat.* Grantham, UK: Woodland Trust.

Tickell, Oliver, and World Wildlife Fund for Nature. 2000. *Certification: A Future for the World's Forests.* Godalming, Surrey, UK: WWF Forests for Life Campaign.

Tilhill Economic Forestry. 2002. About Tilhill. http://www.tef-forestry.co.uk/welcome.htm.

Timber Grower. 1994a. News: Industry gives support to 'Woodmark'. *Timber Grower* 131 (Summer): 5.

——. 1994b. News: WWF certification seminar falls apart. *Timber Grower* 131 (Summer): 5.

——. 1996. Woodmark extends its family. *Timber Grower* 141 (Winter): 17.

——. 1998. Standards debate: FSC in strategy rethink for small owners. *Timber Grower* 146 (Spring): 5.

Timbmet Group. 2000. *Timbmet Group.* Oxford: Timbmet Group.

Tollefson, Chris, ed. 1998. *The Wealth of Forests: Markets, Regulation, and Sustainable Forestry.* 2 vols. Vancouver: University of British Columbia Press.

Tovey, H. 1997. Ford, environmentalism and rural sociology: On the organic farming movement in Ireland. *Sociologia Ruralis* 37 (1): 21–37.

Transfair USA. 2000. *Better or Bitter.* Seattle: TransFair USA.

——. 2000. Peet's Coffee Company launches Fair Trade Certified coffee. Press release. September 20. http://www.transfairusa.org/content/news/nws_story.jsp?type=prs&artid=009020.

——. 2000l. Starbucks Coffee Company brings Fair Trade Certified coffee to retail stores through TransFair USA alliance. Press release. http://www.transfairusa.org/content/news/nws_story.jsp?type=prs&artid=00925.

Turner, Roger. 1991. *Forests for the Future: Integrating Forestry and the Environment.* Edinburgh: Royal Society for the Protection of Birds.

——. 1992. Comment and analysis: The birds and the trees. *Timber Grower* 122 (Spring): 10.

UKWAS Steering Group. 2000. *Introduction to the UK Woodland Assurance Scheme.* Edinburgh: UKWAS Steering Group.

United Kingdom. 2001. *Forestry Devolution Review: Interdepartmental Group Report.* http://www.forestry.gov.uk/website/pdf.nsf/pdf/fdr7Efull7Efinal.pdf/$FILE/fdr7E full7E final.pdf.

United States Department of Agriculture Committee of Scientists. 1999. *Sustaining the People's Lands: Recommendations for Stewardship of the National Forests and Grasslands into the Next Century.* Washington, DC: US Department of Agriculture.

United States Forest Service. 2000. *1997 RPA Assessment: The United States Forest Resources Current Situation (Final Statistics).* Washington, DC: United States Department of Agriculture.

Upton, Christopher, and Stephen Bass. 1996. *The Forest Certification Handbook.* Delray Beach, FL: St. Lucie Press.

Van Evera, Stephen. 1997. *Guide to Methods for Students of Political Science.* Ithaca: Cornell University Press.

van Kooten, G.C., Bill Wilson, and Ilan Vertinsky. 1999. Sweden. In *Forest Policy: International Case Studies,* edited by B. Wilson, G. C. v. Kooten, I. Vertinsky, and L. Arthur. Oxon, UK: CABI Publishing.

Verband Deutscher Papierfabriken. 1998. Statements und Stellungnahmen zum Thema Zertifizierung: Verband Deutscher Papierfabriken. *Forst und Holz* 53 (14): 452–453.

——. 2000. Papier 2000: Ein Leistungsbericht. Bonn: Verband Deutscher Papierfabriken e.V.

Verband Deutscher Zeitschriftenverleger. 2001. http://www.vdz.de.

Viana, Virgilio, Jamison Ervin, Richard Donovan, Chris Elliott, and Henry Gholz. 1996. *Certification of Forest Products: Issues and Perspectives.* Island Press.

Vlosky, R. P. 2000. *Certification: Perceptions of Non-Industrial Private Forestland Owners in Louisiana.* Baton Rouge: Louisiana State University Agricultural Center.

Vlosky, Richard, Shoana Humphries, and Carter Douglas. 2001. Certified wood products merchants in the United States: A comparison between 1995 and 1998. *Forest Products Journal* 51 (6): 32–39.

Vlosky, Richard P., and Lucie K Ozanne. 1997. Forest products certification: The business customer perspective. *Wood and Fiber Science* 29 (2): 195–208.

——. 1998. Environmental certification of wood products: The US manufacturers' perspective. *Forest Products Journal* 48 (9): 21–26.

Vogel, David. 1995. *Trading Up: Consumer and Environmental Regulation in a Global Economy.* Cambridge, MA: Harvard University Press.

——. 2001. Environmental Regulation and Economic Integration. In *Regulatory Competition and Economic Integration: Comparative Perspectives,* edited by D. C. Esty and D. Geradin. Oxford: Oxford University Press.

Voigt, Paul C., and Frederick Cubbage. 2000. An analysis of the costs and returns of development, construction, and operation of a wetlands mitigation bank in North Carolina. Paper read at Proceedings of the 28th Annual Southern Forest Economics Workshop, Biloxi, MS.

Volz, K.-R., and A. Bieling. 1998. Zur Soziologie des Kleinprivatwaldes. *Forst und Holz* 3:67–71.

Wallinger, Scott. 1995. A commitment to the future: AF&PA's sustainable forestry initiative. *Journal of Forestry* 93 (1): 16–19.

Washington Environmental Council. 1999. *Washington Environmental Council's Decision Not to Return to Timber, Fish and Wildlife (TFW) Forum at this Time.* Seattle, Washington, July 19.

Washington State. 1987. *The Timber, Fish and Wildlife Agreement: A Better Future in Our Streams and Woods.* Olympia: Washington State Department of Natural Resources, Wildlife, Ecology, Fisheries, Labor and Industries.

Webb, Kernaghan, ed. 2002. *Voluntary Codes: Private Governance, the Public Interest and Innovation.* Ottawa: Carleton University Research Unit Innovation, Science and the Environment.

Weber, Norbert. 1999. Neue akteure in der forstpolitischen Arena. *Forst und Holz* 54 (12): 355–359.

Webster, Robin. 2000. A policy network analysis of the Forest Stewardship Council (UK): A case study of an NGO/business alliance in conservation. Master of Science thesis in Conservation, University College London.

Wenban-Smith, Matthew. 2001. *Soil Association Woodmark.* Bristol, UK: Soil Association.

Wenban-Smith, Matthew, and Chris Elliot. 1998. The Forest Stewardship Council and the International Organization for Standardization. *Newsletter of the Forest Stewardship Council* 1 (3): 6,7,12.

Western Forest Products Limited. 2000. Forest certification in British Columbia, Canada. In *Information Brochure.* Vancouver: Western Forest Products Limited.

Williams, Ronnie. 1995. Progress over timber certification. *Timber Grower* 135 (Summer): 14.

Wilson, Jeremy. 1990. Wilderness politics in B.C.: The business dominated state and the containment of environmentalism. In *Policy Communities in Canada: A Structural Approach,* edited by W. D. Coleman and G. Skogstad. Mississauga, ON: Copp Clark Pitman.

——. 1998. *Talk and Log: Wilderness Politics in British Columbia.* Vancouver: University of British Columbia Press.

Wilson, Peter. 1996. Comment and analysis: Certified madness for silly season. *Timber Grower* 140 (Autumn): 9.

——. 1998. Comment and analysis: Tune in for the latest episode of certification. *Timber Grower* 148 (Autumn): 9.

Wolfe, Charles, Jr. 1988. *Market or Government: Choosing Between Imperfect Alternatives.* Cambridge, MA: MIT Press.

World Wildlife Fund. 1998. *WWF's Global Annual Forest Report: Forests for Life.* Washington, DC: WWF.

——. 2002. *Forestry and Wood Certification.* Newsletter. World Wildlife Fund European Policy Office. http://www.panda.org/resources/programmes/epo/publications/NewsletterApril200 2.doc.

World Wildlife Fund for Nature. 1999. *WWF's Global Forests and Trade Initiative.* Washington, DC: World Wildlife Fund United States.

World Wildlife Fund Germany. 2000a. Faktenservice Wald- und Holzzertifizierung, vol. 5 Frankfurt, Germany: World Wide Fund for Nature.

——2000b. Faktenservice Wald- und Holzzertifizierung, vol. 2. Frankfurt, Germany: World Wide Fund for Nature.

————. 2000c. Landesregierungen greifen mit Einschüchterungsversuchen in den Holz-markt ein. Frankfurt.

World Wildlife Fund United Kingdom. n.d. *Guidance Notes for Contractors: Indepen-dent Certification Schemes*, edited by R. Shotton. Matlock, UK: WWF-UK.

————. n.d. *The Local Authorities: Suggested Timber Purchasing Policy for Local Au-thorities*, edited by R. Shotton. Matlock, UK: WWF-UK.

————. 1997. *1995+Group Update*. Godalming, Surrey, UK: World Wildlife Fund UK.

————. 2000. *WWF 95+ Group*. Surrey, UK: World Wildlife Fund for Nature.

————. 2001a. *Guidance for Implementing the Suggested Timber Purchasing Policy Writ-ten for Local Authorities by WWF*, edited by R. Shotton. Matlock, UK: WWF-UK.

————. 2001b. History of the WWF 95+ group. http://www.wwf-uk.org/95+group/1286 History.html.

————. 2001c. *The Local Authorities Project*. UK: WWF.

Yaffee, Steven Lewis. 1994. *The Wisdom of the Spotted Owl: Policy Lessons for a New Century*. Covelo, CA: Island Press.

Yull, Len. 2001. Comment and analysis: Certification progress steady. *Timber Growers* 158 (Spring): 10.

Zhang, Daowei, and Peter H. Pearse. 1996. Differences in silviculture investment under various types of forest tenure in British Columbia. *Forest Science* 42 (4): 442–449.

Zietsma, Charlene, and Ilan B. Vertinsky. 1999–2001. Shades of green: Cognitive fram-ing and the dynamics of corporate environmental response. *Journal of Business Ad-ministration and Policy Analysis* 27–29:217–247.

Index

Accreditation Service (UKAS) (United Kingdom), 280

Alaska, Department of Fish and Game, 21

American Forest and Paper Association, 88, 262, 265, 273

American Forest Council, 271

American Paper Institute, 101, 271

American Society of Association Executives, 273

American Tree Farm System, 14, 89, 114, 246, 284, 290–91

Andersen Corporation (United States), 107, 112

Annual Allowable Cut (British Columbia), 59, 64–65, 84, 268

Arkansas Forestry Association, 123

Asian financial crisis, 62

AssiDomän (Sweden), 130, 147, 191, 194–95, 201–3, 205, 208, 214, 283

Association of German Magazine Publishers (*Verband Deutscher Zeitschriftenverleger*), 167, 171, 175, 182

Association of German Paper Producers (*Verband Deutscher Papierfabriken*), 162, 167, 171, 175, 255

Association of Municipalities of Rheinland-Pfalz (*Gemeinde- und Städtebund Rheinland-Pfalz*) (Germany), 175–77

Association of Professional Foresters (United Kingdom), 155–56, 256, 274, 277

Associational system cohesion, 46, 49, 220, 232

— British Columbia (Canada), 60, 64, 66, 78, 226

— Germany, 163, 166, 226

— Sweden, 215, 226–67

— United Kingdom, 130, 135, 158, 226

— United States, 90, 93, 95

auditing, 14–15, 26–27, 30, 102, 113, 117, 124, 136, 143, 150–53, 156, 182, 205, 207, 242, 268, 271, 275, 279–80

authority, 3–5, 7–8, 11, 13–15, 17, 20–

authority (*cont.*)
23, 25, 27–29, 42, 71, 111, 130, 136–
37, 139–41, 144, 147, 149, 180, 194,
196, 217, 219–22, 224, 226, 228, 230,
234, 236–40, 242–45, 263, 265–66,
268, 275, 278, 286, 288
Axel Springer–Verlag (Germany), 74,
173, 182

B&Q (United Kingdom), 11, 74, 77–78,
129, 138, 142, 144, 147–48, 202, 206,
208, 211, 239, 254, 275, 278
Baltic states exports (Latvia & Estonia to
United Kingdom), 148
Best Management Practices (United
States), 10, 13, 19, 108, 117
Big Creek Lumber (United States), 105
Boise (United States), 120, 271
Boots the Chemists (United Kingdom),
278
Bowater (United States), 120, 271
boycott campaigns, 10, 23, 41, 63, 67–
68, 74, 98, 138, 169–70, 190, 275
British Broadcasting Corporation Maga-
zine (United Kingdom), 74, 77, 269
British Columbia, Pulp and Paper Asso-
ciation, 268
Bureau of Land Management, United
States (listed under Department of the
Interior)

California Forestry Association, 123
The Campbell Group (United States), 122
Canada-U.S. Soft wood lumber agree-
ment, 63, 77, 267
Canada-U.S. trade relations, 63, 76, 267
Canadian Council of Forest Ministers,
14, 70
Canadian Forest Service, 70
Canadian Pulp and Paper Association
(Forest Products Association of Can-
ada), 60, 259, 261, 268
Canadian Standards Association (CSA),
— origins, 13, 14

— description, 13–16, 18, 19, 26, 60–61,
70–76, 78–79, 81–82, 182, 225, 245–
46, 249–53, 268–69, 285
Canadian Sustainable Forestry Certification Coalition (CSFCC), 13, 66, 70,
72, 268–69
Canfor (Canadian Forest Products), 62–
63, 81
Centex Corporation (United States), 73,
107, 112, 118
certification
— first party, 16, 113
— second party, 16, 113
— third party, 15–16, 26, 70, 102, 106,
108, 111, 113, 115, 117, 120, 122,
124, 144, 152, 175, 181, 246, 268
Certified Forest Products Council (US),
33, 77, 105–6, 111, 115, 258, 269
chain of custody, 15–16, 24, 45, 65, 82,
93, 100, 102, 112, 114–15, 117, 121,
124, 130, 141, 143, 147–48, 152–53,
156, 174, 182, 184, 196, 206–7, 210,
213, 225, 264, 284
chemical use (forest management), 15,
18, 150–51, 204, 249–50, 252
Church of Sweden, 196, 201, 255, 257
citizens, 3, 44, 125, 199
civil society, 9–10, 15, 17, 20, 29, 36, 44,
286–87
Clayoquot Sound (British Columbia), 68,
82
Clean Water Act (United States), 96
clearcutting, 66, 68, 71, 197, 236
Collins Pine (United States), 105, 260
concentration of industry (vertical and
horizontal integration), 45
— British Columbia (Canada), 64–65,
78, 90, 164–65
— Germany, 161, 164, 184
— Sweden, 164, 190, 192–93
— United Kingdom, 130, 153, 157
— United States, 93–94, 125, 164
conceptions (of certification), 8–9, 11,
15, 17, 25, 34, 238

The Conservation Fund, 111, 118
Conservation International, 107, 118
core audiences (of programs), 24, 34, 36, 69, 76, 82, 87, 116, 119, 124–25, 157, 187, 235, 240–41
Council of Forest Industries (British Columbia), 61–62, 64, 66, 77, 258, 268–69
County Forestry Boards (Sweden), 191, 197–99, 283

David Suzuki Foundation, 82, 258
Department of Agriculture, Forest Service (Northern Ireland), 134–35, 152, 275
Department of Agriculture, Forest Service (United States), 39, 94–96, 109–10, 261
Department of Energy (United States), 274
Department of the Interior, Bureau of Land Management (United States), 94, 261
Do It All (United Kingdom), 139, 278
Dogwood Alliance, 119
Domtar Industries (Canada and the United States), 119
durability (of programs), 30, 32, 37, 240

Earth First!, 141–42
eco-label, 16, 268, 276
Eco-Management and Audit Scheme (European Union), 26, 205, 207
Eco-timber (Germany), 172
EcoTrust, 258–59, 261
end consumers, 144, 287. *See also* citizens
endangered forests. *See* old growth forests)
Endangered Species Act (United States), 96, 271
Environmental Advantage (United States), 104, 272
Environmental Investigation Agency, 278–79

Environmental Protection Agency (United States), 10, 91, 110
evaluations (of programs), 4, 20, 22–23, 28, 32–33, 35–36, 38, 40, 42–43, 84, 123, 144, 159, 169, 191, 220–23, 227, 230–31, 235–37, 243, 265, 282, 287–88
external audiences (of programs), 20, 23, 28, 32–35, 286–87
European Process on Forests (Helsinki, 1993 and Lisbon, 1998), 14, 139, 145, 179, 207, 282

Fair Trade, 5, 24–25, 264, 298, 311, 313
Fauna and Flora International, 278
Fern, 156, 257, 278
FICGB Woodmark, 141–43, 145, 256, 277
—description, 142–46, 274, 277
—origins, 135, 141, 144
Florida Forestry Association, 123
Forest Alliance of British Columbia, 66, 70–71, 73, 75, 78–79, 262, 269
Forest convention, 11–12, 99, 131, 138–39, 170
Forest Enterprise Cooperatives (Forstbetriebsgemeinschaften) (Germany), 166
Forest Investment Associates (United States), 122
Forest Leadership Forum, 118
Forest Management Foundation (United Kingdom), 278
forest landownership, 21, 38–39, 45, 51, 235, 263, 265–66
—British Columbia (Canada), 64
—Germany, 161, 163–64, 186
—Sweden, 193, 196
—United Kingdom, 134
—United States, 93, 96, 121
Forest Licenses (British Columbia), 80
Forest Practices Act (British Columbia), 83
Forest Practices Board (British Columbia), 258, 260, 268
Forest Practices Code (British Columbia), 59, 67, 70, 198, 270

Forest Preservation, 172, 175, 179, 197, 244. *See also* Protected Areas

Forest Products Association (United Kingdom), 135, 156, 255, 278

Forest Products Association of Canada. *See* Canadian Pulp and Paper Association

Forest Service, Northern Ireland. *See listing under* Department of Agriculture

Forest Service, United States. *See listing under* Department of Agriculture

Forest Stewards Guild, 260–61

Forest Stewardship Council, 5, 12, 32, 60, 69, 79, 84, 86, 88–89, 103, 109, 112, 129, 146, 153, 155, 160, 177, 182–83, 185, 189, 202, 219, 249, 254, 255–56, 258–62, 268–70, 272, 277–78, 280, 284–85

— Principles and Criteria, 12, 86, 109, 120, 142, 145, 150, 176, 201, 249, 264, 270, 272

-Principle Nine (B.C.), 73, 75–76, 78–79, 82

-Principle Ten (United States, Southeast), 110

-Principle Three (Canada, Sweden), 202

— SLIMF Program, 107, 116

Forest Stewardship Council in British Columbia

— Aboriginal Fourth Chamber in Canada, 69, 72

— Draft 3, of standard, 83–84

— economic assessment, 85

— standards committee, 85, 106

— standards development, 60–61, 79–87

— standards team, 79–80, 83

— steering committee, 79, 83, 270

Forest Stewardship Council in Great Britain (later United Kingdom), 4, 5, 7, 33, 74, 129–59, 256–57, 274–80

— cross reference with UKWAS, 143, 151–56, 158, 212–13, 234, 279–80, 284

— draft standard, 142, 150, 280

— origins, 9, 136, 138, 275

— percentage–based claims, 143, 149, 152–53, 155, 191, 207, 226

Forest Stewardship Council in Germany

— German National Initiative, 170–78, 183–87

Forest Stewardship Council in Sweden

— StockDove, 209, 211–13

— Swedish National Initiative, 256

Forest Stewardship Council in the United States, 88–126

— National indicators, 112

— Pacific Coast regional standards development, 105, 109, 272–73

— Southeast regional standards development, 105–6, 110, 260, 272

The Forestland Group (United States), 122

Forest Systems (United States), 122

ForestEthics (Coastal Rainforest Coalition), 73–74, 105, 115, 119

Forestry Accord (1996, United Kingdom), 145

Forestry Act (Sweden), 197, 199

Forestry and Timber Association (United Kingdom), 156

Forestry Commission (Great Britain), 134–36, 139, 142–46, 148, 150–52, 154, 157, 243, 256–57, 275, 278–79

— Conservancies, 137

— Forest Authority, 137

— Forest Enterprise, 80, 137, 149, 151–2, 154, 275, 278

— Policy and Practices Branch, 137

— Woodland Grant Scheme, 279

Forestry Industry Council of Great Britain (FICGB), 135, 142–45, 150, 274, 277

Forestry Society (Skogssallskapet) (Sweden), 201, 284

Forestry Standard (United Kingdom), 142, 145–46, 150–51, 279

Fountains plc. (United Kingdom), 134, 154

Fraser Papers, Inc. (Canada), 273

Friends of the Earth, 74–75, 201, 284
— Germany (Bund *für Umwelt und Naturschutz Deutschland*), 171,
— UK, 141–42, 151, 156, 275–76, 278–79

Georgia Forestry Association, 123
Georgia-Pacific (United States), 120, 261, 271
German Forestry Association (*Deutscher Forstwirtschaftsrat*), 166, 171–72, 174–75, 178–79, 256, 281
German Society for Nature Conservation (*Naturschutzbund Deutschland*) 172
Global Forest and Trade Network, 33, 77, 85
Good Wood Alliance (Woodworkers Alliance for Rainforest Protection), 77, 99, 272
Good Wood Watch, 82
Governance
— public, 9, 28
— private, 9, 28
Great Bear Rainforest (British Columbia), 68, 77
Green Tag (United States), 89
Greenpeace
— Canada, 67, 71, 75, 261, 268
— Germany, 161, 168, 170–71, 256
— International, 42, 70–71, 74–75, 77, 82, 141–42, 156, 177, 201, 203, 269–70, 277
— Sweden, 205, 242
— United Kingdom, 42, 66–67, 77, 269
Gruppe 98 (Germany), 174, 176, 181, 183–84

Habitat conservation plans (United States), 10
Habitat UK, 278
Haindl (Germany), 74, 182, 255, 257, 268
Hancock Timber Resource Group (United States), 122, 258

Harrods (United Kingdom), 141
Holzabsatzfond, 179, 254, 282
Home Depot (United States), 11, 23, 26, 35, 73, 77, 79, 88, 99, 105–6, 111–13, 118, 123, 224, 234, 239, 260
Homebase (United Kingdom), 144

IKEA (Sweden), 11, 26, 213, 239, 256
illegal logging, 4, 247, 280
Indicative Forestry Strategies (United Kingdom), 278
Indigenous rights, 13, 202
— customary rights, 202, 204
— First Nations (Canada), 76
— Sami people (Sweden), 201–2, 204, 212, 283–84
Industrial Wood and Allied Workers of Canada (IWA), 73, 75, 79, 260
Industry Canada, 259
Initiative Tropenwald (Germany) 170–71
institutional consumers, 16
Interfor (International Forest Products) (Canada), 77, 81
Interior Lumber Manufacturer's Association (British Columbia), 268
International Forest Industry Roundtable, 115
International Paper (United States), 120, 234, 260, 271
International Standardization Organization (ISO), 14, 26, 70, 81, 284–85
International Tropical Timber Agreement, 11
— Target 2000, 138, 276
International Tropical Timber Organization (ITTO), 11, 138, 276
Izaak Walton League of America, 259

J. D. Irving (Canada and the United States), 109, 258, 285
John Dickinson Stationary (United Kingdom), 278

Kaufman & Board Homes Corporation (United States), 73, 107
Kimberly-Clark, 67
Kinnarps AB (Sweden), 214
Korsnäs AB (Sweden), 214

Label of Origin (*Herkunftszeichen*) (Germany), 172–74, 178, 207
label, 11, 16, 38, 82, 107, 117, 140, 142, 144–47, 149, 151–52, 154, 156, 170, 172–74, 178, 182, 191, 207, 210, 213, 250, 264–65, 267–68, 276, 278, 280–81
label of origin, 140, 142, 144–46, 149, 172–74, 178, 207
legitimacy, forms of
— cognitive, 34–35, 37–38, 40, 159, 221, 240, 242–43, 285–86
— moral, 3, 32–38, 40, 43, 82, 87, 92, 157, 161, 169, 171, 175–76, 179, 186, 221, 223, 231, 240–42, 288
— pragmatic, 7, 32–40, 47, 52, 81–82, 179, 214, 221, 231, 236, 240–42, 265, 287–88
legitimacy achievement strategies, 229, 234, 273, 287
— conforming, 32–34, 38, 41, 49, 60–61, 70, 72–74, 82, 87, 89, 95, 98, 104, 106–8, 110, 112–15, 118, 121, 124–25, 129, 140, 142–43, 153, 157–58, 174, 177, 180–82, 184, 200, 205, 207–11, 213, 221, 230, 235, 265, 286
— converting, 10, 32–34, 37–38, 41–46, 49, 60–61, 63, 65, 68–69, 72–77, 84–87, 90, 94, 98, 106–7, 121, 129, 137, 140, 142–44, 146, 151, 157–59, 161, 173–74, 176, 181, 183–84, 186, 190, 192, 196, 204, 208, 214–15, 221, 223, 225–26, 228, 230, 240, 265, 287
— informing, 26, 33–34, 61, 72, 75–77, 107, 110, 115, 145, 174, 180, 183, 209, 230–31, 265, 273, 287
level playing field (Trade), 138

Local Authorities Project (United Kingdom), 143, 147
Local Authorities' wood procurement policies (United Kingdom), 146–47, 275, 278
logo, 15–16, 284
Los Angeles, wood procurement policy (United States), 111
Louisiana Forestry Association, 123
Louisiana-Pacific (United States), 271
Lowe's (United States), 73, 107, 223
84 Lumber (United States), 118

MacMillan Bloedel (Canada), 35, 45, 73, 77–78, 81, 260, 270
Marine Stewardship Council, 21–22, 24–25, 264
Market-based policy instruments, 6, 10, 61, 68, 74, 77–78, 85–86, 98, 129, 140, 147, 157, 170, 176, 186, 189, 192, 214, 223, 225, 244, 279
MASCO (United States), 118
MasterBrand Cabinets (United States), 118
Meadwestvaco (United States), 120
Menards (United States), 118
Mellanskog (Sweden), 215
Michigan Forestry Association, 123
Ministry of Environment (United Kingdom), 152
Ministry of Environment, Lands and Parks (British Columbia), 268
Ministry of Forests (British Columbia), 64, 68, 75, 80, 258–59, 268–69
Missouri Forest Products Association, 123
MoDo Skog AB (Sweden), 214
Molplus Woodland Group (United States), 122
Montreal Process, 70
Mutual recognition, 13–14, 26, 107, 114, 178, 182, 186, 207, 212–13

National Auditing Organisation (United Kingdom), 136, 275, 276, 279–80, 284

National Environmental Policy Act (United States), 96

National Federation of Swedish Forest Owners, 215

National Forest lands (United States), 90–91, 96–97, 109–10

National Forest Management Act (United States), 96

National Forest Products Association (United States), 271

National Hardwood Lumber Association (United States), 123

National Indicators (United States Forest Stewardship Council). *See* Forest Stewardship Council

National Initiatives, of the Forest Stewardship Council. *See* Forest Stewardship Council

National Small Woods Association (United Kingdom), 155

National Woodland Owners Association (United States), 123, 260

Natural Resources Defense Council, 97–98, 271

The Nature Conservancy, 107, 118

Naturland (Germany), 171–72, 174, 281

New Hampshire Timberland Owners Association, 123

Nexfor Ltd. (formerly CSC Forest Products Ltd.) (United Kingdom), 148

non–industrial private forest owners, 14, 24, 39, 46–47, 89–91, 94, 109–10, 114, 189–92, 196, 200, 202, 204–5, 207, 234–35, 245–46

Non–state market driven authority, 4–9, 12, 15, 17, 20–31, 35–36, 40, 50–51, 55, 84, 87, 159, 236–37, 240, 242–43
— characteristics of, 5, 20 (table 1.5), 26, 89, 250, 252

Nordic Forest Certification Program (Sweden), 201, 208

Norm Thompson Outfitters (United States), 118

Norra Skogsägarna (Sweden), 215

Norrbottens Skogsägare (Sweden), 215

Norrskog (Sweden), 215

OBI (Germany), 170, 174, 176, 181–82, 184, 257

Ohio Forestry Association, 123

old growth forests, 12, 33, 45, 60, 64, 67, 76, 78, 82, 119, 183, 197, 236, 251, 269–70, 283

Oregon Forest Industries Council, 123, 259, 261, 271

Oregon, Forest Practices Act (United States), 97, 103

Otto-Versand (Germany), 173, 176

Overseas Development Institute (United Kingdom), 278

Pan European Forest Certification (PEFC)
— description, 13–16, 19, 73, 81–82, 115, 129, 160, 190, 226, 249, 251, 253–54
— origins, 13–16

Pan European Forest Certification in Germany
— established, 174, 179
— support, 160–62, 168–69, 178–87

Pan European Forest Certification in Sweden
— established, 205, 207–8, 210, 215
— support, 190, 192, 209–13

Pan European Forest Certification in the United Kingdom
— established, 143, 155, 156
— support, 129, 131, 143, 154–57

paper campaign (United States, Staples), 119

path dependency, 221–22, 229

Payless Cashways (United States), 118

percentage-based claims (chain of custody), 143, 149, 152–53, 155, 191, 207, 209, 226, 280

percentage-in percentage-out (chain of custody), 15, 207, 210–11, 213, 241 performance-based standards, 12, 16, 26, 100

Pella Windows (United States), 118

The Pinchot Institute, 117, 119, 262

Plantations, 15, 18, 33, 106, 110, 121, 249–50, 252, 272, 275

Plum Creek Timber Company (United States), 120, 258

political "backlash," 157, 162

popular sovereignty, 42–44, 265

Potlatch Corporation (United States), 120, 271

Praktiker (Germany), 176, 281

product tracking, 124, 205

protected areas, 49, 59, 161, 199, 203, 253. *See also* forest preservation

price premium, 4, 23, 116, 264

public forest policy, 220

— British Columbia (Canada), 49, 66–68, 70, 82, 270

— Germany, 167–69, 177

— Sweden, 196–200

— United Kingdom, 136–37, 144, 154, 158

— United States, 96–98

public policy agenda, 40, 47, 50, 60, 66, 69, 86, 89–91, 96, 98, 104, 123, 136–37, 157, 161, 167, 169, 187, 190–91, 196, 200, 220–21, 223–27

public policy networks, 49, 103, 216, 227–28

Putnam, Robert (Two level games), 159

Rainforest Action Network, 35–36, 43, 75, 77, 102, 105–6, 111, 115

Rainforest Alliance, 5, 27, 99, 116, 263

Rainforest Conservation Society, 82

Rayonier (United States), 120

Reforest the Earth (United Kingdom), 278–79

Regional Boards (Sweden), 191, 199

Resource Planning Act (United States), 96

Responsible Care (Chemical industry), 27

Results Based Code (British Columbia), 244

Rettenmeier Holzindustrie (Germany) 182

riparian zones, 15, 19, 59, 83, 270

Riverside Forest Products (Canada), 81

Robin Wood (Germany), 172

routinization, 159, 242

Royal Society for the Protection of Birds (United Kingdom), 136, 149, 257, 275, 278

Royal Swedish Academy of Agriculture and Forestry, 198–99

Sainsbury's (United Kingdom), 202

SCA Skog AB (Sweden), 214

Scandinavian exports (Finland & Sweden to United Kingdom), 19, 132–33, 147, 163–64, 173, 178, 193–94, 211

Scientific Certification Systems (SCS), 99–100, 103, 108, 120, 261

Scott Paper (United Kingdom), 270

Scott Ltd. (Canada), 67

Scottish Landowners Federation, 155

Scottish Woodlands, 134, 153–54, 254, 274, 277

Seven Islands (United States), 105, 260

SGS Qualifor, 72, 142, 150–51, 256, 260, 278–79

Shotton Paper Company (subsidiary of UPM–Kymmene) (United Kingdom), 274

Sierra Club

— of Canada, 66, 258

— of British Columbia, 82, 260, 270

— of the United States, 96, 106, 273–74

Sierra Legal Defence Fund, 260

Skeena Cellulose (Canada), 81

Slocan Forest Products (Canada), 81

Small Business Forest Enterprise Programme (British Columbia), 80, 259

SmartWood, 80, 99–100, 102–3, 106–7, 116, 261, 263

— TREES program, 116, 275

Smurfit-Stone Container Corporation (United States), 120
social license, 28, 47, 67, 223
Södra Skogsägarna (Sweden), 215
Soil Association, 140–43, 147, 151, 256, 277–79
— Local Authorities Project (with WWF), 143, 147
— Responsible Forest Programme (1992, United Kingdom), 140–42
— WoodMark Certification Scheme (1994, United Kingdom), 141–42, 256, 277
Southern Forest Products Association, 123
sovereign authority, 4, 20, 22, 28, 130, 221, 243, 265–56, 286, 288
State and municipal purchasing policies (Germany), 166, 168–70
Stora Skog AB (Sweden), 203, 214
supply chain, 4, 8, 12, 16, 20, 23–27, 34–36, 38, 40, 42–45, 60–61, 74, 84, 89–90, 94, 104, 110–11, 113, 119, 121, 123–24, 129, 134, 137, 139, 147, 153, 157, 161, 164–56, 173, 175–78, 180–81, 183–84, 187, 190–91, 205, 211, 220, 225, 232–34, 236–40, 271, 277, 286, 288
sustainable forest management, 4–6, 8–9, 12, 14–15, 17, 30, 70–71, 74, 81–82, 88, 100, 102–3, 115, 125, 131, 138–39, 144, 146, 149–50, 152, 156, 188–89, 202, 205, 207, 213, 219–20, 234, 240, 243, 245, 249, 276, 282
Sustainable Forestry Initiative (SFI) (United States)
— description, 13–16, 18, 26, 79, 88–89, 91, 96, 101–2, 104–8, 110–18, 120–25, 174, 182, 216, 245–46, 249, 250–53, 265, 273, 284–85
— Expert (now External) Review Panel, 102, 106, 110, 118
— GOAL program, 107, 114
— Interim inconsistent practices reporting protocol, 106, 113
— Licensing program, 111
— origins, 88–89, 91, 96
State Implementation Committees (SIC), 106, 114
Sustainable Forestry Board, 14, 89, 107, 113
sustained yield, 64, 66, 167
Sveaskog (Sweden), 194–96, 256, 282–83
Swedish Forest Industries Association, 190, 195, 205–6, 212–14, 284
Swedish Society for Nature Conservation, 197, 201, 208, 257, 283–84

Taiga Network, 182, 284
Tembec (Canada), 84
Temple-Inland Inc. (United States), 120
Tetra Pak (The Netherlands), 201
Texas Forestry Association, 123
The Campbell Group (United States), 122
The Conservation Fund, 111, 118
The Forestland Group (United States), 122
The Nature Conservancy, 107, 118
The Pinchot Institute, 117, 119, 262
The Wilderness Society, 258, 261, 274
Tilhill Economic Forestry (United Kingdom), 134, 274
Timber Growers Association (Great Britain), 135, 144, 156, 234, 256–57, 274, 277–78
Timber investment management organization (United States), 90, 94, 122
Timber Trade Federation (United Kingdom), 255, 276
Timbmet Ltd. (United Kingdom), 141, 254, 276–77
Time Magazine (United States), 67
Timfor (Canada), 80
Trade, import and export dependence, 41–42, 44
— British Columbia (Canada), 63, 74, 76, 86, 192, 267

— Germany, 74, 170, 182, 192, 201
— Sweden, 189, 192, 196, 201, 208
— United Kingdom, 74, 131–32, 192
— United States, 63, 76, 201, 267
Tree Farm Licenses (British Columbia), 64
Treetop thinning (*Waldsterben* or "forest die-back"), 167, 282
Tropical deforestation, 98–99
Tropical imports, 171
— From Brazil, Indonesia and Malaysia, 132
Two-level games (Putnam, Robert), 159

UBS Timber Investors (United States), 122
Ulster Timber Growers Organisation, 150, 155–56
United Nations Conference on Environment and Development (1992, Earth Summit), 11–12, 98, 131, 139, 145, 168, 170, 198

Vertical and horizontal integration (industry concentration), 45
— British Columbia (Canada), 64–65, 78, 90, 164–65
— Germany, 161, 164, 184
— Sweden, 164, 190, 192–93
— United Kingdom, 130, 153, 157
— United States, 93–94, 125, 164
Virginia Forestry Association, 123

Wachovia (United States), 122
Washington Contract Loggers Association, 123
Washington Farm Forestry Association, 123
Washington Forest Protection Association, 123, 258
Weldwood (subsidiary of International Paper) (Canada), 81
West Coast Environmental Law, 82, 259
West Fraser Timber (Canada), 81

Western Canada Wilderness Committee, 66, 259, 269
Western Forest Products (Canada), 72–73, 77–79, 261
Western Wood Products Association (United States), 123
Westphalian sovereignty, 4, 20–22, 28, 221, 243, 265–66
Weyerhaeuser Company (Canada and the United States), 35, 68, 77, 81, 120, 234, 259, 261
Wickes (United States), 73, 118
The Wilderness Society, 258, 261, 274
Wood Panel Industry Federation (United Kingdom), 274
Wood swapping (Sweden), 195–96, 206, 284
Woodworkers Alliance for Rainforest Protection (Good Wood Alliance), 77, 99, 272
Woodland Assurance Scheme (UKWAS) (United Kingdom), 143, 150–56, 158, 212–13, 234
Woodland Trust (United Kingdom), 257, 278
Working Group of German Forest Owner Associations (*Arbeitsgemeinschaft Deutscher Waldbesitzerverbände*), 166, 172, 178
Working Group on Natural Silviculture (*Arbeitsgemeinschaft* Naturgemäße *Waldwirtschaft*) (Germany), 167, 177, 255
— Commission on Environment and Development (1987, Brundtland Commission), 145
World Bank, 77, 147
World Resources Institute, 261
World Trade Organization (WTO), 70
World Wide Fund for Nature (World Wildlife Fund)
— Canada, 72, 74, 77, 85, 241, 258, 261, 268
— Germany, 174, 177, 254

—Sweden, 200–2, 208–9, 211–12, 284
—United Kingdom, 100, 138–43, 146–47, 151, 156–57, 205, 256–57, 276–78

—United States, 99, 104–5, 116
WWF 95 (later 95 +) group (UK), 74, 77, 100, 141, 146–48, 151, 156–57, 202, 268, 276